AF356832

The Mobilization, Speciation and Transformation of Organic and Inorganic Contaminants in Soil-Groundwater Ecosystems

The Mobilization, Speciation and Transformation of Organic and Inorganic Contaminants in Soil-Groundwater Ecosystems

Editors

Yizhi Sheng
Wanjun Jiang
Min Zhang

Basel • Beijing • Wuhan • Barcelona • Belgrade • Novi Sad • Cluj • Manchester

Editors
Yizhi Sheng
China University of
Geosciences
Beijing
China

Wanjun Jiang
China Geological Survey
Tianjin
China

Min Zhang
Chinese Academy of
Geological Sciences
Shijiazhuang
China

Editorial Office
MDPI AG
Grosspeteranlage 5
4052 Basel, Switzerland

This is a reprint of articles from the Special Issue published online in the open access journal *Applied Sciences* (ISSN 2076-3417) (available at: https://www.mdpi.com/journal/applsci/special_issues/1NW31522SG).

For citation purposes, cite each article independently as indicated on the article page online and as indicated below:

Lastname, A.A.; Lastname, B.B. Article Title. *Journal Name* **Year**, *Volume Number*, Page Range.

ISBN 978-3-7258-1834-1 (Hbk)
ISBN 978-3-7258-1833-4 (PDF)
doi.org/10.3390/books978-3-7258-1833-4

Contents

About the Editors

Yizhi Sheng

Yizhi Sheng is currently a full professor at the China University of Geosciences, Beijing. Yizhi Sheng's research aims to understand how and why microorganisms and geological media (rocks, minerals, water) interact, and how we can study their mutual interactions to understand a range of biogeochemical processes on Earth and beyond.

Wanjun Jiang

Wanjun Jiang is an associate researcher at the China Geological Survey. Wanjun Jiang's research focuses on hydrogeochemistry, isotopic geochemistry, environmental assessment and environmental geochemistry.

Min Zhang

Min Zhang is a full professor at the Chinese Academy of Geological Sciences. Min Zhang's research focuses on the natural attenuation of soil and groundwater pollution, source analysis of groundwater contamination, and biogeochemical processes and modeling.

Preface

This collection of thirteen articles presents cutting-edge research addressing critical challenges related to soil and groundwater management and environmental sustainability. By integrating advanced methodologies and innovative technologies, the articles enhance our understanding and management of complex soil and groundwater environmental and geological issues.

This collection emphasizes evaluating and improving soil and groundwater quality. Key studies explore the dynamics of pollutants in soil and groundwater ecosystems, offering important insights into processes such as toluene mineralization and petroleum contamination self-purification. Some articles employ models such as GIS-based DRASTIC for decision-making in groundwater protection. These findings inform effective remediation and management strategies. Research on low-salinity nanofluids for oil recovery in sandstone reservoirs shows promise for improving extraction efficiency. Analysis of pathogen-microbiota interactions in karst springs provides insights into managing waterborne disease risks, particularly in rocky desertification areas. Urban studies include geo-environmental suitability evaluations and urban engineering construction assessments. Models such as the entropy weight hierarchy-cloud model and combined weighting methods assess urban underground spaces, emphasizing strategic environmental considerations in urban planning. Groundwater vulnerability and protection strategies in coastal areas are highlighted, underscoring the need for effective groundwater management. Integrated site remediation methods tailored to contaminant distribution present innovative solutions for site cleanup, enhancing environmental safety and public health.

This Special Issue is motivated by the urgent need to address challenges posed by environmental degradation and climate change. By presenting innovative research and practical solutions, we aim to contribute to sustainable urban environments and the protection of vital natural resources. This Special Issue is intended for researchers in environmental science, urban planning, geosciences and related disciplines. It provides insights and fosters collaboration among professionals dedicated to advancing environmental sustainability and resilience.

We express our sincere gratitude to all authors for their invaluable contributions and to the reviewers for their rigorous evaluations and constructive feedback. We hope that this collection will serve as a valuable resource and inspiration for ongoing research and practical applications in soil and groundwater management.

Yizhi Sheng, Wanjun Jiang, and Min Zhang
Editors

Editorial

Mobilization, Speciation, and Transformation of Organic and Inorganic Contaminants in Soil–Groundwater Ecosystems

Yizhi Sheng [1,*], Wanjun Jiang [2,*] and Min Zhang [3]

[1] Center for Geomicrobiology and Biogeochemistry Research, State Key Laboratory of Biogeology and Environmental Geology, China University of Geosciences, Beijing 100083, China

[2] Tianjin Center (North China Center for Geoscience Innovation), China Geological Survey, Tianjin 300170, China

[3] Institute of Hydrogeology and Environmental Geology, Chinese Academy of Geological Sciences, Shijiazhuang 050061, China; minzhang205@live.cn

* Correspondence: shengyz@cugb.edu.cn (Y.S.); jiangwanjun0718@163.com (W.J.)

Citation: Sheng, Y.; Jiang, W.; Zhang, M. Mobilization, Speciation, and Transformation of Organic and Inorganic Contaminants in Soil–Groundwater Ecosystems. *Appl. Sci.* **2023**, *13*, 11454. https://doi.org/10.3390/app132011454

Received: 8 October 2023
Accepted: 17 October 2023
Published: 19 October 2023

The delicate balance of our ecosystems is under threat from the unrelenting release of contaminants into the environment. Among the most concerning are organic and inorganic pollutants that infiltrate the soil and permeate into the groundwater, posing significant risks to both environmental health and the wellbeing of humans who use groundwater as drinking water. To address this pressing issue, we present our Special Issue on the topic of "The Mobilization, Speciation, and Transformation of Organic and Inorganic Contaminants in Soil–Groundwater Ecosystems". This collection of research articles and studies seeks to shed light on these critical processes and foster innovative solutions for safeguarding our soil–groundwater ecosystems.

The common types of inorganic pollutants in soil and groundwater environments include inorganic salts, toxic metals, radioactive substances, etc. [1–4]. The common types of organic pollutants in soil and groundwater environments include polycyclic aromatic hydrocarbons (PAHs), volatile organic compounds (VOCs), chlorinated solvents, polychlorinated biphenyls (PCBs), pesticides, and other emerging contaminants [5–8]. Biological contaminants have also raised significant concerns [9]. Both organic and inorganic pollutants are propagated through soil–groundwater systems and eventually enter the food chain, posing health risks to humans. Long-term rock–water interactions, different groundwater recharge patterns, and intensive human activities have resulted in a complicated enrichment of those pollutants [10,11]. The mobilization, speciation, and transformation of these pollutants in the soil and groundwater ecosystem vary greatly depending on the specific hydro-biogeochemical processes and environments.

Understanding the factors influencing the mobilization of contaminants is crucial to implementing effective strategies to prevent their spread and mitigate their impacts. Unforeseen climatic events, human activities, and land-use changes all contribute to the release of harmful substances into subsurface environments. Climate change poses a threat to groundwater by affecting various aspects of the physical, chemical, and biological characteristics of soil and surface water bodies and aquifer recharge patterns. For instance, when heavy precipitation causes flooding, soil erosion occurs, and pollutants such as heavy metals, organic compounds, nutrients, and pathogens are transported from the soil into surface water bodies [12]. In regions where geomorphic units facilitate frequent interactions between surface water and groundwater, pollutants are carried into the groundwater aquifers as surface water infiltrates in large quantities, subsequently deteriorating the groundwater environment [13].

The hydraulic connections and frequent exchanges between surface water and groundwater (SW–GW) constitute a widespread phenomenon [14]. The variations in the water quality and quantity in these bodies of water are significantly influenced by their mutual interactions on both the time and space scales [15–17]. Once one of them becomes polluted, the other is inevitably under serious threat. During the process of pollutants infiltrating and recharging groundwater from surface water, the migration process and flux of pollutants are notably influenced by the local hydrodynamic characteristics and biogeochemical processes within the hyporheic zone [18,19]. Differences in flow pathways, velocities, residence times, and the exchange between surface water and groundwater control the reactive transport of, and affect the flux in, pollutants.

In the process of pollutants from the vadose soil environment entering the groundwater system, their adsorption, migration, and transformation within the geological media are complex and take forms such as surface runoff, leaching, erosion, desorption, and dissolution [19–21]. This complexity encompasses factors including the soil lithology, porosity, moisture content, pollutant type and concentration, oxidation–reduction potential (ORP), pH, organic matter, cation exchange capacity, microbial communities, etc. [22,23]. It is worth noting that certain elements (e.g., phosphorus, iron, fluoride, arsenic, iodine, uranium, and molybdenum,) are naturally abundant in the form of minerals/rocks in soils and groundwater sediments [3,24–26].

Although elevated levels of these inorganic constituents are often attributed to geogenic processes such as weathering, leaching, and water–rock interactions, anthropogenic activities can be expected to introduce and further intensify the contaminations [27–29]. These contaminants are primarily derived from domestic, agricultural, and industrial sources, including the application of fertilizers and pesticides in agricultural production, irrigation with sewage (or reclaimed water), leakage from domestic sewage networks and landfill site wastewater, waste residues generated during mining, or oil and gas field exploitation. [1–8,18–20]. For instance, fluctuations in groundwater levels caused by either natural or anthropogenic processes accelerate the leaching of organic matter, nitrogen, and heavy metals from the soil into shallow groundwater [30] and affect the aquifer redox cycling and microbial activity [31]. Intensive groundwater pumping can draw recently recharged contaminated modern groundwater into deeper aquifer systems [32]. Petroleum and natural gas exploitation causes petroleum hydrocarbon and salinity contamination in the vadose zone [33] and shallow groundwater [34]. Furthermore, the role of metals (e.g., Hg, Zn, Cu, and Cd) in co-selecting antibiotic resistance might be important in the spread of antibiotic resistance genes [20].

The speciation of contaminants adds another layer of complexity to the soil–groundwater environment. As pollutants interact with the soil matrix and microbial communities, they undergo chemical transformations, altering their properties and toxicity. Contaminants can undergo different chemical transformations, resulting in various species with different properties and toxicities. Metals can exist in different oxidation states, and organic contaminants can undergo degradation or transformation into metabolites and greenhouse gases. The speciation of contaminants affects their mobility, bioavailability, and toxicity, and an understanding of these forms is crucial to accurately assessing their environmental impact and risk.

For instance, arsenic (As) primarily exists as As(V) and can transform into more mobile As(III) under reducing conditions, affecting the groundwater quality [35]. Chromium (Cr) exists as both the Cr(VI) and Cr(III) forms, with Cr(VI) becoming soluble and migrating into groundwater under oxidizing conditions [36]. Uranium (U) in soil and sediments is typically found as less mobile U(IV) compounds, while, in groundwater, it exists as more mobile uranyl U(VI) and UO_2^{2+} complexes, which are strongly influenced by various groundwater physicochemical factors [37]. Redox conditions and natural organic matter play key roles in these transformations [24,29,38,39].

Fluoride often exists in nature in the form of F^- and forms soluble salts with monovalent alkali metals in groundwater, such as fluoride salt (NaF) and fluoride potassium salt (KF) [3]. Meanwhile, fluorite (CaF_2), sellaite (MgF_2), and fluoride-bearing silicate minerals such as mica, amphibole, tourmaline, and fluorapatite are commonly found in soils or rocks. Therefore, the weathering of fluoride minerals could be a natural source of fluoride in water [3]. In the groundwater systems, I commonly exists in the form of iodide (I^-), iodate (IO_3^-), and organic iodine (OI), and I^- is the dominant species in groundwater, whereas IO_3^- and OI are the common species in the soil/sediments [29].

Organic contaminants include a range of conventional pollutants such as petroleum hydrocarbon, phenolic compounds, and pesticides, while concerns have also extended to persistent organic pollutants, such as perfluoroalkyl and polyfluoroalkyl substances (PFAS), pharmaceuticals and personal care products (PPCPs), antibiotics, and microplastics [2,5,8]. Perfluorinated compounds (PFCs) including PFAS are persistent organic pollutants that can linger in soil and impact groundwater long after their use [40]. They are highly toxic and include chemicals such as PFOA and PFOS. The antibiotic pollution in soils and groundwater is diverse, with substances such as ciprofloxacin and sulfamethoxazole found in high concentrations [8,20]. Microplastics include small plastic particles including polyethylene and polyvinyl chloride, etc., resulting from the breakdown of larger plastic waste [41,42].

Pollutants in soil and groundwater pose significant risks to both ecosystems and human health [39,43]. These pollutants enter the human body through pathways including bioaccumulation, the food chain, and drinking water, leading to different types of toxicity, including chemical toxicity, radiation toxicity, and carcinogenicity. Common pollutants, such as nitrogen, when present in excess, can cause various cancers and other health issues [21]. Heavy metals, such as arsenic, chromium, and lead, can lead to cancer, organ damage, and nervous system disorders [29,36]. Radioactive isotopes in groundwater can be highly toxic, affecting various bodily systems. Additionally, non-degradable organic pollutants, such as antibiotics and microplastics, pose emerging health risks due to their persistence and limited microbial degradation [20,42].

The transformation of contaminants within soil–groundwater ecosystems can lead to either an exacerbation or attenuation of their effects. The processes of degradation, redox reactions, and other chemical transformations can significantly impact the persistence and fate of pollutants. An in-depth exploration of these mechanisms together is vital to developing sustainable solutions for remediating contaminated sites. The range of strategies for alleviating the impact of contaminants in soil–groundwater systems includes the monitoring and assessment of pollutant levels, the implementation of containment measures, the promotion of natural attenuation through bioremediation, the incorporation of remediation materials, and the employment of engineered solutions such as permeable reactive barriers or pump-and-treat systems to capture and treat contaminated groundwater, as well as phytoremediation to treat contaminants before they seep into aquifers ([44–49]). Before these techniques are used, their cost and resulting environmental impacts should be considered.

Bioremediation emerges as a particularly promising, sustainable, environmentally friendly, and cost-effective strategy. To start with, understanding the assembly of microbial communities, their driving forces (e.g., their pH, salinity, nutrients, and metals) and their role in transforming pollutants in the vadose-zone–groundwater ecosystem is instrumental to unlocking the full potential of bioremediation [22,50–52]. While advanced techniques have been developed for characterizing microbes, significant gaps remain in our understanding of dynamic subsurface microbial communities and the biogeochemical processes in the environments they inhabit [53], which hinder the effective use of biostimulation or bioaugmentation as an in situ remediation strategy. For example, the complex coupled carbon and iron cycling at multiple redox interfaces across subsurface environments can be expected to positively and negatively affect both microbial and extracellular enzyme activity [54]. Dissimilatory nitrate reduction to ammonia is enriched in the downgradient along the groundwater flow path [55]. Sulfate reduction accelerates groundwater arsenic contamination in aquifers with replete iron oxides [56]. The intrusion of produced water enhances the salinity and petroleum hydrocarbon levels in shallow groundwater, causing a transformation in the composition and functionality of bacterial and archaeal communities [34]. Moreover, hydrodynamic disturbance is another possible driving force for microbial community assembly and biogeochemical processes in sedimentary environments [57].

The mobilization, speciation, and transformation of contaminants in soil–groundwater ecosystems represent an urgent call for action in the face of environmental degradation. The task of preserving the integrity of our soil–groundwater ecosystems is a complex one that requires a multidisciplinary approach. There is still a gap in our understanding of the mechanisms that connect mobilization, speciation, health impacts, and microbial processes related to groundwater contaminants. To effectively address soil pollution sources and manage the risks of groundwater pollution, it is essential to comprehend how complex geological and hydrogeological factors in soil–groundwater systems interplay with climate change and human activities. This understanding can significantly inform the development of remediation strategies.

As we present this Special Issue on "The Mobilization, Speciation, and Transformation of Organic and Inorganic Contaminants in Soil–Groundwater Ecosystems", we call upon researchers to come together, share knowledge, and work toward a shared vision of a cleaner and greener future. Through the insightful research and innovative solutions showcased in this Special Issue, we are reminded of our collective responsibility to safeguard the delicate balance of nature.

Funding: This research was funded by National Natural Science Foundation of China (42302299).

Acknowledgments: We would like to thank all the authors who contributed and will contribute to this Special Issue. We would like to also thank all the reviewers for providing their valuable comments, which have greatly improved the quality of the published papers.

Conflicts of Interest: The authors declare no conflict of interest.

References

1. Abascal, E.; Gómez-Coma, L.; Ortiz, I.; Ortiz, A. Global diagnosis of nitrate pollution in groundwater and review of removal technologies. *Sci. Total Environ.* **2022**, *810*, 152233. [CrossRef]
2. Kurwadkar, S.; Kanel, S.R.; Nakarmi, A. Groundwater pollution, occurrence, detection, and remediation of organic and inorganic pollutants. *Water Environ. Res.* **2020**, *92*, 1659–1668. [CrossRef] [PubMed]
3. Jiang, W.J.; Sheng, Y.Z.; Liu, H.W.; Ma, Z.; Song, Y.X.; Liu, F.T.; Chen, S.M. Groundwater quality assessment and hydrogeochemical processes in typical watersheds in Zhangjiakou region, northern China. *Environ. Sci. Pollut. Res.* **2022**, *29*, 3521–3539. [CrossRef] [PubMed]
4. Martinez-Morata, I.; Bostick, B.C.; Conroy-Ben, O.; Duncan, D.T.; Jones, M.R.; Spaur, M.; Patterson, K.P.; Prins, S.J.; Navas-Acien, A.; Nigra, A.E. Nationwide geospatial analysis of county racial and ethnic composition and public drinking water arsenic and uranium. *Nat. Commun.* **2022**, *13*, 7461. [CrossRef] [PubMed]
5. MacLeod, M.; Arp, H.P.H.; Tekman, M.B.; Jahnke, A. The global threat from plastic pollution. *Science* **2021**, *373*, 61–65. [CrossRef]
6. Sheng, Y.; Tian, X.; Wang, G.; Hao, C.; Liu, F. Bacterial diversity and biogeochemical processes of oil-contaminated groundwater, Baoding, North China. *Geomicrobiol. J.* **2016**, *33*, 537–551. [CrossRef]
7. Fernández-Fernández, V.; Ramil, M.; Cela, R.; Rodríguez, I. Occurrence and risk assessment of pesticides and pharmaceuticals in viticulture impacted watersheds from Northwest Spain. *Chemosphere* **2023**, *341*, 140098. [CrossRef]
8. Zainab, S.M.; Junaid, M.; Xu, N.; Malik, R.N. Antibiotics and antibiotic resistant genes (ARGs) in groundwater, A global review on dissemination, sources, interactions, environmental and human health risks. *Water Res.* **2020**, *187*, 116455. [CrossRef]
9. Dong, Y.; Jiang, Z.; Hu, Y.; Jiang, Y.; Tong, L.; Yu, Y.; Cheng, J.; He, Y.; Shi, J.; Wang, Y. Pathogen contamination of groundwater systems and health risks. *Crit. Rev. Environ. Sci. Technol.* **2023**, 1–23. [CrossRef]
10. Jiang, W.J.; Wang, G.C.; Sheng, Y.Z.; Shi, Z.M.; Zhang, H. Isotopes in groundwater (^{2}H, ^{18}O, ^{14}C) revealed the climate and groundwater recharge in the Northern China. *Sci. Total Environ.* **2019**, *666*, 298–307. [CrossRef]
11. Jiang, W.J.; Sheng, Y.Z.; Wang, G.C.; Shi, Z.M.; Liu, F.T.; Zhang, J.; Chen, D.L. Cl, Br, B, Li, and noble gases isotopes to study the origin and evolution of deep groundwater in sedimentary basins, a review. *Environ. Chem. Lett.* **2022**, *20*, 1497–1528. [CrossRef]
12. Griebler, C.; Avramov, M. Groundwater ecosystem services: A review. *Freshw. Sci.* **2015**, *34*, 355–367. [CrossRef]
13. Hartmann, A.; Jasechko, S.; Gleeson, T.; Wada, Y.; Andreo, B.; Barberá, J.A.; Brielmann, H.; Bouchaou, L.; Charlier, J.B.; Darling, W.G.; et al. Risk of groundwater contamination widely underestimated because of fast flow into aquifers. *Proc. Natl. Acad. Sci. USA* **2021**, *118*, e2024492118. [CrossRef] [PubMed]
14. Sophocleous, M. Interactions between groundwater and surface water: The state of the science. *Hydrogeol. J.* **2022**, *10*, 52–67. [CrossRef]

15. Zhou, P.P.; Wang, G.C.; Mao, H.R.; Liao, F.; Shi, Z.M.; Huang, H.X. Numerical modeling for the temporal variations of the water interchange between groundwater and surface water in a regional great lake (Poyang Lake, China). *J. Hydrol.* **2022**, *610*, 127827. [CrossRef]
16. Adyasari, D.; Dimova, N.T.; Dulai, H.; Gilfedder, B.S.; Cartwright, I.; McKenzie, T.; Fuleky, P. Radon-222 as a groundwater discharge tracer to surface waters. *Earth-Sci. Rev.* **2023**, *238*, 104321. [CrossRef]
17. Liao, F.; Cardenas, M.B.; Ferencz, S.B.; Chen, X.B.; Wang, G.C. Tracing Bank Storage and Hyporheic Exchange Dynamics Using 222Rn, Virtual and Field Tests and Comparison with Other Tracers. *Water Resour. Res.* **2021**, *57*, e2020WR028960. [CrossRef]
18. Liu, Y.; Wang, P.; Gojenko, B.; Yu, J.J.; Wei, L.Z.; Luo, D.G.; Xiao, T.F. A review of water pollution arising from agriculture and mining activities in Central Asia, Facts, causes and effects. *Environ. Pollut.* **2021**, *291*, 118209. [CrossRef]
19. Gandhi, T.P.; Sampath, P.V.; Maliyekkal, S.M. A critical review of uranium contamination in groundwater, treatment and sludge disposal. *Sci. Total Environ.* **2022**, *825*, 153947. [CrossRef]
20. Amarasiri, M.; Sano, D.; Suzuki, S. Understanding human health risks caused by antibiotic resistant bacteria (ARB) and antibiotic resistance genes (ARG) in water environments, Current knowledge and questions to be answered. *Crit. Rev. Environ. Sci. Technol.* **2020**, *50*, 2016–2059. [CrossRef]
21. Picetti, R.; Deeney, M.; Pastorino, S.; Miller, M.R.; Shah, A.; Leon, D.A.; Dangour, A.D.; Green, R. Nitrate and nitrite contamination in drinking water and cancer risk, A systematic review with meta-analysis. *Environ. Res.* **2022**, *210*, 112988. [CrossRef] [PubMed]
22. Sheng, Y.; Li, G.; Dong, H.; Liu, Y.; Ma, L.; Yang, M.; Liu, Y.; Liu, J.; Deng, S.; Zhang, D. Distinct assembly processes shape bacterial communities along unsaturated, groundwater fluctuated, and saturated zones. *Sci. Total Environ.* **2021**, *761*, 143303. [CrossRef] [PubMed]
23. Gnesda, W.R.; Draxler, E.F.; Tinjum, J.; Zahasky, C. Adsorption of PFAAs in the vadose zone and implications for long-term groundwater contamination. *Environ. Sci. Technol.* **2022**, *56*, 16748–16758. [CrossRef]
24. Dong, H.L.; Huang, L.Q.; Zhao, L.D.; Zeng, Q.; Liu, X.L.; Sheng, Y.Z.; Shi, L.; Wu, G.; Jiang, H.C.; Li, F.R.; et al. A critical review of mineral-microbe interaction and coevolution, mechanisms and applications. *Natl. Sci. Rev.* **2022**, *9*, nwac128. [CrossRef] [PubMed]
25. Jia, Y.; Xi, B.; Jiang, Y.; Guo, H.; Yang, Y.; Lian, X.; Han, S. Distribution, formation and human-induced evolution of geogenic contaminated groundwater in China: A review. *Sci. Total Environ.* **2018**, *643*, 967–993. [CrossRef]
26. Sheng, Y.; Baars, O.; Guo, D.; Whitham, J.; Srivastava, S.; Dong, H. Mineral-bound trace metals as cofactors for anaerobic biological nitrogen fixation. *Environ. Sci. Technol.* **2023**, *57*, 7206–7216. [CrossRef]
27. Podgorski, J.; Berg, M. Global analysis and prediction of fluoride in groundwater. *Nat. Commun.* **2022**, *13*, 4232. [CrossRef]
28. Podgorski, J.; Berg, M. Global Threat of Arsenic in Groundwater. *Science* **2020**, *368*, 845–850. [CrossRef]
29. Wang, Y.X.; Li, J.X.; Ma, T.; Xie, X.J.; Deng, Y.M.; Gan, Y.Q. Genesis of geogenic contaminated groundwater, As, F and I. *Crit. Rev. Environ. Sci. Technol.* **2020**, *51*, 2895–2933. [CrossRef]
30. Vázquez-Suñé, E.; Sánchez-Vila, X.; Carrera, J. Introductory review of specific factors influencing urban groundwater, an emerging branch of hydrogeology, with reference to Barcelona, Spain. *Hydrogeol. J.* **2005**, *13*, 522–533. [CrossRef]
31. Sheng, Y.; Dong, H.; Coffin, E.; Myrold, D.; Kleber, M. Inhibition of Extracellular Enzyme Activity by Reactive Oxygen Species upon Oxygenation of Reduced Iron-Bearing Minerals. *Environ. Sci. Technol.* **2023**, *57*, 3425–3433. [CrossRef] [PubMed]
32. Thaw, M.; GebreEgziabher, M.; Villafañe-Pagán, J.Y.; Jasechko, S. Modern groundwater reaches deeper depths in heavily pumped aquifer systems. *Nat. Commun.* **2022**, *13*, 5263. [CrossRef] [PubMed]
33. Sheng, Y.; Liu, Y.; Yang, J.; Dong, H.; Liu, B.; Zhang, H.; Li, A.; Wei, Y.; Li, G.; Zhang, D. History of petroleum disturbance triggering the depth-resolved assembly process of microbial communities in the vadose zone. *J. Hazard. Mater.* **2021**, *402*, 124060. [CrossRef] [PubMed]
34. Chen, X.L.; Sheng, Y.Z.; Wang, G.; Guo, L.; Zhang, H.; Zhang, F.; Yang, T.; Huang, D.D.; Han, X.; Zhou, L. Microbial compositional and functional traits of BTEX and salinity co-contaminated shallow groundwater by produced water. *Water Res.* **2022**, *15*, 118277. [CrossRef]
35. Guo, H.; Wen, D.; Liu, Z.; Jia, Y.; Guo, Q. A review of high arsenic groundwater in Mainland and Taiwan, China, distribution, characteristics and geochemical processes. *Appl. Geochem.* **2014**, *41*, 196–217. [CrossRef]
36. Perraki, M.; Vasileiou, E.; Bartzas, G. Tracing the origin of chromium in groundwater: Current and new perspectives. *Curr. Opin. Environ. Sci. Health* **2021**, *22*, 100267. [CrossRef]
37. Vengosh, A.; Coyte, R.M.; Podgorski, J.; Johnson, T.M. A critical review on the occurrence and distribution of the uranium- and thorium-decay nuclides and their effect on the quality of groundwater. *Sci. Total Environ.* **2022**, *808*, 151914. [CrossRef]
38. Zhang, Z.; Guo, H.M.; Han, S.B.; Gao, Z.P.; Niu, X. Controls of Geochemical and Hydrogeochemical Factors on Arsenic Mobility in the Hetao Basin, China. *Groundwater* **2022**, *61*, 44–55. [CrossRef]
39. Xie, X.; Shi, J.; Pi, K.; Deng, Y.; Yan, B.; Tong, L.; Yao, L.; Dong, Y.; Li, J.; Ma, L.; et al. Groundwater Quality and Public Health. *Annu. Rev. Environ. Resour.* **2023**, *48*, 12.1–12.24. [CrossRef]
40. Xiang, L.; Qiu, J.; Chen, Q.Q.; Yu, P.F.; Liu, B.L.; Zhao, H.M.; Li, Y.W.; Feng, N.X.; Cai, Q.Y.; Mo, C.H.; et al. Development, Evaluation, and Application of Machine Learning Models for Accurate Prediction of Root Uptake of Per- and Polyfluoroalkyl Substances. *Environ. Sci. Technol.* **2023**. [CrossRef]
41. Law, K.L.; Rochman, C.M. Large-scale collaborations uncover global extent of plastic pollution. *Nature* **2023**, *619*, 254–255. [CrossRef] [PubMed]

42. Rochman, C.M.; Hoellein, T. The global odyssey of plastic pollution. *Science* **2020**, *368*, 1184–1185. [CrossRef]
43. Li, P.; Karunanidhi, D.; Subramani, T.; Srinivasamoorthy, K. Sources and consequences of groundwater contamination. *Arch. Environ. Contam. Toxicol.* **2021**, *80*, 1–10. [CrossRef]
44. Hou, D.; Al-Tabbaa, A.; O'Connor, D.; Hu, Q.; Zhu, Y.G.; Wang, L.; Kirkwood, N.; Ok, Y.S.; Tsang, D.C.; Bolan, N.S.; et al. Sustainable remediation and redevelopment of brownfield sites. *Nat. Rev. Earth Environ.* **2023**, *4*, 271–286. [CrossRef]
45. Hashim, M.A. Soumyadeep Mukhopadhyay, Jaya Narayan Sahu, and Bhaskar Sengupta. Remediation technologies for heavy metal contaminated groundwater. *J. Environ. Manag.* **2011**, *92*, 2355–2388. [CrossRef]
46. Sheng, Y.; Zhang, X.; Zhai, X.; Zhang, F.; Li, G.; Zhang, D. A mobile, modular and rapidly-acting treatment system for optimizing and improving the removal of non-aqueous phase liquids (NAPLs) in groundwater. *J. Hazard. Mater.* **2018**, *360*, 639–650. [CrossRef] [PubMed]
47. Ossai, I.C.; Ahmed, A.; Hassan, A.; Hamid, F.S. Remediation of soil and water contaminated with petroleum hydrocarbon: A review. *Environ. Technol. Innov.* **2020**, *17*, 100526.
48. Vangronsveld, J.; Herzig, R.; Weyens, N.; Boulet, J.; Adriaensen, K.; Ruttens, A.; Mench, M. Phytoremediation of contaminated soils and groundwater, lessons from the field. *Environ. Sci. Pollut. Res.* **2009**, *16*, 765–794. [CrossRef]
49. Dong, H.; Coffin, E.S.; Sheng, Y.; Duley, M.L.; Khalifa, Y.M. Microbial reduction of Fe (III) in nontronite: Role of biochar as a redox mediator. *Geochim. Cosmochim. Acta* **2023**, *345*, 102–116. [CrossRef]
50. Sheng, Y.; Bibby, K.; Grettenberger, C.; Kaley, B.; Macalady, J.L.; Wang, G.; Burgos, W.D. Geochemical and temporal influences on the enrichment of acidophilic iron-oxidizing bacterial communities. *Appl. Environ. Microbiol.* **2016**, *82*, 3611–3621. [CrossRef]
51. Sheng, Y.; Wang, G.; Zhao, D.; Hao, C.; Liu, C.; Cui, L. Groundwater microbial communities along a generalized flowpath in confined aquifers in the Qaidam Basin, China. *Groundwater* **2018**, *56*, 719–731. [CrossRef] [PubMed]
52. Ruff, S.E.; Humez, P.; de Angelis, I.H.; Diao, M.; Nightingale, M.; Cho, S.; Connors, L.; Kuloyo, O.O.; Seltzer, A.; Bowman, S.; et al. Hydrogen and dark oxygen drive microbial productivity in diverse groundwater ecosystems. *Nat. Commun.* **2023**, *14*, 3194. [CrossRef] [PubMed]
53. Kaur, G.; Kaur, G.; Krol, M.; Brar, S.K. Unraveling the mystery of subsurface microorganisms in bioremediation. *Curr. Res. Biotechnol.* **2022**, *4*, 302–308. [CrossRef]
54. Dong, H.; Zeng, Q.; Sheng, Y.; Chen, C.; Yu, G.; Kappler, A. Coupled Iron Redox Cycling and Organic Matter Transformation Across Multiple Interfaces. *Nat. Rev. Earth Environ.* **2023**, *4*, 659–673. [CrossRef]
55. Guo, L.; Xie, Q.; Sheng, Y.Z.; Wang, G.C.; Jiang, W.J.; Tong, X.X.; Xu, Q.Y.; Hao, C.B. Co-variation of hydrochemistry, inorganic nitrogen, and microbial community composition along groundwater flowpath. *Appl. Geochem.* **2022**, *140*, 105296. [CrossRef]
56. Nghiem, A.A.; Prommer, H.; Mozumder, M.R.H.; Siade, A.; Jamieson, J.; Ahmed, K.M.; van Geen, A.; Bostick, B.C. Sulfate reduction accelerates groundwater arsenic contamination even in aquifers with abundant iron oxides. *Nat. Water* **2023**, *1*, 151–165. [CrossRef] [PubMed]
57. Chen, Y.J.; Leung, P.M.; Cook, P.L.; Wong, W.W.; Hutchinson, T.; Eate, V.; Kessler, A.J.; Greening, C. Hydrodynamic disturbance controls microbial community assembly and biogeochemical processes in coastal sediments. *ISME J.* **2022**, *16*, 750–763. [CrossRef]

Article

Development and Application of an Integrated Site Remediation Technology Mix Method Based on Site Contaminant Distribution Characteristics

Min Zhang [1,2,†], Shuai Yang [3,†], Zhifei Zhang [4], Caijuan Guo [1,2], Yan Xie [3], Xinzhe Wang [3], Lin Sun [1,2] and Zhuo Ning [1,2,*]

[1] Institute of Hydrogeology and Environmental Geology, Chinese Academy of Geological Sciences, Shijiazhuang 050061, China; minzhang205@live.cn (M.Z.); caizidongdong@163.com (C.G.); sunlin@mail.cgs.gov.cn (L.S.)

[2] Key Laboratory of Groundwater Remediation of Hebei Province & China Geological Survey, Shijiazhuang 050061, China

[3] SINOPEC Research Institute of Safety Engineering Co., Ltd., Qingdao 266071, China; ysyang901029@126.com (S.Y.); xiey.qday@sinopec.com (Y.X.); wangxz.qday@sinopec.com (X.W.)

[4] Hebei Geological Environment Monitoring Institute, Shijiazhuang 050000, China; hkyzzf@163.com

* Correspondence: ningzhuozhuo@163.com; Tel.: +86-0311-67598605

† These authors contributed equally to this work.

Abstract: Millions of contaminated sites worldwide need to be remediated to protect the environment and human health. Although numerous remediation technologies have been developed, selecting optimal technologies is challenging. Several multiple criteria decision-making methods for screening the optimal remediation technology have been proposed, but they mostly focus on a specific area rather than the whole contaminated site. In recent years, the "contamination source control—process blocking—in situ remediation" technology mix model has gradually gained high appreciation. Nevertheless, the screening of technologies within each chain of this model relies heavily on arbitrary personal experience. To avoid such arbitrariness, a petroleum-contaminated site containing light non-aqueous phase liquids (LNAPLs) was used as an example, and a scientific screening and combination procedure was developed in this study by considering the distribution characteristics of contaminants. Through the procedure, a technology mix, which includes institutional control, risk monitoring, emergency response, multiphase extraction, interception ditch, monitoring of natural attenuation, hydrodynamic control, as well as some alternative technologies, was found, aiming at different locations and strata. The clear spatial relationship concept promises to enhance the effectiveness of contaminated site remediation. The proposed method only gave us a technical framework and should be tested and enriched in future studies.

Keywords: contaminated site; contaminants distribution; remediation technology; integrated mix method

Citation: Zhang, M.; Yang, S.; Zhang, Z.; Guo, C.; Xie, Y.; Wang, X.; Sun, L.; Ning, Z. Development and Application of an Integrated Site Remediation Technology Mix Method Based on Site Contaminant Distribution Characteristics. *Appl. Sci.* **2023**, *13*, 11076. https://doi.org/10.3390/app131911076

Academic Editor: Alessio Adamiano

Received: 12 September 2023
Revised: 1 October 2023
Accepted: 7 October 2023
Published: 8 October 2023

1. Introduction

Contaminated sites and the environmental problems they bring are issues of major global concern. Taking petrochemical sites as an example, Europe and the US have approximately 342,000 and 200,000 petrochemical-contaminated sites, respectively [1]. China has approximately 120,000 filling stations, 300 refineries, and bulk factories occupying 78 million m^3 in our previous statistics in 2020. The soil and groundwater at these sites are all laden with potential contaminants. The US EPA and other national environmental agencies emphasize that contaminated sites must be remediated in order to restore and protect groundwater resources, and thereby create a safe living or working environment [2,3].

There are many types of contaminated site remediation technologies. For instance, nearly 30 soil and groundwater contamination remediation and control technologies have been used in the remediation of the 1468 Superfund sites in the US [4–6]. The List of Contaminated Site Remediation Technologies (first batch) issued by China in 2014 lists 15 types of remediation and risk control technologies [7]. In most cases, these techniques have commonly been evaluated by experienced experts, but different technologies are often selected because they are familiar but not because they are the most applicable or cost-effective for a given site [8].

To emilite the arbitrary technology selection, several multiple criteria decision-making methods, including simple additive weighting (SAW), ordered weighted average (OWA), analytic hierarchy process (AHP), preference ranking organization method for enrichment evaluation (PROMETHEE), and technique for order preference by similarity to ideal solution (TOPSIS), have been used to screen the optimal technology [8–13].

These screening processes for remediation methods mainly focus on specific areas of contaminated sites, with technologies primarily selected or ruled out based on the type of contaminants, stratigraphy, and petrology [14–17]. However, contaminants in the soil and aquifers can migrate with groundwater flow [18], leading to variations in contaminants over time and space. Depending on the phase and concentration distribution of contaminants in different areas, contaminated areas or strata at a site can be classified as the contamination source zone and contamination plume [19]. The undifferentiated treatment of different strata and vague spatial delineation of remediation technology application tend to overlook the inhomogeneous and dynamic distribution of contaminants across sites, probably result in the excessive remediation of certain areas or post-remediation reappearance of contamination, and are not consistent with a precise and scientific control system.

In the currently recognized remediation framework, the ultimate goal is to prevent potential receptors from being contaminated or restore receptor risks within acceptable limits [20]. Therefore, within this framework, we only need to protect the receptors without considering non-receptors. However, in order to protect the receptors effectively, every link that contributes to receptor contamination should be taken into account. This means that when selecting remediation technologies, we should consider the entire chain of contamination—source zone, plume, and receptors—as a whole. Based on this understanding, researchers have recently proposed and lent widespread support to the "contamination source control—process blocking—in situ remediation" technology mix model [21–23]. However, the screening of technologies under the technology mix model is largely reliant on empirical methods, and the model lacks a scientific screening method of the combinatorial system.

To develop a valid site remediation technology mix method matching the "contamination source control—process blocking—in situ remediation" technology mix model, in the study, a petrochemical-contaminated site was set as an example. Based on the site contaminant distribution characteristics, and clear-cut "contamination source zone—plume—potentially contaminated area" linkage, an integrated site remediation technology mix method with a clear spatial relationship concept was established.

2. Materials and Methods

2.1. Technology Mix Method

Based on the site's contaminant distribution characteristics, we adopt the "contamination source control—process blocking—in situ remediation" technology mix concept and then establish a list of remediation technologies classified by area in keeping with the standards and guidelines issued by various countries' environmental agencies. After performing an assessment and ranking of each technology from technological, environmental, and social perspectives, we optimize the technology mix in accordance with ranking results, which yields a solution with an optimized technology mix. The specific steps include:

(1)　Clarification of the contamination source zone, plume area, and formation process

Based on a site survey or collection of data, determine the site's chief contaminant types and their distribution characteristics, analyze the hydrological and geological data including the site's strata, lithology, and water table, and establish a conceptual model of the contaminant's hydrological and geological characteristics. Identify the contamination source (including site contamination source and groundwater contamination source), contamination plume, and potentially contaminated area within the site and the peripheral area based on contaminant concentration or phase, and predict the contamination pathways.

Taking a site contaminated with non-aqueous phase liquids (NAPLs) as an example, the groundwater contamination source can be defined as the area containing NAPLs, the contamination plume can be defined as the area in which contaminants exist as a soluble phase [24], and the potentially contaminated area can be defined as the area of concern, i.e., the area which has not yet been contaminated, but which may be contaminated in the future.

(2)   Establishment of a list of classified remediation technologies

In accordance with the contaminated source zone, plume area, potentially contaminated area, and contamination pathways identified in Step 1, analyze each area's contamination control and prevention needs on the basis of remediation requirements and assess whether the migration of contaminants between different areas must be controlled. Determine the technologies that can be used for site remediation, including but not limited to source control, process blocking, contamination plume in situ remediation, and other assisting technologies. Generally, in the source zone, source control technologies can be selected; in the plume area, in-suit technologies can be selected; process blocking technologies are mainly used to block the contaminants migrating from the source zone to the plume area, and the plume area to the potentially contaminated area. Then, collect specific feasible contamination remediation techniques based on the foregoing technology types. It should also be noted that a certain specific remediation technique can be associated with multiple technologies.

(3)   Assessment of the classified remediation technologies

Use assessment methods such as decision-making support system [25] and multicriteria decision system [26] to assess and rank specific contaminated source control, process blocking, and contamination plume in situ remediation technologies in accordance with technological, environmental, and social indicators.

Technological indicators chiefly consider technological feasibility. Assessment is generally performed on the basis of a hydrological and geological conceptual model of site contamination and reflects technological principles and applicable conditions. The applicability of site contaminant characteristics, stratigraphy, lithology, contamination strata, and water table conditions is then assigned points, where full applicability is assigned 3 points, basic applicability 2 points, conditional applicability 1 point, and no applicability 0 points. A sample site condition assessment matrix is shown in Table 1. The specific number of points assigned for each condition will reflect the actual contaminated site and contaminant types.

Environmental indicators chiefly consider the impact of the remediation technologies on peripheral safety and the environment during the design, construction, operation, and maintenance processes. These indicators consist of safety, environmental friendliness, low resource/energy consumption, and sustainability. See Table 2 for the content and assessment standards for each indicator. High, moderate, and low are assigned respective scores of 3 points, 2 points, and 1 point.

Social indicators chiefly consider the feasibility of technology implementation. The indicators shown below have been adapted and revised from the guidelines in the contaminated site remediation technology screening matrix issued by the US Department of Defense Environmental Technology Transfer Committee (DOD ETTC) [27], and chiefly consist of the five aspects, including technological maturity, social acceptability, technology complexity, time frame, and capital investment. The content and scoring standards of each indicator are shown in Table 3.

Table 1. Meaning of technological indicators for site contamination control technologies and scoring standards.

Indicator	Meaning	High (3 Points)	Moderate (2 Points)	Low (1 Point)	Not Applicable (0 Point)
Contaminant characteristics	Whether the technology is able to remove the contaminant	Applicable to all contaminants	Applicable to most contaminants or applicable to all contaminants in conjunction with other technologies	Applicable to some contaminants in conjunction with other technologies	The method cannot remove the contaminants
Stratigraphy and lithology	Lithology can satisfy the technology's application requirements	Fully satisfies requirements	Basically satisfies requirements, but may extend remediation time or increase remediation cost	Requires other accompanying technologies to basically satisfy requirements	The method is not applicable to stratigraphic conditions
Target strata	Applicability of the technology to the target strata (vadose zone, water table fluctuation zone, aquifer, impermeable layers)	Applicable to target strata	Basically applicable to target strata	Applicable to the target strata in a minority of situations	The technology is not applicable to target strata
Other conditions	Conditions such as deployment location, water table, and depth of target strata	Fully satisfies the conditions	A condition is near the critical value	Two or more conditions are near the critical value	A condition is below the critical value

Table 2. Meaning of environmental indicators for site contamination control technologies and scoring standards.

Indicator	Meaning	High (3 Points)	Moderate (2 Points)	Low (1 Point)
Safety	Whether it will affect the safety of personnel at the site and peripheral areas during the project implementation process	No significant impact	Possibly some impact	Significant impact
Environmental friendliness	Whether it will worsen contamination in some areas or cause secondary contamination	Will not worsen contamination in some areas or cause secondary contamination	May worsen contamination in some areas or cause secondary contamination	Will worsen contamination in some areas or cause secondary contamination
Low resource/energy consumption	Cost expenditure and energy consumption	Relatively little energy consumption compared to other methods	Moderate energy consumption compared to other methods	High overall energy consumption compared to other methods
Sustainability	Whether it will affect land, soil, or groundwater ecology and reuse	Will not affect land, soil, or groundwater ecology or reuse	May affect land, soil, or groundwater ecology or reuse	Will affect land, soil, or groundwater ecology or reuse

Table 3. Meaning of social indicators for site contamination control technologies and scoring standards [27].

Indicator	Meaning		High (3 Points)	Moderate (2 Points)	Low (1 Point)
Technological maturity	Status and scale of the usable technology		Has been used in multiple site-scale applications as a part of remediation projects and there are complete and clear records and descriptions	Has been used in site-scale applications, but still requires further improvement and verification	Has not yet been used in site-scale applications, but has been used in pilot projects or laboratory test equipment experiments (such as sandbox or soil column experiments), and laboratory-scale testing (such as static micro-scale testing), and has been verified to have application potential
Technological complexity	Independence	Whether it can be independently used as a remediation technology	Independent technology (not complex in terms of material/number of technologies, can serve as a "routine" technology)	Relatively simple (around two technologies), easy to master, widely used	Complex (relatively complex technologies/materials, produces relatively more waste products)
	Implementability	Number of suppliers able to design, construct, and maintain the technology	More than 4 suppliers	2–4 suppliers	Fewer than 2 suppliers
	Ease of operation/maintenance	Intensity of operation and maintenance work	Low operation/ maintenance intensity	Average operation/ maintenance intensity	High operation/ maintenance intensity
Social acceptability	Whether the technology can be accepted by surrounding residents		Will be accepted by surrounding residents	May be accepted by surrounding residents	Not readily accepted by surrounding residents
Economic cost	Cost of design, construction, operation, and maintenance		Low overall cost when compared with other technologies with equivalent effectiveness	Equivalent cost when compared with other technologies with equivalent effectiveness	High overall cost when compared with other technologies with equivalent effectiveness
Time frame	Time necessary for the implementation	Soil	Fewer than 1 year	1–3 years	More than 3 years
		Groundwater	Fewer than 3 years	3–10 years	More than 10 years

Among the technological, environmental, and social indicators, technological indicators are the most fundamental; if a technology is completely infeasible (a score of 0), no further assessment will be conducted.

In the assessment standard tables, the technology's resource/energy consumption among environmental indicators and economic cost among social indicators should be "equivalent to that of other comparable technologies". As a consequence, the same remediation technology may have different final scores in different contaminated areas. For example, although multiple extraction technologies can be applied to contamination source and contamination plume areas when compared with such other technologies as natural attenuation and enhanced bioremediation, its economic cost is relatively high when applied to the contamination plume, giving it a lower score, but its economic cost is relatively low when applied to contamination source areas, giving it a higher score.

(4)   Preferred technology mix: Follow the principle of "optimize each item, seek mutual compatibility; if not compatible, score again".

The implementation process involves the selection of optimal technologies from source control, process blocking, and in situ remediation technologies, and the assessment of their mutual compatibility; if they are compatible, they constitute an optimal solution. A compatible technology mix implies that the implementation of these technologies in tandem will not affect the original scores of these technologies in Step 3; if the implementation of any one technology causes changes in environmental conditions, which makes it impossible to satisfy the conditions for use of another remediation technology, or if it causes the technological, environmental, or social suitability of another technology in the mix to decrease, then the implemented technology does not meet the requirement for compatibility. For instance, the implementation of certain chemical oxidation technologies may affect the activity of microbes within an aquifer [28], which will cause the suitability of natural attenuation and microbial augmentation methods to decline. This implies that chemical oxidation technology is incompatible with the natural attenuation and microbial augmentation methods.

If optimal technologies are not compatible, continued screening should be performed via the following two approaches: (1) sequentially substitute suboptimal technologies in a certain item and re-assess compatibility, until mutual compatibility has been achieved; then calculate the total score for the compatible technologies; (2) re-assess the score of each technology of other types in Step 3 after changing the implementation conditions of an incompatible technology, select the optimal mix, and calculate the total score. Compare the total scores in (1) and (2) and take the mix with the highest score as the optimal solution.

2.2. Site Situation

The case site consists of the gasoline and diesel tank area at a certain in-production refining enterprise on the mid-stream section of China's Yangtze River. The tank area rests on a 1–2 m layer of fill; an approximately 5 m layer of silty clay lies below the fill, with a layer of fine sand in some places; a roughly 20 m layer of fine sand lies below this layer; the fine sand layer is underlain by clay. The water table lies 2–3 m below the surface. The aquifer has a classic floodplain binary structure, with the upper portion consisting of a silty clay phreatic aquifer and the lower portion consisting of an underlying fine sand micro-confined aquifer, where the clay layer below the aquifer can be considered an impermeable layer. The main groundwater flow direction is from south to north. Contaminants at the site chiefly consist of petroleum hydrocarbons, and contaminants consisting of light non-aqueous phase liquids (LNAPLs) in the form of gasoline and diesel components are present near the water table. A conceptual hydrological and geological model of contamination at the site is shown in Figure 1.

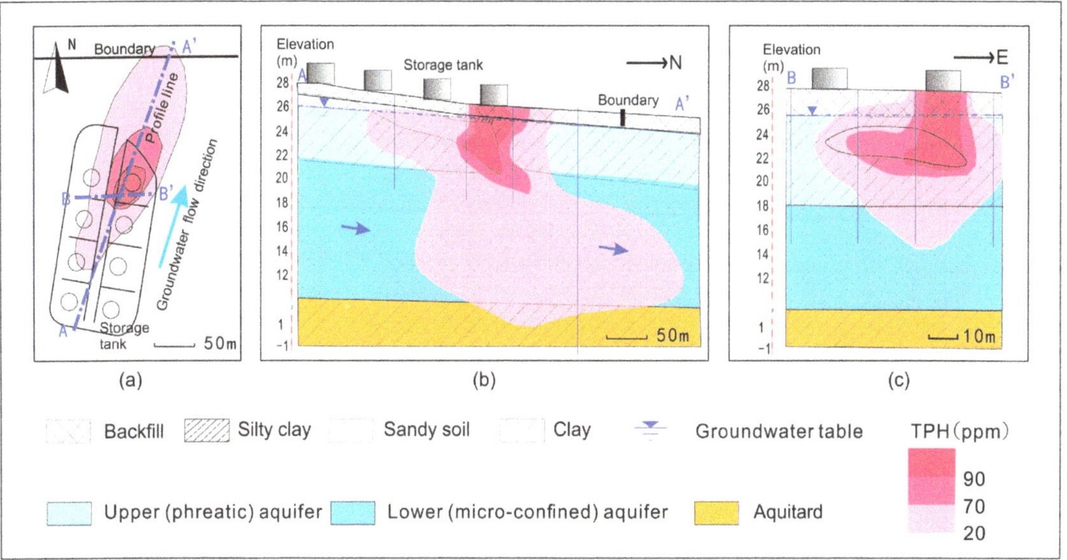

Figure 1. Conceptual hydrological and geological model of contamination at the case site ((**a**) is a plan, (**b**) is a cross-section along A-A′, and (**c**) is a cross-section along B-B′).

3. Results

3.1. Identification of Contamination Source, Contamination Plume, and Contamination Pathways

It can be seen from the foregoing conceptual model that the source of contamination consists of oil storage tanks, and the source of groundwater contamination consists of the non-aqueous phase petroleum hydrocarbons leaking into the ground from the oil storage tanks. The source area is chiefly located in the clay phreatic aquifer constituting the upper portion of the aquifer, and the contamination plume is largely distributed in the downstream micro-confined fine sand aquifer constituting the lower portion. The potentially contaminated area, namely, the acceptor, chiefly consists of soil and groundwater outside the boundaries of the plant and also includes downstream surface water bodies. The chief contaminant migration pathway consists of the lower micro-confined aquifer.

3.2. Establishment of a Classified Remediation Technology List

In accordance with the results of the assessment in Step 1, since a source of contamination and source of groundwater contamination existed at the site, it was necessary to eliminate the source in order to prevent the continued leakage of contamination; the pathway by which contamination would migrate into the potentially contaminated area could be obstructed by process blocking, and the lower micro-confined aquifer is the chief migration strata of contaminant migration which requires key point control; if the risk that the contamination plume would still contaminate the potentially contaminated area continue to exist after eliminating the source and blocking pathways, in situ remediation of the contamination plume could be performed. Since the contaminant chiefly migrated vertically in the vadose zone, and the migration of contaminant into the groundwater aquifer might cease after the site's contamination source was eliminated, so that contaminant would have almost no effect on the potentially contaminated area, contamination of the vadose zone could be ignored for the time being. Based on this understanding, remediation of the site could employ the following approaches: site source control, groundwater contamination source elimination, process blocking (of contaminant migration in the lower micro-confined aquifer), and in situ remediation of the contamination plume. We consequently gathered various usable technological methods (Table 4).

Table 4. List of petrochemical site contamination control technologies.

Type		Technology
Source control	Site contamination source	Institutional control [29], risk monitoring (of production installation leaks) [30], emergency response [31]
	Groundwater contamination source	Natural source zone depletion [32], soil gas phase extraction, multiphase extraction, pump and treat, enhanced soil solubilization, in situ thermal desorption, vitrification immobilization
Process blocking		Containment barrier, interception ditch, permeable reactive barrier, hydrodynamic control
In situ remediation of contamination plume		Monitoring of natural attenuation, soil gas phase extraction, multiphase extraction, pump and treat, enhanced soil solubilization, in situ thermal desorption, vitrification immobilization, bioventing, enhanced bioremediation, phytoremediation, chemical oxidation, aeration, well stripping
Assisting technologies		Pressure fracturing, directional wells

3.3. Assessment of the Classified Remediation Technologies

We used a decision-making support system (Figure 2) to assess contaminated source control, process blocking, and in situ remediation technologies in different strata in accordance with technological, environmental, and social indicators (Tables 1–3). Addressing this petrochemical site, we referred to the applicable conditions for remediation technologies in the US EPA's Underground Storage Tank Sites [17]; the resulting technology score matrix is shown in Table 5.

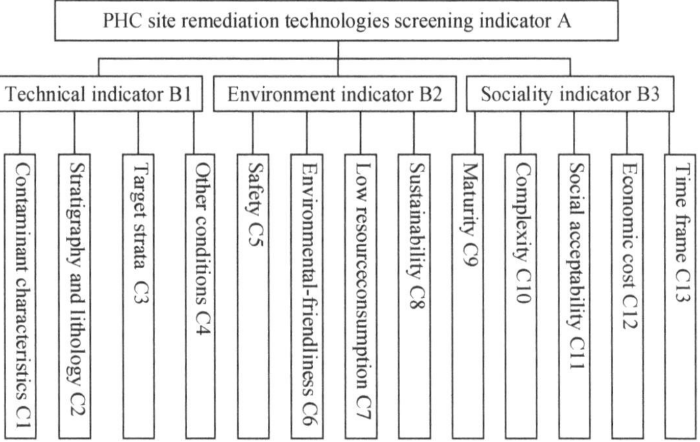

Figure 2. Technology screening decision-making support system model.

It is assumed that in this case that technological indicators, environmental indicators, and social indicators are equally important, and the secondary indicators and various links within the technological indicators, environmental indicators, and social indicators are also equally important. If the technological indicator score is 0, it is assumed that the technology is not suitable for this site, and no further assessment will be performed. A summary of assessment results and ranking of contamination source control—process blocking—in situ remediation technologies is shown in Table 6.

Table 5. Scoring standards for petrochemical site contamination control technologies.

Technology	Contaminant C1					Stratigraphy and Lithology C2								Target Strata C3				Other Conditions C4
	Gasoline	Kerosene	Diesel	Fuel Oil	Lubricating Oil	Gravel and Cobble	Coarse Sand	Medium Sand	Fine Sand	Silt	Silty Soil	Silty Clay	Clay	Vadose Zone	Water Table Fluctuation Zone	Aquifer	Impermeable Layer	
Institutional control	3	3	3	3	3	3	3	3	3	3	3	3	3	3	3	3	3	
Risk monitoring	3	3	3	3	3	3	3	3	3	3	3	3	3	3	3	3	3	
Emergency response	3	3	3	3	3	3	3	3	3	3	3	3	3	3	3	3	3	
Natural source zone depletion	3	3	3	3	3	3	3	3	3	3	3	3	3	3	3	3	3	
Soil gas phase extraction	3	2	2	2	0	2	2	3	3	2	2	1	0	3	1	0	0	Generally, requires water table > 3 m below surface
Multiphase extraction	3	3	3	3	2	3	3	3	3	3	3	3	2	2	3	3	0	
Pump and treat	3	3	3	3	2	3	3	3	3	3	2	1	0	0	2	3	0	
Enhanced soil solubilization	3	3	3	3	2	1	2	2	3	3	3	2	1	3	3	3	0	
In situ thermal desorption	3	3	3	3	2	3	3	3	3	3	2	2	0	3	3	3	3	
Vitrification immobilization	3	3	3	3	3	3	3	3	3	3	3	3	3	3	2	0	0	
Containment barrier	3	3	3	3	3	3	3	3	3	3	3	2	2	3	3	3	2	Generally, installed between the source and source zone, or between the source zone and plume zone
Interception ditch	3	3	3	3	3	3	3	3	3	3	3	3	3	3	3	3	3	Generally, installed between the source zone and the plume zone
Permeable reactive barrier	3	3	3	3	2	3	3	3	3	3	2	2	1	0	1	3	0	Generally, has a depth of <15 m, installed between the source zone and the plume zone

Table 5. *Cont.*

Technology	Contaminant C1					Stratigraphy and Lithology C2								Target Strata C3				Other Conditions C4
	Gasoline	Kerosene	Diesel	Fuel Oil	Lubricating Oil	Gravel and Cobble	Coarse Sand	Medium Sand	Fine Sand	Silt	Silty Soil	Silty Clay	Clay	Vadose Zone	Water Table Fluctuation Zone	Aquifer	Impermeable Layer	
Hydrodynamic control	3	3	3	3	3	3	3	3	3	3	2	1	0	0	2	3	0	Can be installed between the source and source zone, between the source zone and plume zone, and between the plume and receiving body
Monitoring of natural attenuation	3	3	3	3	3	3	3	3	3	3	3	3	3	3	3	3	3	
Bioventing	3	3	3	3	2	3	3	3	3	2	2	2	0	3	2	0	0	Generally, requires water table > 3 m below surface
Enhanced bioremediation	3	3	3	3	2	3	3	3	3	3	2	2	0	3	3	3	2	
Phytoremediation	3	3	3	3	2	3	3	3	3	3	3	3	2	3	3	3	3	Contamination within the scope of plant root systems
Chemical oxidation	3	3	3	3	3	3	3	3	3	3	2	2	0	3	3	3	2	
Aeration	3	2	2	2	0	3	3	3	3	3	2	2	0	2	2	3	2	
Well stripping	3	2	2	2	0	3	3	3	3	3	2	2	0	0	2	3	2	
Directional wells	3	3	3	3	3	3	3	3	3	3	3	3	2	3	3	3	2	
Pressure fracturing	3	3	3	3	3	0	0	0	2	2	2	3	3	3	3	3	3	

If fully feasible, 3 points; if feasible under most situations, 2 points; if feasible under some situations, 1 point; if completely infeasible, 0 points. C4 is scored according to this standard.

Table 6. Ranking of petrochemical site contamination remediation technologies.

Type		Upper (Phreatic) Aquifer			Lower (Micro-Confined) Aquifer		
		Remediation Technology	Score	Rank	Remediation Technology	Score	Rank
Source control	Site contamination source control	Institutional control	3.00	1	Institutional control	3.00	1
		Risk source monitoring	3.00	1	Risk source monitoring	3.00	1
		Emergency response	3.00	1	Emergency response	3.00	1
	Groundwater contamination source control	Multiphase extraction	2.68	1	Multiphase extraction	2.68	1
		Natural source zone depletion	2.52	2	Pump and treat	2.54	2
		In situ thermal desorption	2.51	3	Natural source zone depletion	2.52	3
		Enhanced soil solubilization	2.51	3	In situ thermal desorption	2.51	3
		Pump and treat	2.38	5	Enhanced soil solubilization	2.42	5
		Soil gas phase extraction	0.00	/ *	Soil gas phase extraction	0.00	/
		Vitrification immobilization	0.00	/	Vitrification immobilization	0.00	/
Process blocking		Interception ditch	2.59	1	Hydrodynamic control	2.80	1
		Hydrodynamic control	2.55	2	Interception ditch	0.00	/
		Permeable reactive barrier	2.53	3	Permeable reactive barrier	0.00	/
		Underground barrier	2.31	4	Underground barrier	0.00	/
In situ remediation of contamination plume		Monitoring of natural attenuation	2.82	1	Monitoring of natural attenuation	2.82	1
		Enhanced bioremediation	2.72	2	Enhanced bioremediation	2.63	2
		Phytoremediation	2.69	3	Multiphase extraction	2.61	3
		Multiphase extraction	2.61	4	Chemical oxidation	2.57	4
		Chemical oxidation	2.49	5	Pump and treat	2.54	5
		In situ thermal desorption	2.44	6	Aeration	2.51	6
		Enhanced soil solubilization	2.44	6	In situ thermal desorption	2.44	7
		Aeration	2.43	8	Enhanced soil solubilization	2.36	8
		Pump and treat	2.38	9	Well stripping	2.27	9
		Well stripping	2.19	10	Phytoremediation	0.00	/
		Soil gas phase extraction	0.00	/	Soil gas phase extraction	0.00	/
		Vitrification immobilization	0.00	/	Vitrification immobilization	0.00	/
		Bioventing	0.00	/	Bioventing	0.00	/
Assisting technologies		Directional wells	0.00	/	Directional wells	0.00	/
		Pressure fracturing	0.00	/	Pressure fracturing	0.00	/

* "/" indicates that the technology is not suitable for the site, and scored 0.00, and will not be ranked.

From the table, institutional control, risk source monitoring, and emergency response all gained the full mark. That is because the site is located in an in-production plan, and all the site contamination source control procedures should be carried out. Assisting technologies gained scores of zero, for the remediation work condition was good and no additional assisting technologies were needed.

For other types of technologies, many factors affected the final scores. For example, for process blocking, the upper (phreatic) aquifer, both interception ditch and hydrodynamic control gained higher scores. In environmental factors, they obtained the same score, 2.85 (See Table S1). In social factors, the interception ditch obtained a lower score (2.40) than hydrodynamic control (2.53), for the lower development status of the interception ditch (See Table S2). In technical factors, the interception ditch obtained a higher score (2.50) than the hydrodynamic control (2.25), for the hydrodynamic control is hardly used in clay stratum (See Table S3).

3.4. Preferred Technology Mix

The mix when selecting the technologies ranked first in Table 6 is as follows: site contamination source control technologies consist of institutional control, risk monitoring, and emergency response; for upper phreatic aquifer remediation, the groundwater contamination source control technology consists of multiphase extraction, process blocking technology consists of interception ditch, and in situ plume remediation technology consists of monitoring of natural attenuation; for lower micro-confined aquifer remediation, the groundwater contamination source control technology consists of multiphase extraction, process blocking technology consists of hydrodynamic control, and the in situ plume remediation technology consists of monitoring of natural attenuation. These technologies are mutually compatible, and, therefore, form an optimal technology mix for this site.

4. Discussion

4.1. The Screened Technologies for the Case Site

The screened technologies include institutional control, risk monitoring, emergency response, multiphase extraction, interception ditch, monitoring of natural attenuation, and hydrodynamic control. These technologies have been selected and widely used in managing various contaminated sites [33–36], but combining these technologies together to remediate one site is seldom performed.

Previous studies always aimed at a certain layer of the aquifer or considered several layers as a whole [37–39]. Therefore, there was always only one optimal technology for each location. In the present study site, two aquifer layers have been contaminated. According to the present developed method, the different layers should be treated discriminately, which resulted in the two optimal technologies at one location. For example, for blocking the transportation of contaminates, interception ditch and hydrodynamic control were both selected, but the former is for the upper (phreatic) aquifer and the latter is for the lower (micro-confined) aquifer. The interception ditch is hardly operated when the depth is greater than 15 m and, therefore, was not suitable for the lower aquifer, while in the shallow aquifer, it is easily developed, operated, and maintained [40–42], and, therefore, was selected for upper aquifer in the study. The other technology, hydrodynamic control, is not suitable for the clay medium [43] and, therefore, was not been selected for the upper aquifer, but is suitable for the sandy lower (micro-confined) aquifer. Separately treating for different characteristic stratum or contaminant concentrations is the precondition for precise remediation and can avoid ineffective remediation.

In practice, combining more than one technology to remediate contaminated soil and aquifer is not rare [44]. For example, Fenton processes and biotreatment were combined for soil remediation [45]; microbial combined methods, including microbe-biochar, microbe–nutrition, and microbe–plant technologies were used to remediate the petroleum-contaminated soils [46]; a plant–microbial combined bioremediation of polychlorinated naphthalene-contaminated soil was established using the intelligent integration of analytic hierarchy process and formula evaluation methods [47]; a sustainability assessment methodology for prioritizing the technologies of groundwater contamination remediation was established [48]. In comparison, combining multiple techniques to remedy a contaminated site is rare. A classic case is combining five technologies to remediate a PHC site, which was a former gasoline service station located in southwest Ohio [49]. At first, the remediation plan consisted of a combined groundwater pump and treat and soil vapor extraction. However, accompanied by the remediation processes, to meet the requirements, the plan had to be adjusted, and natural attenuation needed to be monitored, so surfactant-based soil washing and catalytic oxidation had been added empirically. The case represents most remediation projects that choosing the techniques is experiential and lacking systematicness. In our study, the contaminated soil and groundwater, as well as the receptors, were deemed as a system, and the techniques can be chosen systematically. Combining with the assessment methods and compatibility evaluation, an effective and reasonable approach is expected to be found.

4.2. Implications

Protecting the receptor is the terminal goal, and, therefore, the screened optimal technology mix does not necessarily include the site's ultimate remediation technologies; these technologies can be selected after reviewing them in the order of their scores during subsequent effectiveness modeling [50,51]. For instance, in this case study, the preferred in situ contamination plume remediation technology consists of natural attenuation monitoring; if it is found during modeling that natural attenuation monitoring will not yield the desired remediation effects during the period of validity, enhanced bioremediation can be assessed. If funding permits, different types of technology can be employed jointly; the technologies employed jointly can be selected in accordance with their ranking order. This principle can be also used to adjust the technologies during the course of a remediation project, especially when the initially selected technologies may not yield the desired results.

During the technology screening process, attention must be paid to the spatial distribution characteristics of contamination and contamination migration processes in order to clarify the spatial locations and relationships of the contamination source, contamination plume, and potential receiving bodies. In order to enhance the specificity of technology screening, each remediation technology must be assessed on the basis of different standards when used in different contaminated areas or for different remediation links.

Compatibility assessment must be performed for technologies in the preferred technology mix, and a dynamic score adjustment and re-screening plan must be made when technologies are incompatible. The goal is to enhance the compatibility of the remediation mix as a whole and realize an integrated remediation solution.

With regard to the establishment of an easy-to-implement technological and environmental indicators system, the technology score matrix can be used directly in the case of most petrochemical-contaminated sites; in the case of environmental indicators, the safety, environmental friendliness, resource/energy consumption, and sustainability of remediation technologies during the four stages of design, construction, operation, and maintenance can be assessed.

By providing ideas and principles concerning technology screening via classified assessment, the technology screening mix method presented in this paper offers powerful expandability. For example, during the classified assessment of remediation technologies in Step 3, various types of decision-making support systems can be employed, and other screening techniques can be used as well. These methods can be used in a flexible manner in light of the actual situation. The proposed procedure is not only suitable for hydrocarbon-contaminated unconsolidated sediment sites but also suitable for other contaminants and lithologies. When the contaminated sites or conditions vary, we simply need to follow the procedure and construct a new mix.

5. Conclusions

This study proposed a method of screening the optimal technologies for site remediation. The method is rooted in the "contamination source control—process blocking—in situ remediation" site remediation principles and the distribution characteristics of contaminants, as well as relationships among different areas including the source zone, plume area, and potentially contaminated area. The clear spatial relationship concept promises to enhance the effectiveness of contaminated site remediation. When using the procedure in an in-production petrochemical-contaminated site where LNAPLs are present, a technology mix, which includes institutional control, risk monitoring, emergency response, multiphase extraction, interception ditch, monitoring of natural attenuation, hydrodynamic control, as well as some alternative technologies, was found, aiming at different locations and strata. The technologies in the mix may not be the final ones and should be tested using computer simulation or pilot scale tests before carrying out the remediation work. The proposed method only gave us a technical framework and should be tested and enriched in future studies.

Supplementary Materials: The following supporting information can be downloaded at: https://www.mdpi.com/article/10.3390/app131911076/s1, Table S1: The scores in environmental factors; Table S2: The scores in social factors; Table S3: The scores in technical factors.

Author Contributions: Methodology, Z.Z., C.G., L.S. and Z.N.; Software, Y.X. and X.W.; Validation, Y.X. and X.W.; Investigation, S.Y., C.G., L.S. and Z.N.; Resources, Z.Z., C.G. and Z.N.; Data curation, Z.Z., C.G., L.S. and Z.N.; Writing—original draft, M.Z.; Writing—review & editing, M.Z., Z.Z., L.S. and Z.N. All authors have read and agreed to the published version of the manuscript.

Funding: This research was funded by the National Natural Science Foundation of China, grant number 42007171; the Sinopec Science Department Project, grant number 322082; the Hebei Natural Science Foundation, D2022504009; the CAGS Research Fund, grant number SK202207.

Institutional Review Board Statement: Not applicable.

Informed Consent Statement: Not applicable.

Data Availability Statement: The data presented in this study are available on request from the corresponding author.

Conflicts of Interest: Author Shuai Yang, Yan Xie and Xinzhe Wang were employed by the company SINOPEC Research Institute of Safety Engineering Co., Ltd. Author Zhifei Zhang was employed by the company Hebei Geological Environment Monitoring Institute. The remaining authors declare that the research was conducted in the absence of any commercial or financial relationships that could be construed as a potential conflict of interest. The authors declare that this study received funding from Sinopec Science Department Project. The funder was not involved in the study design, collection, analysis, interpretation of data, the writing of this article or the decision to submit it for publication.

References

1. U.S. Environmental Protection Agency. Petroleum Brownfields. Available online: https://www.epa.gov/ust/petroleum-brownfields (accessed on 12 July 2023).
2. U.S. Environmental Protection Agency. *Releases from Underground Storage Tanks*; U.S. Environmental Protection Agency: Washington, DC, USA, 2023.
3. Mitab, B.T.; Hamdoon, R.M.; Sayl, K.N. Assessing potential landfill sites using gis and remote sensing techniques: A case study in kirkuk, iraq. *Int. J. Des. Nat. Ecodynamics* **2023**, *18*, 643–652.
4. Bai, L.P.; Luo, Y.; Liu, L.; Zhou, Y.Y.; Yan, Z.G.; Li, F.S. Research on the screening method of soil remediation technology at contaminated sites and lts application. *Huanjing Kexue* **2015**, *36*, 4218–4224.
5. Khan, F.I.; Husain, T.; Hejazi, R. An overview and analysis of site remediation technologies. *J. Environ. Manag.* **2004**, *71*, 95–122.
6. Hao, G.; Yong, Q.; Yuan, G.X.; Wang, C.X. Research progress on the soil vapor extraction. *J. Groundw. Sci. Eng.* **2020**, *8*, 57.
7. Ministry of Ecology and Environment of the People's Republic of China. *List of Remediation Techniques for Contaminated Sites (First Batch)*; Ministry of Ecology and Environment of the People's Republic of China: Beijing, China, 2014.
8. Tian, J.; Huo, Z.; Ma, F.; Gao, X.; Wu, Y. Application and selection of remediation technology for ocps-contaminated sites by decision-making methods. *Int. J. Environ. Res. Public Health* **2019**, *16*, 1888.
9. Verta, M.; Kiviranta, H.; Salo, S.; Malve, O.; Korhonen, M.; Verkasalo, P.K.; Ruokojärvi, P.; Rossi, E.; Hanski, A.; Päätalo, K. A decision framework for possible remediation of contaminated sediments in the river kymijoki, finland. *Environ. Sci. Pollut. Res.* **2009**, *16*, 95–105.
10. Meng, X.S.; Chen, H.H.; He, Y.-P.; Zheng, C.Q.; Yue, X. Establishment of the environmental indexes in selection of remediation schemes: A case study of an abandoned coking site. *Environ. Eng.* **2021**, *39*, 7.
11. Luo, C.; Yi, A.; Zhang, Z.; Zhao, N.; Wang, Q.; Huang, Q. Remediation technology selection for pops contaminated sites. *Chin. J. Environ. Eng.* **2008**, *2*, 569–573.
12. Bai, L.; Luo, Y.; Shi, D.; Xie, X.; Liu, L.; Zhou, Y.; Yan, Z.; Li, F. Topsis-based screening method of soil remediation technology for contaminated sites and its application. *Soil Sediment Contam. Int. J.* **2015**, *24*, 386–397.
13. Chen, R.; Teng, Y.; Chen, H.; Yue, W.; Su, X.; Liu, Y.; Zhang, Q. A coupled optimization of groundwater remediation alternatives screening under health risk assessment: An application to a petroleum-contaminated site in a typical cold industrial region in northeastern china. *J. Hazard. Mater.* **2021**, *407*, 124796.
14. Mulligan, C.; Yong, R.; Gibbs, B. Remediation technologies for metal-contaminated soils and groundwater: An evaluation. *Eng. Geol.* **2001**, *60*, 193–207.
15. Bhandari, A.; Surampalli, R.; Champagne, P.; Tyagi, R.; Ong, S.K.; Lo, I. *Remediation Technologies for Soils and Groundwater*; ASCE: Preston, VA, USA, 2007.
16. Bao, Q.; Dong, J.; Dong, Z.; Yang, M. A review on ionizing radiation-based technologies for the remediation of contaminated groundwaters and soils. *Chem. Eng. J.* **2022**, *446*, 136964.
17. U.S. Environmental Protection Agency. *How to Evaluate Alternative Cleanup Technologies for Underground Storage Tank Sites*; U.S. Environmental Protection Agency: Washington, DC, USA, 2017.
18. Jiang, W.; Sheng, Y.; Wang, G.; Shi, Z.; Liu, F.; Zhang, J.; Chen, D. Cl, br, b, li, and noble gases isotopes to study the origin and evolution of deep groundwater in sedimentary basins: A review. *Environ. Chem. Lett.* **2022**, *20*, 1497–1528.

19. Engelmann, C.; Händel, F.; Binder, M.; Yadav, P.K.; Dietrich, P.; Liedl, R.; Walther, M. The fate of dnapl contaminants in non-consolidated subsurface systems–discussion on the relevance of effective source zone geometries for plume propagation. *J. Hazard. Mater.* **2019**, *375*, 233–240.

20. Critto, A.; Cantarella, L.; Carlon, C.; Giove, S.; Petruzzelli, G.; Marcomini, A. Decision support–oriented selection of remediation technologies to rehabilitate contaminated sites. *Integr. Environ. Assess. Manag.* **2006**, *2*, 273–285.

21. Luo, Y.M.; Teng, Y. Research progresses and prospects on soil pollution and remediation in china. *Acta Pedofil* **2020**, *57*, 1137–1142.

22. Li, P.; Cheng, X.; Zhou, W.; Luo, C.; Tan, F.; Ren, Z.; Zheng, L.; Zhu, X.; Wu, D. Application of sodium percarbonate activated with Fe(II) for mitigating ultrafiltration membrane fouling by natural organic matter in drinking water treatment. *J. Clean. Prod.* **2020**, *269*, 122228.

23. Deng, M.; Zhu, Y.; Duan, L.; Shen, J.; Feng, Y. Analysis on integrated remediation model of "phytoremediation coupled with agro-production" for heavy metal pollution in farmland soil. *J. Zhejiang Univ. (Agric. Life Sci.)* **2020**, *46*, 135–150.

24. Suk, H.; Zheng, K.-W.; Liao, Z.-Y.; Liang, C.-P.; Wang, S.-W.; Chen, J.-S. A new analytical model for transport of multiple contaminants considering remediation of both napl source and downgradient contaminant plume in groundwater. *Adv. Water Resour.* **2022**, *167*, 104290.

25. Mysiak, J.; Giupponi, C.; Rosato, P. Towards the development of a decision support system for water resource management. *Environ. Model. Softw.* **2005**, *20*, 203–214.

26. Mohammed, O.; Sayl, K. *A Gis-Based Multicriteria Decision for Groundwater Potential Zone in the West Desert of Iraq*; IOP Conference Series: Earth and Environmental Science, 2021; IOP Publishing: Bristol, UK, 2021; p. 012049.

27. Marks, P.J.; Wujcik, W.J.; Loncar, A.F. *Remediation Technologies Screening Matrix and Reference Guide*, 2nd ed.; DOD Environmental Technology Transfer Committee: Washington, DC, USA, 1994.

28. Sahl, J.; Munakata-Marr, J. The effects of in situ chemical oxidation on microbiological processes: A review. *Remediat. J. J. Environ. Cleanup Costs Technol. Tech.* **2006**, *16*, 57–70.

29. U.S. Environmental Protection Agency. *Institutional Controls: A Site Manager's Guide to Identifying, Evaluating and Selecting Institutional Controls at Superfund and Rcra Corrective Action Cleanups*; U.S. Environmental Protection Agency: Washington, DC, USA, 2000.

30. Ikwan, F.; Sanders, D.; Hassan, M. Safety evaluation of leak in a storage tank using fault tree analysis and risk matrix analysis. *J. Loss Prev. Process Ind.* **2021**, *73*, 104597.

31. Chenhao, J.; Yupeng, X. *Risk Analysis and Emergency Response to Marine Oil Spill Environmental Pollution*; IOP Conference Series: Earth and Environmental Science, 2021; IOP Publishing: Bristol, UK, 2021; p. 012070.

32. Kulkarni, P.R.; Walker, K.L.; Newell, C.J.; Askarani, K.K.; Li, Y.; McHugh, T.E. Natural source zone depletion (NSZD) insights from over 15 years of research and measurements: A multi-site study. *Water Res.* **2022**, *225*, 119170. [PubMed]

33. Zhang, N.; Yang, Y.; Wu, J.; Xu, C.; Ma, Y.; Zhang, Y.; Zhu, L. Efficient remediation of soils contaminated with petroleum hydrocarbons using sustainable plant-derived surfactants. *Environ. Pollut.* **2023**, *337*, 122566. [PubMed]

34. Oladeji, O.; Nigeria, O. An overview of aquifer pollution by petroleum hydrocarbons and possible remediation techniques. *Int. J. Appl. Sci. Eng. Res.* **2013**, *2*.

35. Lv, H.; Su, X.; Wang, Y.; Dai, Z.; Liu, M. Effectiveness and mechanism of natural attenuation at a petroleum-hydrocarbon contaminated site. *Chemosphere* **2018**, *206*, 293–301. [PubMed]

36. Guo, Y.; Wen, Z.; Zhang, C.; Jakada, H. Contamination and natural attenuation characteristics of petroleum hydrocarbons in a fractured karst aquifer, north china. *Environ. Sci. Pollut. Res.* **2020**, *27*, 22780–22794.

37. Ossai, I.C.; Ahmed, A.; Hassan, A.; Hamid, F.S. Remediation of soil and water contaminated with petroleum hydrocarbon: A review. *Environ. Technol. Innov.* **2020**, *17*, 100526.

38. Logeshwaran, P.; Megharaj, M.; Chadalavada, S.; Bowman, M.; Naidu, R. Petroleum hydrocarbons (PH) in groundwater aquifers: An overview of environmental fate, toxicity, microbial degradation and risk-based remediation approaches. *Environ. Technol. Innov.* **2018**, *10*, 175–193.

39. Zhang, S.; Su, X.; Lin, X.; Zhang, Y.; Zhang, Y. Experimental study on the multi-media prb reactor for the remediation of petroleum-contaminated groundwater. *Environ. Earth Sci.* **2015**, *73*, 5611–5618.

40. Zheng, C.; Wang, H.; Anderson, M.; Bradbury, K. Analysis of interceptor ditches for control of groundwater pollution. *J. Hydrol.* **1988**, *98*, 67–81.

41. Canter, L.W. *Ground Water Pollution Control*; CRC Press: Boca Raton, FL, USA, 2020.

42. Zhang, F.; Wang, W.; Liu, H. Application of interceptor trench in risk control and remediation of contaminated groundwater. *Environ. Prot. Sci.* **2020**, *46*, 167–172.

43. Xie, M.; Jarrett, B.A.; Da Silva-Cadoux, C.; Fetters, K.J.; Burton, G.A., Jr.; Gaillard, J.-F.O.; Packman, A.I. Coupled effects of hydrodynamics and biogeochemistry on zn mobility and speciation in highly contaminated sediments. *Environ. Sci. Technol.* **2015**, *49*, 5346–5353.

44. Aparicio, J.D.; Raimondo, E.E.; Saez, J.M.; Costa-Gutierrez, S.B.; Alvarez, A.; Benimeli, C.S.; Polti, M.A. The current approach to soil remediation: A review of physicochemical and biological technologies, and the potential of their strategic combination. *J. Environ. Chem. Eng.* **2022**, *10*, 107141.

45. Huang, D.; Hu, C.; Zeng, G.; Cheng, M.; Xu, P.; Gong, X.; Wang, R.; Xue, W. Combination of fenton processes and biotreatment for wastewater treatment and soil remediation. *Sci. Total Environ.* **2017**, *574*, 1599–1610.

46. Sui, X.; Wang, X.; Li, Y.; Ji, H. Remediation of petroleum-contaminated soils with microbial and microbial combined methods: Advances, mechanisms, and challenges. *Sustainability* **2021**, *13*, 9267.
47. Gu, W.; Li, X.; Li, Q.; Hou, Y.; Zheng, M.; Li, Y. Combined remediation of polychlorinated naphthalene-contaminated soil under multiple scenarios: An integrated method of genetic engineering and environmental remediation technology. *J. Hazard. Mater.* **2021**, *405*, 124139.
48. An, D.; Xi, B.; Wang, Y.; Xu, D.; Tang, J.; Dong, L.; Ren, J.; Pang, C. A sustainability assessment methodology for prioritizing the technologies of groundwater contamination remediation. *J. Clean. Prod.* **2016**, *112*, 4647–4656.
49. Hartsough, S.D. *Total Solutions: Combining Multiple Technologies to Create a Complete Remediation Package*; WEFTEC 2001, 2001; Water Environment Federation: Alexandria, VI, USA, 2001; pp. 439–461.
50. Sprocati, R.; Rolle, M. Integrating process-based reactive transport modeling and machine learning for electrokinetic remediation of contaminated groundwater. *Water Resour. Res.* **2021**, *57*, e2021WR029959.
51. Zhang, C.M.; Guo, X.N.; Richard, H.; James, D. Groundwater modelling to help diagnose contamination problems. *J. Groundw. Sci. Eng.* **2015**, *3*, 285. [CrossRef]

applied sciences

Article

A Conceptual Model for Depicting the Relationships between Toluene Degradation and Fe(III) Reduction with Different Fe(III) Phases as Terminal Electron Acceptors

He Di [1,2,3,4], Min Zhang [1,3,*], Zhuo Ning [1,3], Ze He [1,3], Changli Liu [1,3,4] and Jiajia Song [1]

[1] Institute of Hydrogeology and Environmental Geology, Chinese Academy of Geological Sciences, Shijiazhuang 050061, China; 3020190008@email.cugb.edu.cn (H.D.); ningzhuozhuo@163.com (Z.N.)
[2] School of Chinese Academy of Geological Sciences, China University of Geosciences (Beijing), Beijing 100086, China
[3] Key Laboratory of Groundwater Remediation of Hebei Province & China Geological Survey, Zhengding 050083, China
[4] Key Laboratory of Water Cycle and Ecological Geological Processes, Xiamen 361021, China
* Correspondence: zhangmin@mail.cgs.gov.cn; Tel.: +86-0311-67598605

Abstract: Iron reduction is one of the most crucial biogeochemical processes in groundwater for organic contaminants biodegradation, especially in the iron-rich aquifers. Previous research has posited that the reduction of iron and the biodegradation of organic substances occur synchronously, with their processes adhering to specific quantitative relationships. However, discrepancies between the observed values of iron reduction and organic compound degradation during the reaction and their theoretical counterparts have been noted. To find out the relationship between organic substance biodegradation and iron reduction, this study conducted batch experiments utilizing toluene as a typical organic compound and electron donor, with various iron minerals serving as electron acceptors. Results indicate that toluene degradation follows first-order kinetic equations with different degradation rate constants under different iron minerals, but the generation of the iron reduction product Fe(II) was not uniform. Based on these dynamic relationships, a conceptual model was developed, which categorizes the reactions into two phases: the transformation of toluene to an intermediate-state dominated phase and the mineralization of the intermediate-state dominated phase. This model revealed the relationships between toluene oxidation and Fe(II) formation in the toluene biodegradation through iron reduction. The coupling mechanism of toluene degradation and iron reduction was revealed, which is expected to improve our ability to accurately assess the attenuation of organic contaminants in groundwater.

Keywords: toluene; organic contaminants; iron reduction; biodegradation; groundwater; attenuation

Citation: Di, H.; Zhang, M.; Ning, Z.; He, Z.; Liu, C.; Song, J. A Conceptual Model for Depicting the Relationships between Toluene Degradation and Fe(III) Reduction with Different Fe(III) Phases as Terminal Electron Acceptors. *Appl. Sci.* **2024**, *14*, 5017. https://doi.org/10.3390/app14125017

Academic Editor: Francesco Liberato Cappiello

Received: 7 May 2024
Revised: 24 May 2024
Accepted: 31 May 2024
Published: 8 June 2024

1. Introduction

Organic contaminants, particularly benzene compounds, are widespread in the groundwater of sites contaminated with substances such as petroleum, posing significant adverse effects on both the environment and the economy [1–6]. Due to the prevalent anaerobic conditions in the groundwater of polluted sites [7,8], anaerobic biodegradation has become a crucial approach for addressing organic pollution in site remediation. Organic contaminants undergo biological degradation through processes such as nitrate reduction, iron reduction, manganese reduction, sulfate reduction, and methane production [9–11].

Iron, as one of the most abundant elements in the Earth's crust, constitutes approximately 6% of its surface. It is one of the most abundant electron acceptors, and the dissimilatory iron reduction process stands out as a primary mechanism in anaerobic biodegradation [12]. Research on the iron reduction of organic compounds, especially aromatic hydrocarbons, is particularly crucial [13].

Over the past three decades, researchers have extensively studied dissimilatory iron-reducing bacteria and the reduction of iron-containing minerals. These minerals, including ferrihydrite, goethite, amorphous iron minerals, magnetite, iron-bearing clay minerals, jarosite, and schwertmannite, among others [14–21], serve as electron acceptors in dissimilatory iron reduction reactions, while certain compounds serve as electron donors in these reactions.

$$CH_3COOH + 8Fe^{3+} + 4H_2O \rightarrow 8Fe^{2+} + 2HCO_3- + 10H^+ \; \Delta H = 651.98 \text{ kJ/mol} \tag{1}$$

$$C_6H_6 + 30Fe^{3+} + 18H_2O \rightarrow 30Fe^{2+} + 6HCO_3- + 36H^+ \; \Delta H = -475.92 \text{ kJ/mol} \tag{2}$$

$$C_7H_8 + 36Fe^{3+} + 21H_2O \rightarrow 36Fe^{2+} + 7HCO_3^- + 43H^+ \; \Delta H = -550.39 \text{ kJ/mol} \tag{3}$$

$$C_6H_5OH + 28Fe^{3+} + 17H_2O \rightarrow 28Fe^{2+} + 6HCO_3^- + 34H^+ \; \Delta H = -468.99 \text{ kJ/mol} \tag{4}$$

$$C_8H_{10} + 42Fe^{3+} + 24H_2O \rightarrow 42Fe^{2+} + 8HCO_3^- + 50H^+ \; \Delta H = -637.7 \text{ kJ/mol} \tag{5}$$

In actual pure cultivation experiments, iron-reducing bacteria utilize trivalent iron compounds as electron acceptors for the biological degradation of compounds such as acetate, benzene, toluene, phenol, ethylbenzene, and xylene isomers [22,23]. In these reactions, the electron donor–substrate and the electron acceptor–trivalent iron react in a specific ratio. However, the actual ratio often deviates from the theoretical ratio.

Taking the extensively studied organic pollutant, toluene, as an example, the theoretical molar ratio of toluene oxidation to the generated reduced iron in the reaction is 1/36. However, previous studies [24–27] have found that the ratio of these two components often exceeds the theoretical value of 1/36. The following two perspectives explain this phenomenon: firstly, it is suggested that some organic carbon may be bound to bacterial cells, ultimately promoting biomass growth [28]. On the other hand, it is possible that toluene has undergone partial loss over time. Whether it is due to the loss of synthesized biomass or the adsorption and volatilization loss of toluene over time, these processes are functions of time. However, as of now, no coupled analysis has been conducted regarding the temporal evolution between toluene oxidation and iron reduction.

In order to delve into the stoichiometric relationship of the microbial degradation reaction of toluene with iron under pure cultivation conditions, we designed and conducted an experiment. Four prevalent and extensively studied iron oxides commonly found in geological formations—amorphous iron hydroxide, hematite, magnetite, and goethite—were selected as electron acceptors. Toluene served as the electron donor in this experiment. Based on the above study, a dynamic model for characterizing the relationships between toluene oxidation and Fe(II) formation in toluene biodegradation through iron reduction was developed.

Understanding the coupling mechanism between toluene degradation and ferrous reduction is crucial for accurately quantifying the degradation of pollutants such as toluene, offering new insights and directions. If the degradation amount is underestimated according to the production of iron, the pollution source leakage estimate will be insufficient, and the assessment will be misleading, resulting in insufficient risk understanding.

2. Materials and Methods

2.1. Preparation of Fe(III) Oxide

Amorphous iron hydroxide was synthesized through gradually neutralizing a solution containing 0.4 mol/L $FeCl_3$ with 10 mol/L NaOH until the pH reached 7.0. The metal oxide suspensions underwent three centrifugation and washing cycles before being resuspended in distilled water. Subsequently, the metal oxides were suspended in a basal medium [29]. This process resulted in the formation of brick-red mineral particles with no apparent crystal structure and an amorphous morphology, characterized by a high surface area.

Hematite and magnetite were procured from Guoming Mineral Resources. In order to closely mimic natural conditions, naturally occurring ores were carefully selected. To achieve the highest surface area, the minerals were ground to particles of 350 mesh size. The chosen hematite was a reddish-brown powder with a primary composition of Fe_2O_3, while magnetite was pure black with a primary composition of Fe_3O_4.

Goethite was prepared by dissolving unhydrolyzed $Fe(NO_3)_3 \cdot 9H_2O$ in distilled water to create a fresh 1 mol/L $Fe(NO_3)_3$ solution. Taking 100 mL of the 1 mol/L $Fe(NO_3)_3$ solution in a 2 L polyethylene bottle, 180 mL of the 5 mol/L KOH solution was rapidly added via stirring, resulting in the formation of brownish-red hydrated iron ore precipitation. The suspension was promptly diluted to 2 L with distilled water and incubated at 70 °C for 60 h in a sealed polyethylene bottle [30]. This process yielded fibrous and needle-like crystals of yellow mineral powder.

2.2. Inoculum Source

The site is located in a petroleum and chemical industrial area in Northwest China, near the gas condensate storage tanks at a purification plant. The water table depth ranges from 3.2 to 4.7 m below the ground surface. The vadose and saturated zones are primarily composed of fine sands. Groundwater flows generally from north to south. Due to gas condensate releases from the storage tanks and groundwater flow, contaminants spread nearly across the site. These contaminants are largely light-end petroleum hydrocarbons, such as benzenes and other small volatile hydrocarbons (C6–C9).

In areas where the benzene and iron content jointly exceed contamination standards, soil samples were taken for bacterial inoculation. Under anaerobic conditions, 100 g of soil was concentrated in a 600 mL glass bottle with a medium containing 250 mg NH_4Cl per liter, 600 mg NaH_2PO4, 100 mg KCl, 0.1 mmol toluene, 3.6 mmol amorphous ferric hydroxide, 10 mL mineral solution, and 10 mL vitamin solution [31]. The contents of trace elements and vitamin solutions are listed in Tables 1 and 2. After 20 days of dark incubation at 25 °C, 60 mL of the culture medium was transferred to a new sterile bottle with 500 mL of fresh medium. This process was repeated 10 times over 200 days, enriching the bacterial strains with iron-reducing capabilities.

Table 1. The components of the vitamin solution.

Components	Concentration (mg/L)
Biotin	2.0
Folic acid	2.0
Pyridoxine HCL	10.0
Riboflavin	5.0
Thiamine	5.0
Nicotinic acid	5.0
Pantothenic acid	5.0
B-12	0.1
p-Aminobenzoic acid	5.0
Thioctic acid	5.0

Table 2. The components of the mineral solution.

Components	Concentration (g/L)
Trisodium nitrilotriacetic acid	1.5
$MgSO_4.$	3
$MnSO_4 \cdot H_2O$	0.5
NaCl	1
$FeSO_4 \cdot 7H_2O$	0.1
$CaCl_2 \cdot 2H_2O$	0.1
$CoCl_2 \cdot 6H_2O$	0.1
$ZnCl_2$	0.13
$CuSO_4 \cdot 5H_2O$	0.01
$AlK(SO_4)_2 \cdot 12H_2O$	0.01
H_3BO_3	0.01
Na_2MoO_4	0.025
$NiCl_2 \cdot 6H_2O$	0.024
$Na_2WO_4 \cdot 2H_2O$	0.025

2.3. Anaerobic Incubation with Different Iron Oxides

The experiments followed the same procedure as the enrichment process, using the same medium but replacing the amorphous ferric hydroxide with four different types of iron minerals, as shown in Figure 1. Negative control groups were constructed in the same manner, with the bacterial solution being replaced by purified water. Each type of iron mineral had three parallel experimental groups and three parallel control groups. On days 1, 3, 5, 7, 12, 18, 22, 28, 32, and 40, water samples were collected from each microcosm for analysis. The concentrations of toluene, total ferrous, and total iron in the solution were measured. The sampling process was conducted under strict anaerobic conditions.

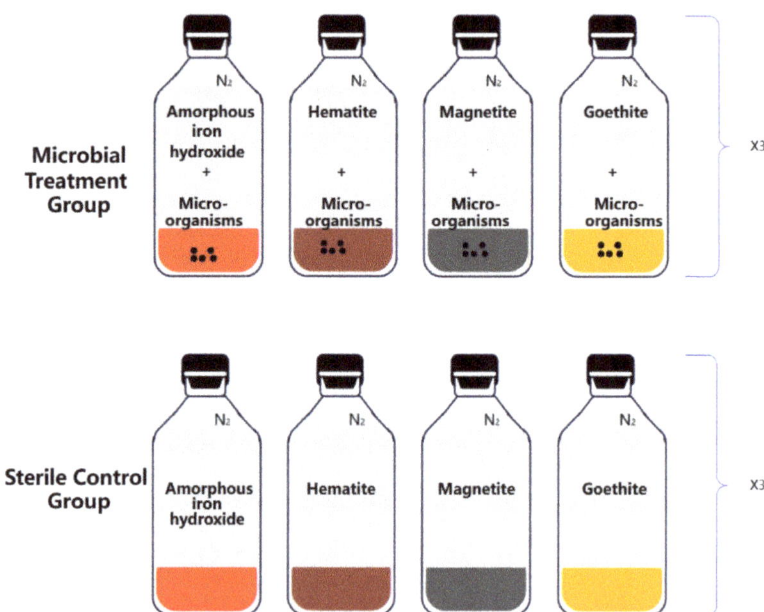

Figure 1. Microbial experiment design.

2.4. Analytical Method

Toluene Analysis Method: Toluene samples were analyzed using headspace-based gas chromatography (Nexis GC-2030, Shimadzu, Kyoto, Japan) with an automated headspace sampler (HS-10, Shimadzu, Kyoto, Japan). The headspace operating conditions were as follows: low shaking of the tested sample solution at 35 °C for 2 min, a GC cycle time of 25 min, and a vial pressurization time of 0.1 min. The gas chromatograph was equipped with a capillary column (HP-5, Shimadzu, Kyoto, Japan) and a flame ionization detector (FID). The injector, detector, and column temperatures were held at 150, 200, and 100 °C, respectively. Air and hydrogen served as fuel gases for the FID. Nitrogen served as a carrier gas, and the flow rate was 40.1 mL/min.

Ferrous Iron Analysis Method: The ortho-phenanthroline colorimetric method is utilized [32]. Before conducting the ferrous iron test, a uniformly mixed culture suspension of 0.5 milliliters is injected into 4.3 milliliters of 0.5 mol/L hydrochloric acid and subjected to extraction at 30 °C for 24 h. Subsequently, 2.4 milliliters of the extraction solution are taken, and 0.32 milliliters of 6 mol/L hydrochloric acid, 0.32 milliliters of 2 mmol/L ammonium fluoride, 0.32 milliliters of 1% o-phenanthroline, 0.48 milliliters of ammonium acetate buffer, and 0.16 milliliters of distilled water are added. The absorbance of the solution is measured at 510 nm using a UV spectrophotometer (UV2550). Then, the total dissolved ferrous iron concentration is calculated according to the standard curve formula, $c\left(Fe^{2+}\right) = c \times n$, where c represents the ferrous iron concentration in the water sample as measured by the spectrophotometer in mg/L, and n is the dilution factor. As the dissolution of iron minerals gradually decreases during the reaction, the experiment employs the method of multiplying the proportion of dissolved ferrous iron values to total iron by the amount of total iron minerals added in order to determine the total ferrous iron generated during the reaction.

2.5. Data Statistical Processing

The toluene and ferrous iron concentration data were presented as mean $\pm$ S.D. Duncan's multiple range test, conducted using SPSS Statistics 22.0, was employed to assess the differences among the three systems, with a significance level set at $p = 0.05$.

To investigate the degradation kinetics of toluene in various systems with different iron oxides as electron acceptors, we assumed first-order kinetics for the reactions. To determine the degradation rate constants, we conducted a series of experiments where the concentration of toluene was monitored over time. The concentration data were then processed through calculating the natural logarithm of the ratio of the remaining concentration at time t (C_t) to the initial concentration (C_0). This relationship is represented as $\ln(C_t/C_0)$.

Subsequently, we plotted $\ln(C_t/C_0)$ against time (t) for each system. According to the first-order reaction kinetics equation, $\ln(C_t/C_0) = -kt.$, where k is the degradation rate constant. The slope of each line corresponds to the negative of the first-order degradation rate constant (k). Through determining these slopes, we were able to calculate the degradation rate constants for toluene in each system with different iron oxides.

The mean absolute percentage error (MAPE) is a metric used for assessing the accuracy of a predictive model. The calculation process is as follows:

For each observed value y_i and its corresponding predicted value $\hat{y}_i$, calculate the absolute percentage error (APE):

$$\text{APE}_I = \left| \frac{y_i - \hat{y}_i}{y_i} \right| \times 100 \tag{6}$$

Compute the average of all APEs, resulting in the mean absolute percentage error (MAPE):

$$\text{MAPE} = \frac{1}{n} \sum_{i=1}^{n} \text{APE}_I \tag{7}$$

Here, $n = 3$, represents the number of observed values.

3. Results

3.1. The Relationship of Toluene over Time under Different Iron Mineral Conditions

Toluene degradation follows first-order kinetics, and the corresponding first-order reaction kinetics equations for toluene under four different electron acceptors are shown in Table 3. The reaction rate constant graphs for the microbial treatment and sterile control groups in various electron acceptor systems are shown in Figure 2.

Table 3. Four Fe(III) oxides and their kinetic equations and parameters for toluene degradation.

Group	First-Order Kinetic Equation	R^2	Reaction Rate Constant
Amorphous iron hydroxide group	$C_{AT} = C_0 e^{-0.128t}$	0.98	0.128
Hematite group	$C_{HT} = C_0 e^{-0.027t}$	0.97	0.027
Magnetite group	$C_{MT} = C_0 e^{-0.025t}$	0.99	0.025
Goethite group	$C_{GT} = C_0 e^{-0.073t}$	0.99	0.073

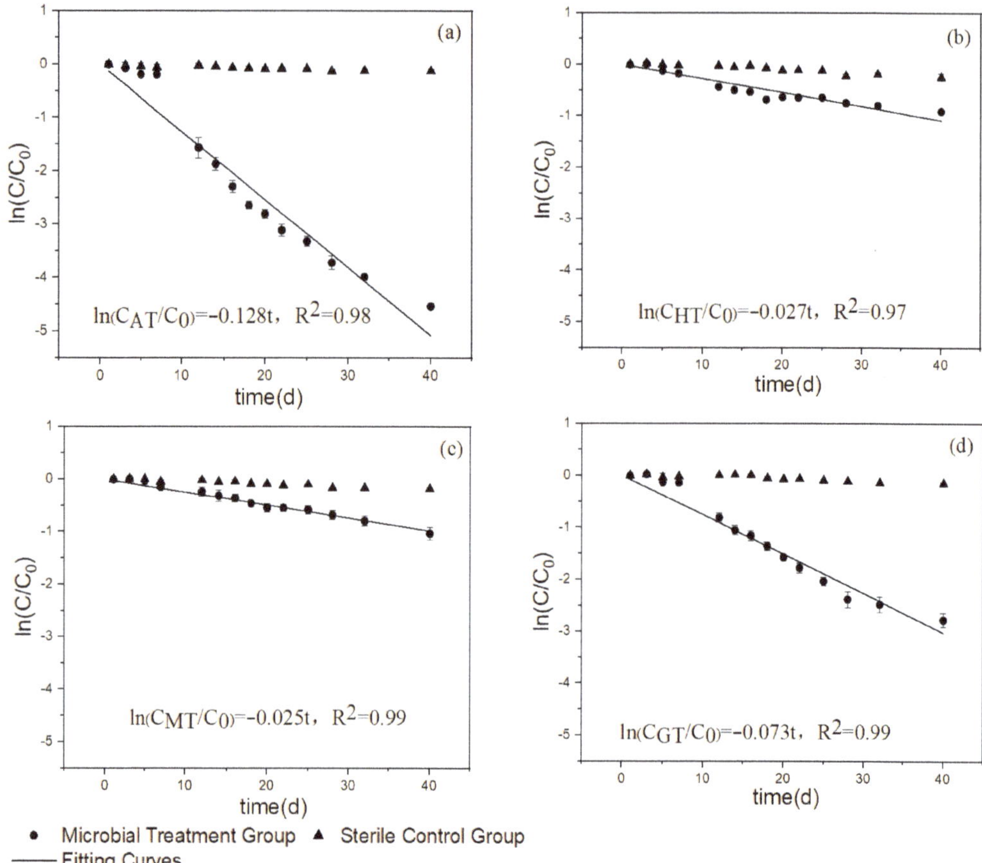

Figure 2. Toluene degradation kinetics fitting curves under the influence of four Fe(III) oxides. (a–d) represent the first-—order reaction kinetics fitting curves for toluene degradation in the presence of four iron minerals (Fe(III) initial concentration of 3.6 mmol/L) acting as electron acceptors, facilitated by iron-reducing bacteria. The initial concentration of toluene added to the culture bottle is denoted as C_0 (0.1 mmol/L). C_{AT} represents the toluene concentration at time t in the amorphous iron hydroxide group, C_{HT} represents the toluene concentration at time t in the hematite group, C_{MT} represents the toluene concentration at time t in the magnetite group, and C_{GT} represents the toluene concentration at time t in the goethite group.

3.2. The Relationship of Ferrous Products over Time under Different Iron Mineral Conditions

In the case of the four electron acceptors, assuming that all electrons from toluene mineralization are used to reduce Fe(III), iron is finally recovered in the form of Fe(II). The theoretical formula for the real-time concentration of Fe(II) is obtained through converting the fitted toluene degradation kinetics formula. According to reaction Equation (3), The theoretical formula for the Fe(II) concentration is obtained using $C_{Fe(II)} = 36C_0(1 − e^{−kt})$, where k is the degradation rate constant of toluene. The theoretical and actual Fe(II) concentration graphs in the four different electron acceptors are shown in Figure 3.

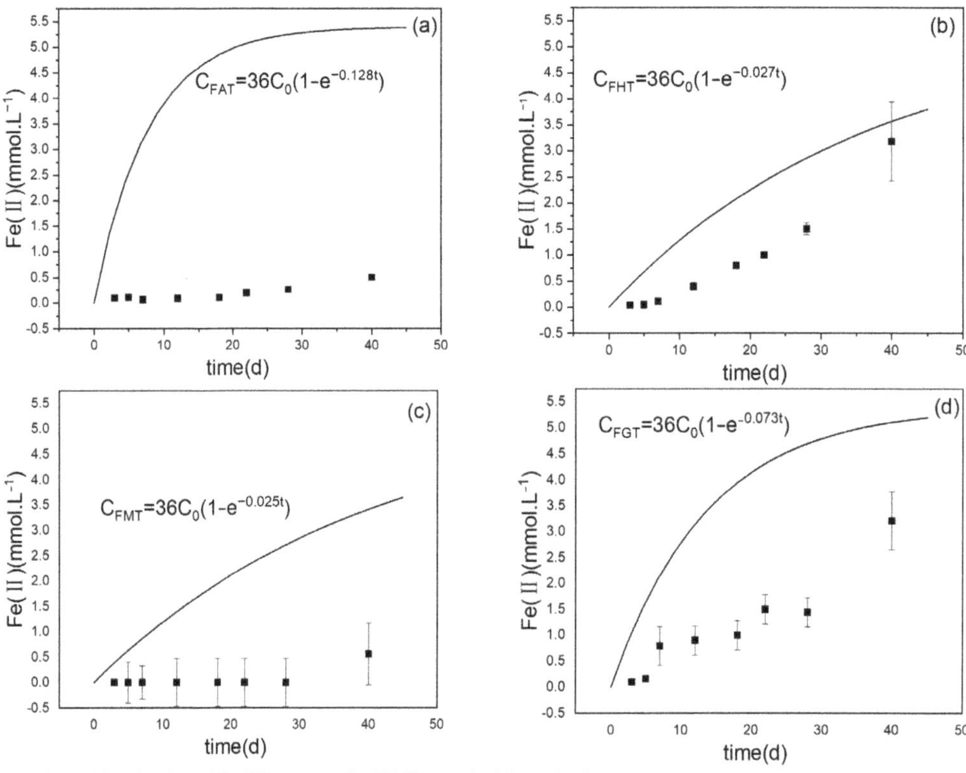

Figure 3. A comparison between the theoretical Fe(II) concentration curves and the actual Fe(II) concentration curves. (**a–d**), respectively, represent the four iron minerals, namely, amorphous iron hydroxide, hematite, magnetite, and goethite, respectively (initial Fe(III) concentration of 3.6 mmol/L) as electron acceptors. The initial concentration of toluene (C_0) was 0.1 mmol/L. The theoretical Fe(II) concentrations (C_{FAT}, C_{FHT}, C_{FMT}, and C_{FGT}) correspond to the mineralization of toluene to Fe(II) at time t for amorphous iron hydroxide, hematite, magnetite, and goethite, respectively.

To elucidate the relationship between theoretical predictions and experimental observations of ferrous iron (Fe(II)) generation from toluene oxidation, and to verify the accuracy and reliability of theoretical models in predicting Fe(II) generation under different conditions, we performed MAPE (mean absolute percentage error) analysis for different groups. The MAPE for four different iron oxides (amorphous iron hydroxide, hematite group, magnetite group, goethite group) are 2864%, 457%, 510%, and 381%, respectively.

3.3. The Relationship between Toluene and Ferrous Products under Different Iron Mineral Conditions

In order to investigate the proportional relationship between the consumption of toluene and the generation of ferrous iron during the reaction process, we measured the toluene consumption (vertical axis) and ferrous iron generation (horizontal axis) at specific time points (3rd, 5th, 7th, 12th, 18th, 22nd, 28th, and 40th days) in the biodegradation of toluene. These findings are visually depicted in the two-dimensional graph shown in Figure 4.

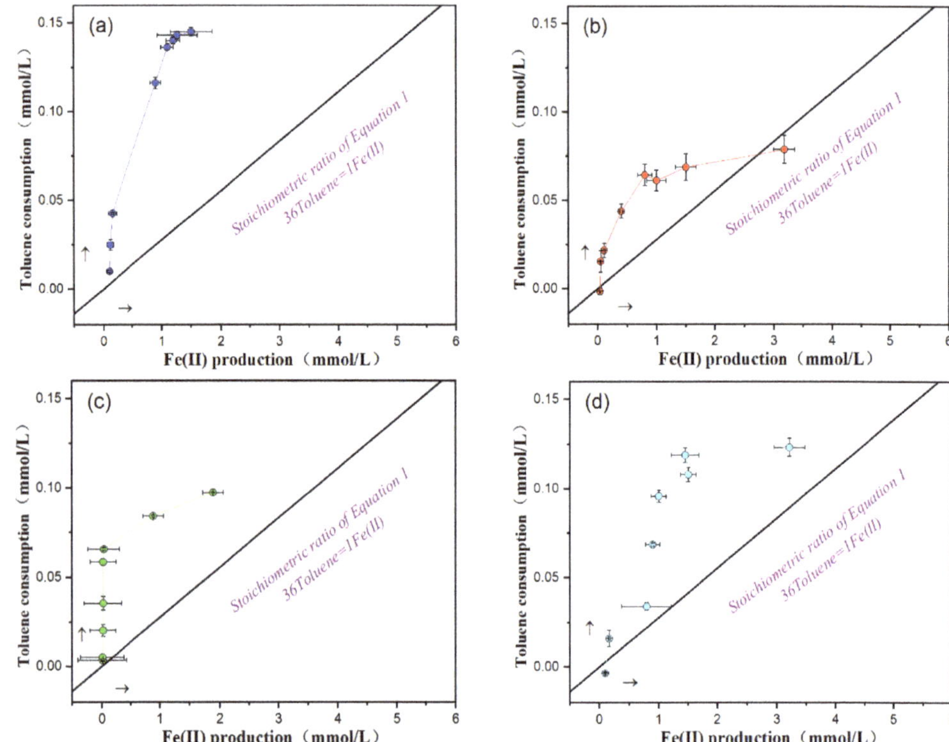

Figure 4. The mutual relationship between the degradation of toluene and iron reduction is depicted in the figure, where (**a–d**), respectively, represent the four iron minerals, namely, amorphous iron hydroxide, hematite, magnetite, and goethite, with an initial Fe(III) concentration of 3.6 mmol/L. The figure illustrates the real-time consumption of toluene (initial concentration of 0.1 mmol/L) and the real-time generation of ferrous ions under the influence of iron-reducing bacteria. For comparison, a reference line y = 1/36x is plotted, representing the theoretical ratio of 1/36 between toluene consumption and ferrous ion generation. This line serves as a reference, indicating that, in the theoretical reaction, the complete oxidation of 1 mmol of toluene would theoretically yield 36 mmol of ferrous ions. The direction of the arrow indicates the direction of the reaction.

According to the graph, the vertical axis, from bottom to top, represents the increasing amount of toluene consumption, while the horizontal axis, from left to right, represents the gradually increasing ferrous iron production, reflecting the progression of the reaction over time.

Hematite group: At the beginning of the reaction, the toluene consumption increases slowly from 0 mmol/L, and the ferrous iron production also starts at 0 mmol/L and increases gradually. If the initial toluene consumption and ferrous iron production follow the theoretical ratio of 1/36, the curve would tend toward the direction of the theoretical curve. However, as observed in the graph, from the first point to the second point, the curve deviates from the theoretical curve toward the Y-axis, indicating higher toluene consumption and relatively insufficient ferrous iron production. From the first point to the sixth point, the slope of the curve decreases, gradually approaching the slope of the theoretical curve, indicating a gradual reduction in toluene consumption and a relative increase in ferrous iron production from 0 to 22 days, reaching a stage where toluene and iron approach the theoretical 1/36 ratio. From the sixth point to the seventh point, a temporary equilibrium is reached, indicating that at some point between 22 and 28 days, toluene consumption and iron production are in a theoretical 1/36 ratio. From the seventh point to the eighth point, the slope of the curve is less than the slope of the theoretical curve, indicating that from 28 to 40 days, the ratio of toluene to iron is less than the theoretical 1/36 ratio. Similar analyses were conducted for the fitting relationship between toluene consumption and ferrous iron production in the hematite, magnetite, and goethite groups, as shown in Table 4.

Table 4. Phases of the coupled relationship between toluene consumption and ferrous iron reduction for four Fe(III) oxides.

Group	Time Periods with Slopes Greater than 1/36	Time Periods with Slopes Approximately Equal to 1/36	Time Periods with Slopes Less than 1/36
Amorphous iron hydroxide Group	0–22	22–28	28–40
Hematite Group	0–12	12–28	28–40
Magnetite Group	0–22	22–28	28–40
Goethite Needle Group	0–18	18–22	22–40

4. Discussion

4.1. Toluene Degradation under Different Iron Mineral Conditions

Under anaerobic conditions mediated by dissimilatory iron-reducing microorganisms, the degradation reaction between toluene and various iron minerals undergoes a series of slow and sequential steps. Initially, toluene enters the microbial cells from the aqueous solution, followed by a series of intracellular reactions leading to gradual degradation within the microbial cells. Ultimately, electrons are transferred from the microbial cells to the iron minerals attached to the dissimilatory iron-reducing microorganisms. Throughout this process, Fe(III) in the iron minerals undergoes reduction [33].

The degradation of toluene follows first-order kinetics, consistent with previous studies by researchers [34,35]. During the preliminary degradation of toluene under four types of iron minerals, the first-order kinetic reaction rate constants are, respectively, as follows: amorphous iron hydroxide group, 0.128 d^{-1}; hematite group, 0.027 d^{-1}; magnetite group, 0.025 d^{-1}; and goethite group, 0.073 d^{-1}. The reaction rate constant depends on the temperature, activation energy, and concentration, and the activation energy depends on the structure (mineral type) and its interaction with the toluene, in this case. In the experiment, amorphous iron hydroxide appears as aggregates of amorphous irregularly shaped particles, while hematite exhibits long needle-like crystals. Hematite primarily exists as granular crystals, and magnetite exists in the form of inverse spinel. Amorphous iron hydroxide with a lower crystallinity and a slightly higher crystallinity of hematite demonstrate higher reaction rate constants. In contrast, hematite with a higher crystallinity exhibits lower reaction rate constants. We observed that the crystallinity of iron minerals affects their interaction with reactants such as toluene. This confirms the existence of a certain relationship between the crystallinity and reaction rate constants, although this relationship may be influenced by various factors [36].

Crystals with a higher crystallinity often have a more complete and compact crystal structure, resulting in a smaller surface area and lower exposure of active sites. In contrast, crystals with a lower crystallinity may have more grain boundaries and defects, leading to an increased surface area and more exposed active sites, thereby promoting the adsorption and conversion of reactants and increasing the reaction rate constants.

Internally, crystals with a higher crystallinity typically have a more complete and ordered structure, limiting the diffusion of reactants within the crystal and affecting the rate of reaction. In contrast, crystals with a lower crystallinity may have more grain boundaries and defects, allowing reactants to diffuse more easily within the crystal, thereby increasing the reaction rate constants. Crystals with a higher crystallinity often have higher bond energy and higher activation energy, increasing the energy required for the reaction to occur and consequently reducing the reaction rate constants. In contrast, crystals with a lower crystallinity may have lower bond energy and lower activation energy, making the reaction easier to occur and thus increasing the reaction rate constants.

It should be noted that the relationship between the crystallinity and reaction rate constants is influenced by specific reaction systems, reaction conditions, and crystal structures, and is not an absolute rule. The order of rate constants is amorphous iron hydroxide > goethite > hematite > magnetite. In this study, we focused on exploring the effect of the reactant structure on reaction rate constants, particularly for iron minerals with different crystallinities, providing valuable insights into the reaction kinetics between iron minerals and organic pollutants.

During the degradation process of toluene, the theoretical amount of iron reduction is much higher than the actual reduction amount. This suggests that the complete oxidation of toluene and the reduction of iron do not occur synchronously. This discrepancy may be due to the degradation process of toluene not being a one-step mineralization, but rather involving intermediate states, leading to the incomplete reduction of ferrous iron and poor model fitting [28,37].

4.2. The Conceptual Model of the Interrelationship between Toluene Degradation and Iron Reduction Processes

Observation reveals a deviation between the experimental curve and the theoretical straight line. Initially, the slope of the experimental curve exceeds that of the theoretical curve. As the reaction progresses, the slope gradually decreases, eventually becoming less than that of the theoretical curve. This suggests the occurrence of electron-delayed ferrous iron reduction during the reaction. After analysis, three potential reasons have been identified. Firstly, during the extracellular anaerobic respiration of microorganisms with solid electron acceptors, electrons are typically transported by electron shuttles, leading to the storage of some electrons on these shuttles. This storage mechanism may cause a deviation in the stoichiometric ratio between toluene and iron during the reaction process [38]. Secondly, the synthetic metabolism and assimilation by microorganisms may result in the partial assimilation of toluene as a carbon source, promoting microbial biomass growth. With an increase in the microbial population, the amount of assimilated toluene also increases, leading to the accumulation of un-reduced toluene inside microbial cells. This accumulation prevents the detection of toluene, resulting in delayed electron transfer, which can be released upon microbial decay, achieving delayed electron release [39]. Thirdly, toluene may be completely degraded into multiple intermediate products within a short period of time. These intermediate products may accumulate during the degradation pathway, causing delayed electron transfer. For example, under the catalysis of benzoyl-CoA synthase, fumarate is added to the methyl group of toluene to form benzoyl succinate. Subsequently, a series of modified β-oxidation reactions occur after the addition reaction, converting phenylsuccinic acid into benzoyl-CoA, which is an intermediate product of the anaerobic degradation of aromatic compounds. Then, benzoyl-CoA is reduced by benzoyl-CoA reductase to form non-aromatic products, with the final products being CO_2 and H_2O [40]. Only one suspected intermediate product was found in the gas chromatography analysis. It is speculated that many intermediates have their specificity, and a specific treatment is required for each intermediate product to achieve the identification of the product. The existence cycle of each intermediate is short, and the identification and characterization of intermediate products can be specialized in the later stage.

Under the influence of microorganisms, all intermediate states can undergo mineralization. For instance, electron shuttles can transfer all their electrons to an electron acceptor, transitioning from a reduced state to an oxidized state, thereby achieving the reconversion of the electron shuttle storage state [41]. Similarly, biomass storage states can be consumed through microbial growth and decay, resulting in the depletion of the electron storage state [39]. Intermediate products, such as benzyl-succinate, benzyl-succinyl-CoA, E-phenylitaconyl-CoA, (hydroxymethylphenyl)-succinyl-CoA, benzoylsuccinyl-CoA, benzoyl-CoA, and succinyl-CoA, are present throughout the reaction process. The electron storage states of these intermediates can be gradually consumed over time [42–44]. Consequently, the degradation capability of intermediate states, whether in the form of an electron shuttle storage state, biomass storage state, or intermediate product storage state, can release electrons for utilization via the iron electron acceptor, leading to the reduction of trivalent iron to divalent iron.

In summary, there exists an intermediate state of electron storage. Here, it is referred to as the intermediary state (electron shuttle storage state, biomass storage state, intermediate product storage state), as shown in Figure 5. The first-order degradation of toluene mentioned earlier might be toluene converting into an intermediate state. Due to the presence of this intermediate state, this reaction is more complex than a single-step reaction. Here, we discuss a relatively simpler continuous reaction process divided into two stages. Let us assume that toluene undergoes a series of processes, first generating the intermediate state B, which then further transforms into the final product C. The entire series of reactions can be represented as follows:

$$\text{toluene} \xrightarrow{v_1} B \xrightarrow{v_2} C \tag{8}$$

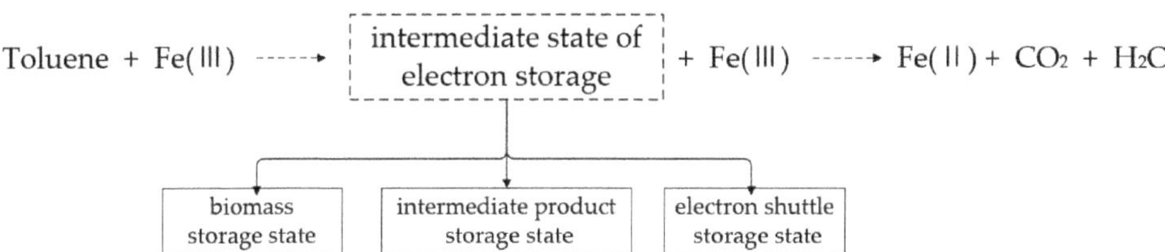

Figure 5. Schematic chemical proposal for the implied mechanism during the toluene degradation.

In the first stage, the conversion of toluene into an intermediate state dominates the reaction process. In the initial phase, the rate of toluene conversion is high, indicating a significant rate of intermediate state generation, while the rate of intermediate state consumption is low ($v_1 > v_2$). Primarily, two types of reactions occur as follows: the conversion of toluene into the intermediate state and the subsequent conversion of the intermediate state into the final product. However, the conversion of toluene into the intermediate state is the predominant reaction, leading to the accumulation of the intermediate state. This indicates that this stage of the reaction is primarily driven by the conversion of toluene into the intermediate state. The initial first-order kinetics equation for toluene consumption represents the initial conversion of toluene, but the conversion of toluene is not completely mineralized, resulting in the generation of ferrous iron not following first-order kinetics. This suggests that there is still an intermediate state carrying some electrons. The ratio of toluene consumption to iron reduction is greater than 1/36.

As the reaction progresses, the rate of toluene conversion gradually decreases along with the decreasing concentration of toluene. The rate of intermediate state generation diminishes, resulting in an accumulation of the intermediate state. Meanwhile, the rate of transformation of the intermediate state also increases. When the generation rate of the intermediate state equals its consumption rate ($v_1 = v_2$), the accumulation of the intermediate state ceases, reaching its maximum value. At this point, a balance is achieved between the conversion of toluene into the intermediate state and the transformation of the intermediate state into the final product, with the ratio of toluene conversion to iron reduction equaling 1/36.

The rate of toluene conversion continues to decrease, and the generation rate of the intermediate state slows down. The consumption rate of the intermediate state exceeds its generation rate ($v_1 < v_2$), indicating that, in this stage, two reactions primarily occur: toluene conversion to an intermediate state and the subsequent transformation of the intermediate state into the final product. There is minimal generation of the intermediate state but there is significant consumption. The intermediates accumulated in the early stage of this phase begin to gradually deplete. Overall, this stage signifies a predominance of the transformation of the intermediate state. Consequently, in the overall reaction, the electrons obtained from ferrous iron reduction exceed those released from toluene consumption during this period. In other words, the lagged electrons at this point contribute to the reduction of ferrous iron. The ratio of toluene conversion to iron reduction is less than 1/36.

It can be inferred that under microbial mediation, the degradation reaction of toluene using trivalent iron minerals as electron acceptors is likely to proceed in stages rather than in a single step; initially, toluene is transformed into certain intermediate states, and eventually transformed into final products. Therefore, through analysis, we have developed this conceptual model, as shown in Figure 6.

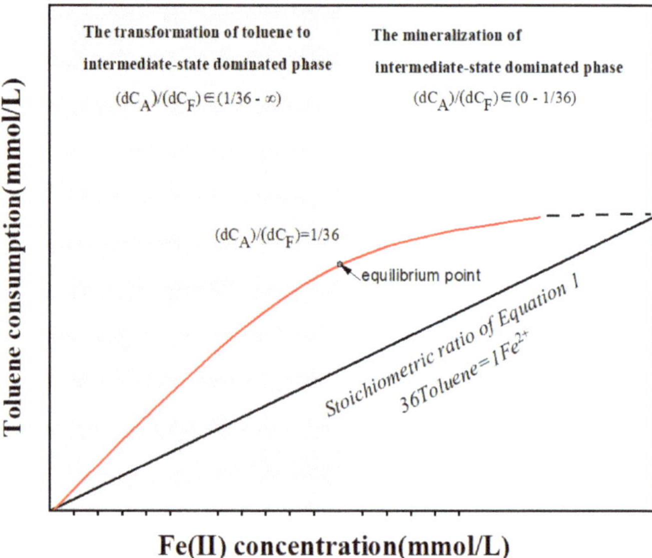

Figure 6. Conceptual model of the interrelationship between toluene degradation and iron reduction processes, where the red curve represents the reaction progress line, and dashed lines indicate reactions that have not occurred. The ○ points denote the points with a slope of 1/36, with the left side representing the transformation of toluene to intermediate-state-dominated phase and the right side representing the mineralization of the intermediate-state-dominated phase. The solid black line, y = 1/36x, represents the reaction progress line according to the theoretical equation.

By integrating experimental data into the model, we can ascertain the predominant stages of various iron minerals. The slope of the regression line, depicted in Figure 4, with the ferrous iron concentration on the x-axis and toluene consumption on the y-axis (representing the ratio of toluene consumption to iron reduction, hereinafter referred to as "the slope"), effectively delineates the phases of different iron minerals within the model. Those with slopes greater than 1/36 denote the dominant phase of toluene transitioning to an intermediate state, while those with slopes less than 1/36 signify the dominant phase of intermediate state mineralization. The phases of the coupled relationship between toluene consumption and ferrous iron reduction for four Fe(III) Oxides represents the division of stages, as illustrated in Table 4.

In the process of toluene transformation to an intermediate-state-dominated phase, the slopes are largest for magnetite and amorphous iron hydroxide, followed by goethite, and then red hematite. A larger slope indicates a faster oxidation rate of toluene relative to the iron reduction rate, with a rapid accumulation of the intermediate state. Active oxidized iron acts as a catalyst, promoting the rapid degradation of toluene. The catalytic effects are in the following order: magnetite > amorphous iron hydroxide > goethite > red hematite.

Through analyzing the duration of toluene transformation to the intermediate-state-dominated phase, the process is longest in amorphous iron hydroxide and magnetite, followed by goethite, and then by red hematite. This suggests that different activities of oxidized iron lead to different accumulation times of the intermediate state. The order of accumulation times is as follows: amorphous iron hydroxide and magnetite > goethite > red hematite.

Arranging based on the slope of the line connecting the endpoint and starting point, the order is amorphous iron hydroxide > magnetite > goethite > red hematite. The slope of the line represents the final ratio of toluene consumption to ferrous iron reduction. This also suggests that higher activity of oxidized iron leads to a lower ratio of toluene consumption to ferrous iron reduction, which might be due to the high activity of oxidized iron affecting the iron reduction in the intermediate state of electron storage, resulting in the accumulation of electrons.

In addition, the microorganisms in the culture system of different minerals may be different, the degradation products may be different, and the ability to degrade intermediate products may be different, resulting in these differences in slope and time [45–48].

4.3. Implication

In the process of practical field application, special attention must be paid to the oxidation of toluene-like compounds and the reduction of iron. When considering the co-existence of toluene-like compounds and iron pollution in a site, it is not sufficient to rely solely on the iron content to assess the original quantity of contaminants. Evaluating the age of contaminants requires a comprehensive consideration of the iron content and cannot simply rely on specific coupling ratios. Additionally, factors such as the type of iron oxides, specific coupling conditions, and the delay in electron transfer must be taken into account. Therefore, in practical applications, it is essential to consider multiple factors comprehensively for a more accurate assessment and treatment of organic and iron pollution on the site.

The integration of our proposed model applies not only to organic contaminants, but also extends to biomass and organic acids. The coupling interaction with iron minerals, leading to the formation of complexes with iron, significantly influences the solubility and mobility of iron. We can further explore the significance of the interaction between carbon and iron in soil ecosystems. Microbes, through metabolic activities such as the decomposition of organic matter and redox reactions, play a regulatory role in the cycling of carbon and iron in the soil. The specific organic substances released by microbes, such as biochar and lignin, may mediate the reduction process of iron, thereby affecting the form and bioavailability of iron in the soil [49]. Our model provides a mechanism to elucidate this influence, particularly in the context of microbial involvement in oxidation–reduction processes.

Based on a comprehensive understanding of toluene degradation and iron reduction processes, the model can reasonably interpret experimental results and possesses a certain degree of universality. However, the model is only in its preliminary qualitative stage and is also subject to limitations imposed by experimental conditions and environmental factors, necessitating further investigation. The model provides important insights into the mechanisms and kinetics of organic degradation and iron reduction processes. Further research could reveal the time lag between toluene degradation and iron generation, providing deeper insights into the mechanism of toluene degradation.

The effect of temperature on the reaction rate is a classic phenomenon in chemical kinetics. Changes in temperature usually do not alter the fundamental mechanistic processes of a reaction [50]. Temperatures may vary the rate constant (k) and may affect the final results of the conceptual model curves, but they do not alter the reaction equations.

Therefore, the presentation may vary at different temperatures, but the established model remains applicable because individual reaction mechanisms remain unchanged.

5. Conclusions

In the presence of iron minerals with varying crystallinity, the degradation of toluene does not coincide with the reduction of iron; the electrons released during toluene degradation fail to fully reduce the iron. Aside from the losses incurred during the toluene experimental process, the assimilation by microbial growth, and the pre-measurement losses of iron, the results indicate the presence of incompletely degraded intermediate states. Consequently, a conceptual model delineating the mutual relationship between toluene degradation and iron reduction processes is established as follows: the heterotrophic iron-reducing microorganisms participate in two reaction stages of toluene and iron mineral heterotrophic iron reduction. The first stage corresponds to the dominant phase of toluene conversion to an inter-mediate state, while the second stage pertains to the dominant phase of intermediate state mineralization.

This study delves into the relationship between toluene degradation and iron reduction, proposing a comprehensive conceptual model. It is helpful to deduce the coupling mechanism of toluene degradation and the ferrous reduction amount. It deepens our understanding of microbial involvement in degradation processes within complex environments, providing a significant theoretical framework and practical guidance for comprehending the degradation of toluene-like contaminants in subsurface environments.

In practical field applications, it is imperative to consider the specific types of iron minerals in the subsurface of the site when evaluating the natural attenuation degree of pollution and the age of the contaminants. Moreover, the delayed electron transfer phenomenon between contaminants and iron also impacts the site assessments. Therefore, this factor should be taken into account during the assessment process.

Supplementary Materials: The following supporting information can be downloaded at https://www.mdpi.com/article/10.3390/app14125017/s1, Figure S1: The gas chromatography spectra comparison between the experimental group and the control group.

Author Contributions: Conceptualization, H.D. and Z.N.; Formal analysis, Z.N. and C.L.; Investigation, H.D., M.Z., Z.N., Z.H., C.L. and J.S.; Methodology, H.D., M.Z. and Z.N.; Supervision, M.Z., Z.H. and C.L.; Validation, H.D.; Visualization, H.D.; Writing—original draft, H.D.; Writing—review and editing, H.D., M.Z., Z.N., Z.H., C.L. and J.S. All authors have read and agreed to the published version of the manuscript.

Funding: The Central Leading Local Science and Technology Development Fund Project, grant number 236Z4204G.

Institutional Review Board Statement: Not applicable.

Informed Consent Statement: Not applicable.

Data Availability Statement: The data presented in this study are available in the article and Supplementary Materials.

Acknowledgments: We appreciate the Nexis GC-2030 gas chromatograph and HS-10 automated headspace sampler manufactured by Shimadzu Corporation for ensuring the accuracy of our research data.

Conflicts of Interest: The authors declare no conflicts of interest.

References

1. Chen, H.; Chen, H.; He, J.; Liu, F.; Shen, Z.; Han, B.; Sun, J. Health-based risk assessment of contaminated sites: Principles and methods. *Earth Sci. Front.* **2006**, *13*, 216–223. [CrossRef]
2. Chen, M. Analytical integration procedures for the derivation of risk-based generic assessment criteria for soil. *Hum. Ecol. Risk Assess. Int. J.* **2010**, *16*, 1295–1317. [CrossRef]
3. Li, C.; Wu, J.; Luo, F. Risk assessment of soil and groundwater for an organic chemical contaminated site. *Soil* **2013**, *45*, 933–939.

4. Dong, M.; Zhang, J.; Luo, F. Health risk assessment of soil and groundwater for a typical organic chemical contaminated site in Southern China. *Soils* **2015**, *47*, 100–106.

5. Zhang, S.; Zhang, C.; He, Z.; Chen, L.; Zhang, F.; Yin, M.; Ning, Z.; Sun, Z.; Zhen, S. Application research of enhanced in-situ micro-ecological remediation of petroleum contaminated soil. *J. Groundwater Sci. Eng.* **2016**, *4*, 157. [CrossRef]

6. Ma, S.; Zhou, J.; Liang, P.; Su, Y. Characteristics-based classification research on typical petroleum contaminants of groundwater. *J. Groundwater Sci. Eng.* **2014**, *2*, 41. [CrossRef]

7. Christensen, T.H.; Kjeldsen, P.; Bjerg, P.L.; Jensen, D.L.; Christensen, J.B.; Baun, A.; Albrechtsen, H.-J.; Heron, G. Biogeochemistry of landfill leachate plumes. *Appl. Geochem.* **2001**, *16*, 659–718. [CrossRef]

8. Alvarez, P.J.J.; Illman, W.A. *Bioremediation and Natural Attenuation: Process Fundamentals and Mathematical Models*; John Wiley & Sons: Hoboken, NJ, USA, 2006.

9. Spormann, A.M.; Widdel, F. Metabolism of alkylbenzenes, alkanes, and other hydrocarbons in anaerobic bacteria. *Biodegradation* **2000**, *11*, 85–105. [CrossRef]

10. Widdel, F.; Rabus, R. Anaerobic biodegradation of saturated and aromatic hydrocarbons. *Curr. Opin. Biotechnol.* **2001**, *12*, 259–276. [CrossRef]

11. Chakraborty, R.; Coates, J. Anaerobic degradation of monoaromatic hydrocarbons. *Appl. Microbiol. Biotechnol.* **2004**, *64*, 437–446. [CrossRef]

12. Dong, H.; Zeng, Q.; Sheng, Y.; Chen, C.; Yu, G.; Kappler, A. Coupled iron cycling and organic matter transformation across redox interfaces. *Nat. Rev. Earth Environ.* **2023**, *4*, 659–673. [CrossRef]

13. Qu, D.; Schnell, S. Microbial Reduction Capacity of Various Iron Oxides in Pure Culture Experiment. *Acta Microbiol. Sin.* **2001**, *41*, 745–749.

14. Weelink, S.A.; Van Doesburg, W.; Saia, F.T.; Rijpstra, W.I.C.; Röling, W.F.; Smidt, H.; Stams, A.J. A strictly anaerobic betaproteobacterium Georgfuchsia toluolica gen. nov., sp. nov. degrades aromatic compounds with Fe (III), Mn (IV) or nitrate as an electron acceptor. *FEMS Microbiol. Ecol.* **2009**, *70*, 575–585. [CrossRef]

15. Botton, S.; Van Harmelen, M.; Braster, M.; Parsons, J.R.; Röling, W.F. Dominance of Geobacteraceae in BTX-degrading enrichments from an iron-reducing aquifer. *FEMS Microbiol. Ecol.* **2007**, *62*, 118–130. [CrossRef] [PubMed]

16. Kane, S.R.; Beller, H.R.; Legler, T.C.; Anderson, R.T. Biochemical and genetic evidence of benzylsuccinate synthase intoluene-degrading, ferric iron-reducing Geobacter metallireducens. *Biodegradation* **2002**, *13*, 149–154. [CrossRef]

17. Shi, Z.; Zachara, J.M.; Wang, Z.; Shi, L.; Fredrickson, J.K. Reductive dissolution of goethite and hematite by reduced flavins. *Geochim. Cosmochim. Acta* **2013**, *121*, 139–154. [CrossRef]

18. Liu, J.; Pearce, C.I.; Shi, L.; Wang, Z.; Shi, Z.; Arenholz, E.; Rosso, K.M. Particle size effect and the mechanism of hematite reduction by the outer membrane cytochrome OmcA of Shewanella oneidensis MR-1. *Geochim. Cosmochim. Acta* **2016**, *193*, 160–175. [CrossRef]

19. Weihe, S.H.; Mangayayam, M.; Sand, K.K.; Tobler, D.J. Hematite crystallization in the presence of organic matter: Impact on crystal properties and bacterial dissolution. *ACS Earth Space Chem.* **2019**, *3*, 510–518. [CrossRef]

20. Kukkadapu, R.K.; Zachara, J.M.; Smith, S.C.; Fredrickson, J.K.; Liu, C. Dissimilatory bacterial reduction of Al-substituted goethite in subsurface sediments. *Geochim. Cosmochim. Acta* **2001**, *65*, 2913–2924. [CrossRef]

21. Rowe, A.R.; Yoshimura, M.; LaRowe, D.E.; Bird, L.J.; Amend, J.P.; Hashimoto, K.; Nealson, K.H.; Okamoto, A. In situ electrochemical enrichment and isolation of a magnetite-reducing bacterium from a high pH serpentinizing spring. *Environ. Microbiol.* **2017**, *19*, 2272–2285. [CrossRef]

22. Botton, S.; Parsons, J.R. Degradation of BTX by dissimilatory iron-reducing cultures. *Biodegradation* **2007**, *18*, 371–381. [CrossRef] [PubMed]

23. Villatoro-Monzón, W.; Mesta-Howard, A.; Razo-Flores, E. Anaerobic biodegradation of BTEX using Mn (IV) and Fe (III) as alternative electron acceptors. *Water Sci. Technol.* **2003**, *48*, 125–131. [CrossRef] [PubMed]

24. Kim, S.J.; Park, S.J.; Cha, I.T.; Min, D.; Kim, J.S.; Chung, W.H.; Chae, J.C.; Jeon, C.O.; Rhee, S.K. Metabolic versatility of toluene-degrading, iron-reducing bacteria in tidal flat sediment, characterized by stable isotope probing-based metagenomic analysis. *Environ. Microbiol.* **2014**, *16*, 189–204. [CrossRef] [PubMed]

25. Frankel, R.B.; Bazylinski, D.A. Biologically induced mineralization by bacteria. *Rev. Mineral. Geochem.* **2003**, *54*, 95–114. [CrossRef]

26. Pilloni, G.; von Netzer, F.; Engel, M.; Lueders, T. Electron acceptor-dependent identification of key anaerobic toluene degraders at a tar-oil-contaminated aquifer by Pyro-SIP. *FEMS Microbiol. Ecol.* **2011**, *78*, 165–175. [CrossRef] [PubMed]

27. Hori, T.; Müller, A.; Igarashi, Y.; Conrad, R.; Friedrich, M.W. Identification of iron-reducing microorganisms in anoxic rice paddy soil by 13C-acetate probing. *ISME J.* **2010**, *4*, 267–278. [CrossRef] [PubMed]

28. Tobler, N.B.; Hofstetter, T.B.; Schwarzenbach, R.P. Carbon and hydrogen isotope fractionation during anaerobic toluene oxidation by Geobacter metallireducens with different Fe (III) phases as terminal electron acceptors. *Environ. Sci. Technol.* **2008**, *42*, 7786–7792. [CrossRef] [PubMed]

29. Langenhoff, A.A.M.; Zehnder, A.J.B.; Schraa, G. Behaviour of toluene, benzene and naphthalene under anaerobic conditions in sediment columns. *Biodegradation* **1996**, *7*, 267–274. [CrossRef]

30. Schwertmann, h.c.U.; Cornell, R.M. Goethite. In *Iron Oxides in the Laboratory*; Wiley: Hoboken, NJ, USA, 2000; pp. 67–92.

31. Lovley, D. *Dissimilatory Fe(III)- and Mn(IV)-Reducing Prokaryotes*; Springer: Berlin/Heidelberg, Germany, 2013.

32. Yao, Q.; Xun, S.; Zhou, Y.; Dong, C. Spectrophotometric Determination of Iron Valency States with 1,10-Phenantbroline. *Chin. J. Health Lab. Technol.* **2000**, *10*, 3–5. [CrossRef]
33. Borch, T.; Kretzschmar, R.; Kappler, A.; Cappellen, P.V.; Ginder-Vogel, M.; Voegelin, A.; Campbell, K. Biogeochemical redox processes and their impact on contaminant dynamics. *Environ. Sci. Technol.* **2010**, *44*, 15–23. [CrossRef]
34. Pavlostathis, S.G.; Giraldo-Gomez, E. Kinetics of anaerobic treatment: A critical review. *Crit. Rev. Environ. Sci. Technol.* **1991**, *21*, 411–490. [CrossRef]
35. Coates, J.D.; Woodward, J.; Allen, J.; Philp, P.; Lovley, D.R. Anaerobic degradation of polycyclic aromatic hydrocarbons and alkanes in petroleum-contaminated marine harbor sediments. *Appl. Environ. Microbiol.* **1997**, *63*, 3589–3593. [CrossRef] [PubMed]
36. Wang, L. Migration and Transformation of Typical Fluoroquinolone Antibiotics during the Crystal Phase Transformation of Ferrihydrite. Ph.D. Thesis, Jilin University, Changchun, China, 2023.
37. Krieger, C.J.; Beller, H.R.; Reinhard, M.; Spormann, A.M. Initial reactions in anaerobic oxidation of m-xylene by the denitrifying bacterium Azoarcus sp. strain T. *J. Bacteriol.* **1999**, *181*, 6403–6410. [CrossRef] [PubMed]
38. Yu, S.-S.; Chen, J.-J.; Cheng, R.-F.; Min, Y.; Yu, H.-Q. Iron Cycle Tuned by Outer-Membrane Cytochromes of Dissimilatory Metal-Reducing Bacteria: Interfacial Dynamics and Mechanisms In Vitro. *Environ. Sci. Technol.* **2021**, *55*, 11424–11433. [CrossRef] [PubMed]
39. Zhu, J.; Liu, D.; Wang, S.; Huang, Z.; Liang, J. Impacts of Soil Nutrients and Stoichiometry on Microbial Carbon Use Efficiency. *J. Guangxi Norm. Univ. Nat. Sci. Ed.* **2022**, *40*, 376–387.
40. Harwood, C.S.; Gerhard, B.; Heidrun, H.; Georg, F. Anaerobic metabolism of aromatic compounds via the benzoyl-CoA pathway. *FEMS Microbiol. Rev.* **1998**, *22*, 439–458. [CrossRef]
41. Ma, J.; Ma, C.; Tang, J.; Zhou, S.; Zhuang, L. Mechanisms and Applications of Electron Shuttle-Mediated Extracellular Electron Transfer. *Prog. Chem.* **2015**, *27*, 1833.
42. Leuthner, B.; Heider, J. Anaerobic toluene catabolism of Thauera aromatica: The bbs operon codes for enzymes of β oxidation of the intermediate benzylsuccinate. *J. Bacteriol.* **2000**, *182*, 272–277. [CrossRef]
43. Leutwein, C.; Heider, J. (R)-Benzylsuccinyl-CoA dehydrogenase of Thauera aromatica, an enzyme of the anaerobic toluene catabolic pathway. *Arch. Microbiol.* **2002**, *178*, 517–524. [CrossRef]
44. von Horsten, S.; Lippert, M.L.; Geisselbrecht, Y.; Schühle, K.; Schall, I.; Essen, L.O.; Heider, J. Inactive pseudoenzyme subunits in heterotetrameric BbsCD, a novel short-chain alcohol dehydrogenase involved in anaerobic toluene degradation. *FEBS J.* **2022**, *289*, 1023–1042. [CrossRef]
45. Fuchs, G.; Boll, M.; Heider, J. Microbial degradation of aromatic compounds—From one strategy to four. *Nat. Rev. Microbiol.* **2011**, *9*, 803. [CrossRef] [PubMed]
46. Forkan, A.; Hasibullah, M.; Jannatul, F.; Nurul, A.M. Microbial Degradation of Petroleum Hydrocarbon. *Bangladesh J. Microbiol.* **2010**, *27*, 10–13.
47. Das, N.; Chandran, P. Microbial degradation of petroleum hydrocarbon contaminants: An overview. *Biotechnol. Res. Int.* **2011**, *2011*, 941810. [CrossRef] [PubMed]
48. Varjani, S.J. Microbial degradation of petroleum hydrocarbons. *Bioresour. Technol.* **2017**, *223*, 277–286. [CrossRef]
49. Sheng, Y.; Dong, H.; Kukkadapu, R.K.; Ni, S.; Zeng, Q.; Hu, J.; Coffin, E.; Zhao, S.; Sommer, A.J.; McCarrick, R.M.; et al. Lignin-enhanced reduction of structural Fe(III) in nontronite: Dual roles of lignin as electron shuttle and donor. *Geochim. Cosmochim. Acta* **2021**, *307*, 1–21. [CrossRef]
50. Atkins, P.W.; Paula, J.C.D.; Keeler, J. *Atkins' Physical Chemistry*; Oxford University Press: Oxford, UK, 2018.

applied sciences

MDPI

Article

The Effects of Toluene Mineralization under Denitrification Conditions on Carbonate Dissolution and Precipitation in Water: Mechanism and Model

Shuang Gan [1,2], Min Zhang [1,3,*], Yahong Zhou [4], Caijuan Guo [1,3], Shuai Yang [5], Yan Xie [5], Xinzhe Wang [5], Lin Sun [1,3] and Zhuo Ning [1,3,*]

[1] Institute of Hydrogeology and Environmental Geology, Chinese Academy of Geological Sciences, Shijiazhuang 050061, China; lin_sun166@163.com (L.S.)
[2] School of Resources and Environmental Engineering, Hefei University of Technology, Hefei 230009, China
[3] Key Laboratory of Groundwater Remediation of Hebei Province & China Geological Survey, Shijiazhuang 050061, China
[4] School of Water Resources and Environment, Hebei GEO University, Shijiazhuang 050031, China
[5] SINOPEC Research Institute of Safety Engineering Co., Ltd., Qingdao 266071, China; yangshu.qday@sinopec.com (S.Y.)
* Correspondence: minzhang205@live.cn (M.Z.); ningzhuozhuo@163.com (Z.N.); Tel.: +86-0311-6759-8605 (M.Z. & Z.N.)

Abstract: The mineralization of benzene, toluene, ethylbenzene, and xylene (BTEX) into inorganic substances by microorganisms may affect the water–rock interaction. However, few studies have quantitatively analyzed the processes. To quantitatively reveal this mechanism, in this study, nitrate and toluene were taken as the typical electron acceptor and BTEX, respectively. Based on hydrogeochemical theory, the mechanism and mathematical model were established. In addition, the model was verified with a toluene mineralization experiment. The mechanism model demonstrated that H^+ was the main factor in the dissolution or precipitation of $CaCO_3$. The mathematical model derived the equations quantitatively between the amount of toluene mineralization, $CaCO_3$, and some biogeochemical indicators, including temperature, microbial consumption, and other major ions in groundwater. According to the model, the amount of dissolved $CaCO_3$ increased with the increasing proportion of completely reduced nitrate. For a complete reaction, the greater the microorganisms' consumption of toluene was, the smaller the precipitation of $CaCO_3$. $CaCO_3$ dissolution was a nonmonotonic function that varied with temperature and the milligram equivalent of other ions. Furthermore, the validation experiments agreed well with the mathematical model, indicating its practicality. The established model provides a tool for assessing the biodegradation of toluene by monitoring the concentration of groundwater ions.

Keywords: BTEX mineralization; toluene; denitrification; carbonate dissolution and precipitation; mechanism; model

Citation: Gan, S.; Zhang, M.; Zhou, Y.; Guo, C.; Yang, S.; Xie, Y.; Wang, X.; Sun, L.; Ning, Z. The Effects of Toluene Mineralization under Denitrification Conditions on Carbonate Dissolution and Precipitation in Water: Mechanism and Model. *Appl. Sci.* **2023**, *13*, 11867. https://doi.org/10.3390/app132111867

Academic Editor: Dibyendu Sarkar

Received: 17 September 2023
Revised: 27 October 2023
Accepted: 28 October 2023
Published: 30 October 2023

1. Introduction

Contamination caused by petroleum leakage occurring during oil production, transportation, refinement, and mishandled storage is widespread [1–5]. Benzene, toluene, ethylbenzene, and xylene (BTEX) are volatile organic compounds in petroleum that result in neurotoxicity, genotoxicity, and reproductive toxicity [6–8]. Due to its mutagenic and high migration, BTEX leads to the deterioration of aquatic and terrestrial environment quality when released into the environment and causes adverse effects on human health when absorbed by breathing, touching, or swallowing [9–11].

Natural attenuation processes, such as biomineralization, adsorption, advection, dispersion, and volatilization, can all decrease the number of contaminants in an aquifer [12–14]. Since biomineralization uses microorganisms to convert BTEX into inorganic substances, it is considered an environmentally friendly technology [15,16]. In the 1960s, scientists found that some bacteria could use BTEX as their source of carbon energy, thus decomposing the BTEX into CO_2 and water, which confirmed that it could be aerobically degraded [17,18]. However, the underground environment is usually anaerobic, and the anaerobic mineralization of BTEX is more feasible and practical. BTEX can be biodegraded with NO_3^-, Mn^{4+}, Fe^{3+}, and SO_4^{2-}, and CO_2 as terminal electron acceptors [19–22]. As NO_3^- dissolves well in water, and denitrification yields the highest energy, NO_3^- can easily be applied for anaerobic mineralization [23–26]. The mineralization mechanism of BTEX under denitrification conditions is as follows: BTEX acts as both a carbon source and energy for the growth and reproduction of bacteria, and NO_3^- acts as an electron acceptor to convert hazardous materials to CO_2 and H_2O, leading to the mineralization of BTEX [27,28].

The BTEX mineralization equations are as follows:

$$C_6H_6 + 6NO_3^- + 6H^+ \rightarrow 6CO_2 + 3N_2 + 6H_2O \tag{1}$$

$$C_7H_8 + 7.2NO_3^- + 7.2H^+ \rightarrow 7CO_2 + 3.6N_2 + 7.6H_2O \tag{2}$$

$$C_8H_{10} + 8.4NO_3^- + 8.4H^+ \rightarrow 8CO_2 + 4.2N_2 + 9.2H_2O \tag{3}$$

It was previously believed that this mineralization could solve the problem of BTEX pollution. However, mineralization impacts the water–rock interaction, especially in the dissolution and precipitation of carbonate rocks. Studies in this area have found that transforming BTEX to inorganic C involves a series of reactions, resulting in a decreased pH and the dissolution of carbonate mineral calcite [29].

Two main aspects influence the dissolution and precipitation of $CaCO_3$ via biomineralization. First, under denitrification conditions, BTEX is converted into neutral or acidic organic compounds, which induce a variety of pH levels [30,31]. Second, BTEX is converted into end-product CO_2 (Equations (1)–(3)), which, together with H^+ from the water, forms H_2CO_3 and causes the pH to decrease (Equations (4)–(8)) [27].

$$CO_2 + H_2O \rightleftharpoons H_2CO_3 \tag{4}$$

$$H_2CO_3 \rightleftharpoons H^+ + HCO_3^- \tag{5}$$

$$HCO_3^- \rightleftharpoons H^+ + CO_3^{2-} \tag{6}$$

$$H_2O \rightleftharpoons H^+ + OH^- \tag{7}$$

$$CaCO_3 \rightleftharpoons Ca^{2+} + CO_3^{2-} \tag{8}$$

Currently, most studies show that the degradation of BTEX promotes the dissolution of calcium carbonate, whether this is demonstrated through indoor column experiments or field data collection [32,33]. Moreover, several studies suggest that the pH could rise slightly as CO_2 is degassed from water and autotrophic microbial metabolism occurs, resulting in carbonate precipitation and removal [34,35]. Until now, few studies have quantitatively analyzed the effect of BTEX mineralization on the dissolution and precipitation of $CaCO_3$.

Therefore, based on hydrogeochemical and organic biomineralization theory, a mechanism model was revealed. To quantitatively evaluate the effect of BTEX mineralization on the water–rock interaction and to infer the impact of BTEX mineralization on downstream water quality, a quantitative model was established, taking nitrate as a typical electron acceptor and toluene, the most abundant species of BTEX, as a typical BTEX pollutant, respectively [36]. Only the toluene mineralization and carbonate equilibrium reaction occurred in the solution. In this case, the difference between the Ca^{2+} and HCO_3^- concentrations in the solution at different periods can be substituted into the mathematical model, and the toluene degradation amount in each period can be roughly determined. Furthermore, a toluene mineralization experiment was conducted to verify the established model.

2. Materials and Methods

2.1. Model Construction

2.1.1. Mechanism Model Construction

This paper integrates relevant hydrogeochemistry and organic matter degradation theories and selects toluene and nitrate as a typical BTEX pollutant and a typical electron acceptor, respectively. According to the theoretical equation of toluene mineralization under denitrification conditions, the changes in H^+, Ca^{2+}, and HCO_3^- concentrations were qualitatively analyzed, and a mechanism model for examining how the mineralization of toluene affected the dissolution and precipitation of calcium carbonate was established.

2.1.2. Mathematical Model Construction

In this study, to quantitatively analyze the effects of toluene mineralization under denitrification conditions on carbonate dissolution and precipitation, based on the above mechanism model and according to mass conservation and charge conservation equations, the relationships between the H^+, Ca^{2+}, and HCO_3^- concentrations and toluene degradation were obtained. To facilitate this analysis, several assumptions were made. Since the main inorganic salt ions in groundwater account for 90–95% of the total inorganic salt ions, the main inorganic salt ions were taken into account by the model [37].

This paper mainly considers the influence of four variables on the model, namely, the influence of the proportion of nitrate consumed by the complete reduction of nitrogen on the total amount consumed; the temperature; the proportion of toluene consumed by microbial growth to total toluene content; and the milligram equivalent of other ions on the dissolution and precipitation of calcium carbonate.

2.2. Model Validation

2.2.1. The Validation Experiment

(1)   Toluene biodegradation enrichment cultures

The enrichment culture microorganisms originated from groundwater in a Chinese coking plant.

Microcosms were set up using 280 mL air-tight bottles. Each microcosm contained 225 mL of sterilized basic culture medium, 25 mL of bacterial solution, and 0.0175 g of $CaCO_3$ to simulate the carbonate environment. The bacterial culture medium was modified from a recipe developed by Zhang [38]. The basic culture medium was prepared according to the following compositions (per liter): KNO_3, 80 mg; KH_2PO_4, 0.50 mg; K_2HPO_4 $3H_2O$, 1.00 mg; $MgSO_4$, 1.00 mg; mineral solution, 0.10 mL; and vitamin solution, 0.20 mL. The components of the vitamin and mineral solutions are shown in Tables 1 and 2. Then, the bottles were autoclaved at 121 °C for 30 min to prevent biological activity. The medium was purged with nitrogen to remove oxygen, and the air-tight bottles were sealed with a halogenated butyl rubber stopper to ensure their air tightness. Furthermore, toluene was added to each bottle to form 0.1 mmol/L toluene solution. Three parallel controls were established for the blank and experimental groups. In the sterile controls, the bacterial solution was replaced with purified water. All of the microcosms were stored in a biochemical incubator at 20 °C. On days 1, 2, 4, and 8, water samples were collected from each microcosm for analysis. Sampling procedures were accomplished under strictly anaerobic conditions. As $CaCO_3$ is an insoluble impurity in water, it was difficult to observe, and there were no other calcium inputs in the system, so the concentration of Ca^{2+} was used to characterize the dissolution and precipitation of calcium carbonate. To evaluate the carbonate dissolution/precipitation via biodegradation, the consumption of Ca^{2+}, HCO_3^-, NO_3^-, and NO_2^- was continuously measured during the entire experimental period.

Table 1. The components of the vitamin solution.

Compounds	Concentration (mg/L)
Biotin	2.00
Folic acid	2.00
Riboflavin	5.00
Nicotinic acid	5.00
p-Aminobenzoic acid	5.00
Thioctic acid	5.00
Calcium D-(+)-pantothenate	5.00
Pyridoxine hydrochloride	10.00

Table 2. The components of the mineral solution.

Compounds	Concentration (g/L)
$FeSO_4\ 4H_2O$	1.80
$CoCl_2\ 6H_2O$	0.25
$CuCl_2\ 2H_2O$	0.01
$NiCl\ 6H_2O$	0.01
$ZnCl_2$	0.10
H_3BO_3	0.50
EDTA	2.50
$MnCl_2\ 4H_2O$	0.70
$(NH_4)_6Mo_7O_{24}\ 4H_2O$	0.01

(2) Physical and chemical parameter analyses

The toluene samples were analyzed by headspace-based gas chromatography (Nexis GC-2030, Shimadzu, Japan) carried out on an automated headspace sampler (HS-10, Shimadzu, Japan). Headspace operating conditions were as follows: low shaking of the tested sample solution at 35 °C for 2 min, GC cycle time of 25 min, and vial pressurization time of 0.1 min. The gas chromatograph was equipped with a capillary column (HP-5, Shimadzu, Japan) and a flame ionization detector (FID). The injector, detector, and column temperatures were held at 150, 200, and 100 °C, respectively. Air and hydrogen served as fuel gases for the FID. Nitrogen served as a carrier gas, and the flow rate was 40.1 mL/min.

The pH value was measured with a portable pH meter (portable pH meter pH30, CLEAN, Sherman Oaks, CA, USA), and the Ca^{2+} and HCO_3^- concentrations were measured via titration. The NO_2^- and NO_3^- concentrations were analyzed according to the International Organization for Standardization ISO 10304-1 [39] with an ultraviolet spectrophotometer (UV-2550, Shimazu, Japan).

2.2.2. The Validation Method

To evaluate the accuracies of the model, residual sum of squares error (SSE), mean square error (MSE), and root mean square (RMS) error were considered as the criteria for the accuracy evaluation and calculated by the following (Equations (9)–(11)).

$$SSE = \sum_{i=1}^{n} \left(\bar{y}_i - y_i \right)^2 \tag{9}$$

$$MSE = \frac{SSE}{n} = \frac{1}{n}\sum_{i=1}^{n} \left(\bar{y}_i - y_i \right)^2 \tag{10}$$

$$RMS = \sqrt{MSE} = \sqrt{\frac{SSE}{n}} = \sqrt{\frac{1}{n}\sum_{i=1}^{n} \left(\bar{y}_i - y_i \right)^2} \tag{11}$$

where $\bar{y}_i$ is the value calculated by the model, y_i is the value of the experimental data, and n is the number of samples. SSE, MSE, and RMS error reflect the reliability and stability of the model. It is commonly accepted that the closer the sum of squares error (SSE) is to 0, the better the model fits the data.

3. Results and Discussion

3.1. Mechanism Model

3.1.1. Response of H^+ to Toluene Mineralization

According to Siobhan M. et al., the nitrification process comprises a two-step reaction during microbially mediated toluene mineralization. The first product of nitrate reduction is nitrite (NO_2^-). Further reduction of nitrite proceeds to nitrogen gas (N_2) [40,41].

Nitrification first step (nitrate reduction):

$$C_7H_8 + 18NO_3^- = 18NO_2^- + 7CO_2 + 4H_2O \tag{12}$$

Nitrification second step (nitrite reduction):

$$C_7H_8 + 12NO_2^- + 12H^+ = 6N_2 + 7CO_2 + 10H_2O \tag{13}$$

According to Song et al., in the two-step nitrification process, nitrite reduction produces hydrogen ions, whereas nitrate reduction does not [42]. During the complete mineralization process of toluene, the generated end products include gaseous carbon dioxide and water [43]. When gaseous carbon dioxide dissolves in water, it reacts and forms carbonic acid. The dissociation of hydrogen ions from the carbonic acid provides bicarbonate, and subsequently, a carbonate ion leads to an increase in H^+ (Figure 1).

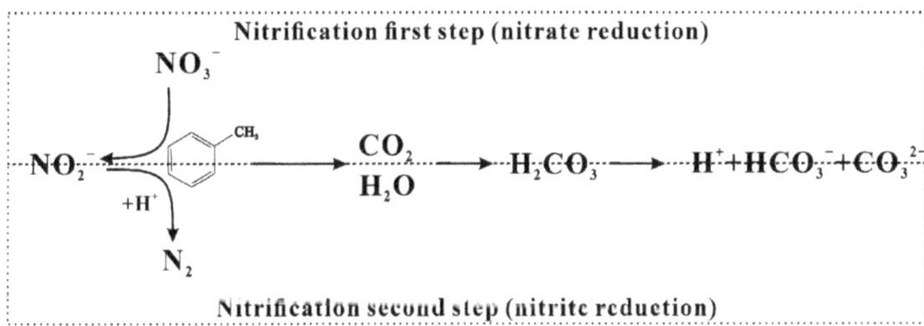

Figure 1. Mechanism modeling of the model.

Under natural practical conditions, nitrate usually degrades directly into nitrogen gas (Equation (2)). Toluene mineralization consumes H^+, and the aqueous pH increases during the reduction process. The end products of mineralization are still carbon dioxide and water, which dissolve into the solution and eventually react with water to form stable carbonate compounds. Due to the complexity of the reactions, changes in the concentration of hydrogen ions in the solution are complicated. Based on the ionization fractions, the ratio of the solute components in the aqueous carbonate system will vary with H^+. As the pH of the solution increases, the concentration of CO_3^{2-} also increases, particularly in an alkaline solution. When the saturation index (SI) reaches a certain value, some of the calcium carbonate in the solution precipitates out.

3.1.2. Response of Ca^{2+} and DIC to Toluene Mineralization

H^+ is one of the critical factors affecting carbonate composition. Along with the change in H^+, the concentration of other ions in the solution will change accordingly. Dissolution and precipitation are more difficult to observe under field conditions; Ca^{2+} and dissolved inorganic carbon (DIC) are the most commonly used to detect $CaCO_3$ dissolution. DIC is composed of three dominant species: dissolved CO_2 (CO_2), bicarbonate (HCO_3^-), and carbonate (CO_3^{2-}).

If the amount of hydrogen ions produced by carbon dioxide dissolved in water and then electrolyzed is greater than the amount of hydrogen ions consumed by the reaction, the $CaCO_3$ dissolves, which leads to an increase in the concentration of Ca^{2+} and CO_3^{2-}, accompanied by a corresponding increase in the concentration of H_2CO_3 and HCO_3^-. On the other hand, when the solution is saturated with $CaCO_3$, then the $CaCO_3$ precipitates, and the concentrations of Ca^{2+}, CO_3^{2-}, and HCO_3^- decrease.

In summary, multiple reactions are involved in toluene mineralization under denitrification conditions, and they are strongly interdependent. Therefore, a mathematical model is required to analyze the impact of toluene mineralization on calcium carbonate in this context.

3.2. Mathematical Model

Groundwater is uniform within an infinite small range. Each ion concentration of groundwater is low, and the temperature of groundwater is generally relatively constant [44]. The aeration zone blocks the exchange of aquifer and atmospheric gas, and the carbonate in an oil-polluted aquifer is often saturated [34].

Therefore, the following assumptions are made to facilitate the establishment of the mathematical model: (1) the components are well mixed, and the pressure, salinity, and ionic strength are uniform, (2) CO_2 is the ideal gas, (3) the temperature is constant, (4) the ionic activity of an ion equals the actual concentration at which the reactions take place in a homogeneous environment, and (5) the solution in the system, which is a closed system, is a calcium carbonate saturated solution.

Based on the mechanism model, the mathematical model was derived from the mass conservation and charge conservation equations in Equations (12) and (13).

In case a carbonate component is not introduced into the system, the Ca^{2+} and dissolved carbonate components come from the degradation of toluene and calcite. As a result, the mass conservation equation can be expressed as follows:

$$7[C_7H_8] + [Ca^{2+}] = [H_2CO_3] + [HCO_3^-] + [CO_3^{2-}] + [CO_2] \qquad (14)$$

where $[C_7H_8]$ is the concentration of toluene involved in the reaction, mol/L. [] is the ionic activity of the ions, mol/L.

The electric charge conservation equation is as follows:

$$[H^+] + [Na^+] + [K^+] + 2[Ca^{2+}] + 2[Mg^{2+}] = [Cl^-] + 2[SO_4^{2-}] + [HCO_3^-] + 2[CO_3^{2-}] + [NO_3^-] + [NO_2^-] + [OH^-] \qquad (15)$$

where [] is the ionic activity of the ions, mol/L.

The equilibrium constant in the system can be expressed as:

$$K_H = \frac{[H_2CO_3]}{P_{co2}} \qquad (16)$$

$$K_{H2CO3} = \frac{[H^+] \times [HCO_3^-]}{[H_2CO_3]} \qquad (17)$$

$$K_{HCO3-} = \frac{[H^+] \times [CO_3^{2-}]}{[HCO_3^-]} \qquad (18)$$

$$K_w = [H^+] \times [OH^-] \qquad (19)$$

$$K_{cal} = [Ca^{2+}] \times [CO_3^{2-}] \qquad (20)$$

According to Equations (16)–(20), the relationship equation between the H^+ concentration and the toluene degradation is established (Equation (21)).

$$
\left\{
\begin{aligned}
14 \times f \times [C_7H_8] &= [H^+] - \frac{K_w}{[H^+]} - [NO_3^-] - [NO_2^-] + E + (2 \times A + B + C) \times \frac{1}{[Ca^{2+}]} \\
A &= \frac{K_{cal} \times [H^+]^2}{K_{H2CO3} \times K_{HCO3-}} \\
B &= \frac{K_{cal} \times [H^+]}{K_{HCO3-}} \\
C &= \frac{K_{cal} \times (V-Va) \times [H^+]^2}{K_{H2CO3} \times K_{HCO3-} \times K_H \times R \times T \times Va} \\
D &= \frac{\left([H^+] - \frac{K_w}{[H^+]} - [NO_3^-] - [NO_2^-] + C\right)^2}{16} - \frac{K_{cal} \times [H^+]}{2 \times K_{HCO3-}} + K_{cal} \\
E &= [Na^+] + [K^+] + 2[Mg^{2+}] - [Cl^-] - 2[SO_4^{2-}] \\
[Ca^{2+}] &= \frac{-[H^+] + \frac{K_w}{[H^+]} + [NO_3^-] + [NO_2^-] - E}{4} + \sqrt{D}
\end{aligned}
\right.
\tag{21}
$$

where [] is the ionic activity of the ions, mol/L. V is the total volume of air and solution, and Va is the volume of solution, L. f is the proportion of toluene consumed by microbial growth to the total toluene content. T is the temperature of the system, °C.

The concentration of H^+ in the system can be calculated from the equation.

d is defined as the proportion of nitrate consumed by the complete reduction of nitrogen to the total amount of nitrate consumed at a certain time. The simplified model can be expressed as:

$$
\left\{
\begin{aligned}
14 \times f \times [C_7H_8] &= [H^+] - \frac{K_w}{[H^+]} - (1+d)[NO_3^-] + E + (2 \times A + B + C) \times \frac{1}{[Ca^{2+}]} \\
A &= \frac{K_{cal} \times [H^+]^2}{K_{H2CO3} \times K_{HCO3-}} \\
B &= \frac{K_{cal} \times [H^+]}{K_{HCO3-}} \\
C &= \frac{K_{cal} \times (V-Va) \times [H^+]^2}{K_{H2CO3} \times K_{HCO3-} \times K_H \times R \times T \times Va} \\
D &= \frac{\left([H^+] - \frac{K_w}{[H^+]} - (1+d)[NO_3^-] + C\right)^2}{16} - \frac{K_{cal} \times [H^+]}{2 \times K_{HCO3-}} + K_{cal} \\
E &= [Na^+] + [K^+] + 2[Mg^{2+}] - [Cl^-] - 2[SO_4^{2-}] \\
[Ca^{2+}] &= \frac{-[H^+] + \frac{K_w}{[H^+]} + (1+d)[NO_3^-] - E}{4} + \sqrt{D}
\end{aligned}
\right.
\tag{22}
$$

Due to the large number of parameters involved in the reaction, the following scenarios are proposed. The system is saturated with calcium carbonate. Based on the data from a contaminated site that we studied, the toluene concentration was 0.1 mmol/L. According to the theoretical stoichiometric relationship, the ratio of the nitrate concentration to the toluene concentration was 7.2:1. The specific discussion is as follows.

3.2.1. The Effect of the Proportion of Nitrate Consumed by the Complete Reduction of Nitrogen to the Total Amount Consumed (d)

It has been reported that microbial growth can consume from 5% to 8% of toluene, so it is assumed that the consumption of microbial growth is 6.5% in this section [45]. The temperature is assumed to be 20 °C, which combines the average groundwater temperature with the culture temperature of the bacteria used in this paper.

If d is 0, there is no nitrite in the solution, and the nitrate entirely transforms into nitrogen (Equation (2)). As shown in Figure 2, carbon dioxide produced by toluene mineralization reacts with water, increasing the amount of hydrogen ions and producing a large amount of bicarbonate. The precipitation rate of calcium carbonate resulting from reaching the SI value is greater than the dissolution rate of calcium carbonate caused by increased hydrogen ions. This result is consistent with the study of Alenezi et al. [10].

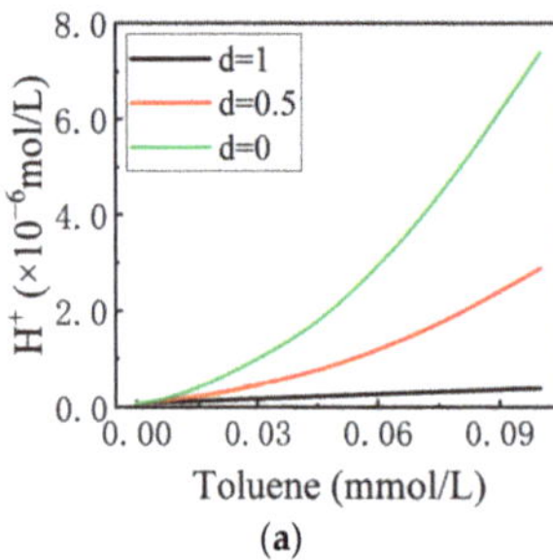 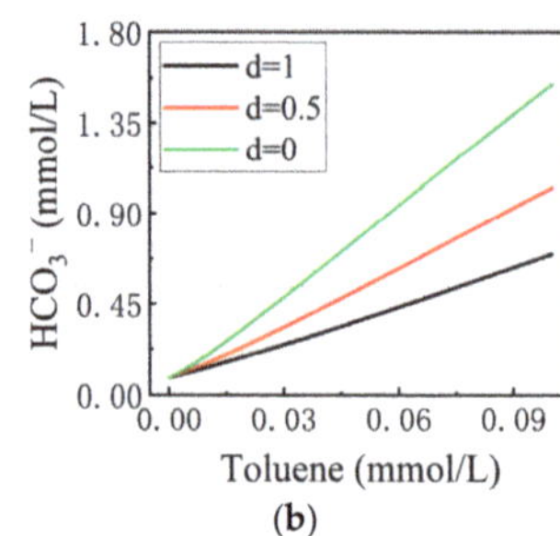 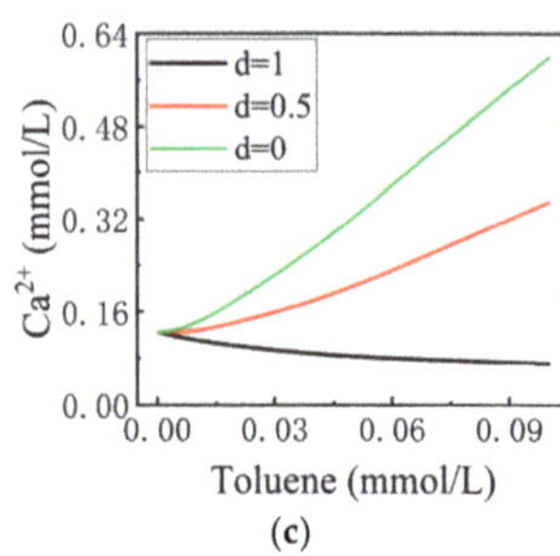

Figure 2. Variations in ion concentrations with the proportion of nitrate consumed by the complete reduction of nitrogen to the total amount consumed. (**a**) Variation in H^+ with toluene concentration; (**b**) variation in HCO_3^- with toluene concentration; (**c**) variation in Ca^{2+} with toluene concentration.

When d equals 0.5, half of the nitrate in the solution is reduced to nitrogen, while the remaining half is reduced to nitrite. The concentration of H^+ in the solution increases due to the production of carbon dioxide (Equations (12) and (13)) and the consumption of OH^- (Equation (13)).

Carbon dioxide produced by toluene mineralization reacts with water to increase the concentration of bicarbonate, which is the same as when d = 0. However, the precipitation rate of calcium carbonate in this state is higher than the formation rate, and the concentration of Ca^{2+} increases.

When d is 1, all the nitrate in the solution is converted to nitrite. The reduction of nitrate to nitrite neither consumes nor produces hydrogen ions. However, the carbon dioxide produced by the reaction increases the concentration of H^+ in the solution. Increased bicarbonate in solution makes the SI always greater than the solubility rate, and the Ca^{2+} concentration decreases.

It can be seen from the figure that when the reaction does not occur, the amount of each element in the solution is the same in the three cases. The mineralization of toluene increases the concentration of hydrogen ions to varying degrees, and the formation of carbon dioxide increases the concentration of bicarbonate. Depending on the reaction mechanism in the solution, the concentration of Ca^{2+} varies with the proportion of nitrate consumed in the two-step reaction.

Under natural practical conditions, the nitrite concentration in the solution is small, the reaction is less likely, and d = 0 occurs more often.

3.2.2. The Effect of the Proportion of Toluene Consumed by Microbial Growth on the Total Toluene Content (*f*)

Mathematical models were created to illustrate the impact of microbial growth on $CaCO_3$, with diagrams presented in Figure 3 for levels of 0%, 4%, and 8%. Under natural practical conditions, nitrate is completely reduced to nitrogen, and it is assumed that d is 0 and the temperature is 20 °C.

It can be seen that f has little effect on concentration, the consumption of toluene increases, and the concentrations of H^+, Ca^{2+}, and HCO_3^- decrease (Figure 3). The influence of microbial growth on calcium carbonate is mainly reflected in the fact that the more significant the consumption of microorganisms is, the smaller the amount of toluene used for the reaction.

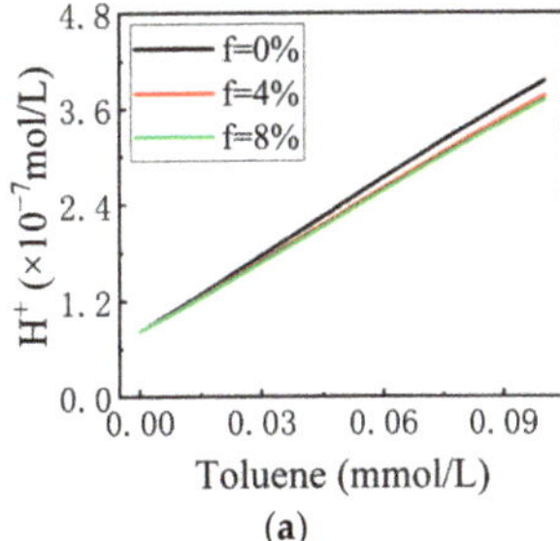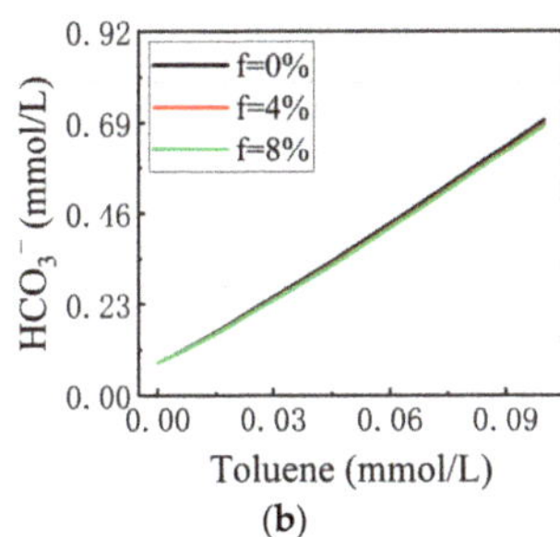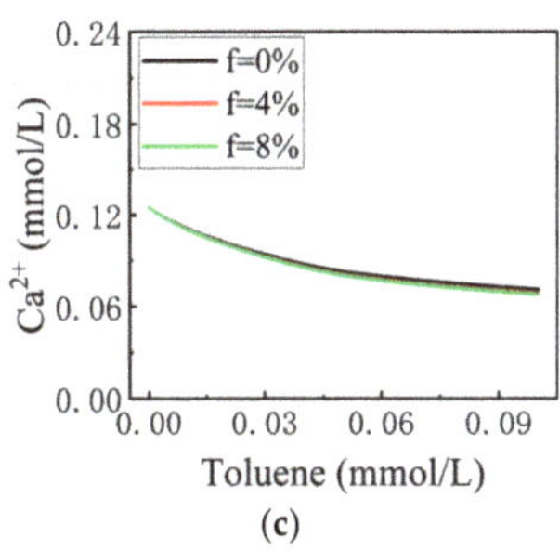

Figure 3. Variation in ion concentration with the proportion of toluene consumed by microbial growth to the total toluene content. (**a**) Variation in H^+ with toluene concentration; (**b**) variation in HCO_3^- with toluene concentration; (**c**) variation in Ca^{2+} with toluene concentration.

3.2.3. The Effect of Temperature (T)

Since the water temperature of aquifers is mainly in the range of 9–20 °C, in order to simulate the relationship between $CaCO_3$ and toluene at different temperatures, mathematical model diagrams of 10 °C, 15 °C, and 20 °C are presented in Figure 4. This paper assumes that the nitrate is entirely transformed into nitrogen and that the consumption of microbial growth is 6.5%.

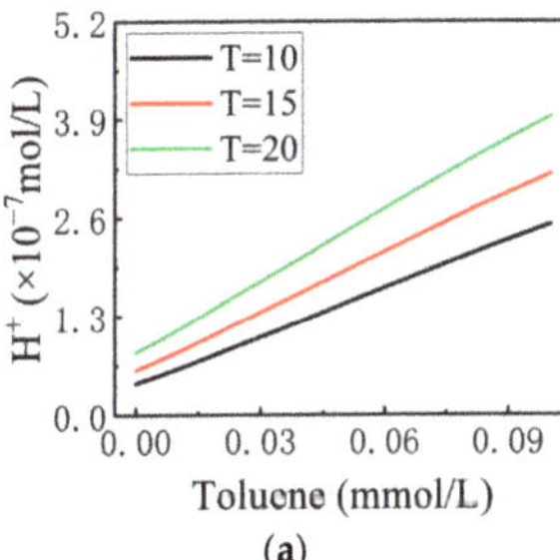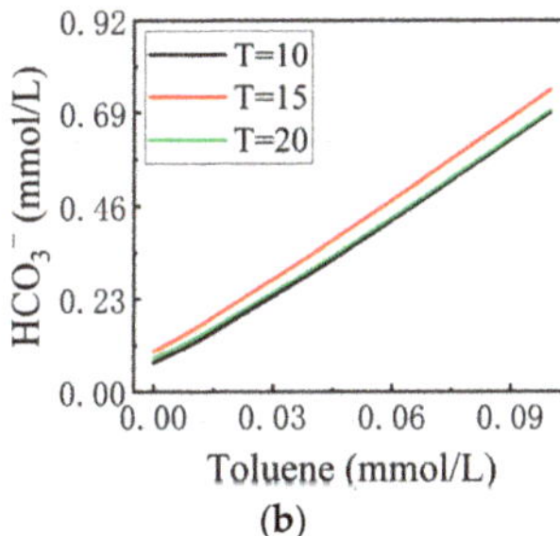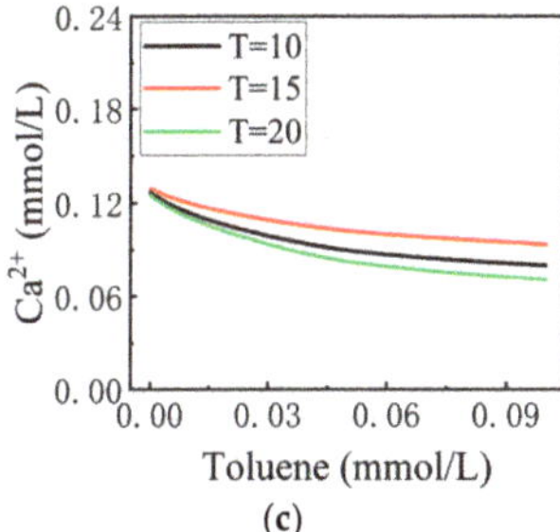

Figure 4. Variation in ion concentration with temperature. (**a**) Variation in H^+ with toluene concentration; (**b**) variation in HCO_3^- with toluene concentration; (**c**) variation in Ca^{2+} with toluene concentration.

As the temperature increases, the activity of ions increases, and the equilibrium constant changes. In addition to directly affecting the temperature assignment in the model, temperature also affects the equilibrium constant of the reaction. As shown in Figure 4, the higher the temperature is, the higher the concentration of H^+. The trend of Ca^{2+} and H^+ changes could be more consistent, possibly due to the combination of temperature and the equilibrium constant. Therefore, the influence of temperature on toluene mineralization is complicated. The Ca^{2+} concentration in the system is not a monotonic function that varies with temperature.

3.2.4. The Effect of the Milligram Equivalent of Other Ions (E)

Figure 5 presents the effects of the milligram equivalent of other ions (E) under toluene mineralization on carbonate dissolution and precipitation. This section assumes that nitrate is entirely transformed into nitrogen, the consumption of microbial growth is 6.5%, and the temperature of the system is 20 °C.

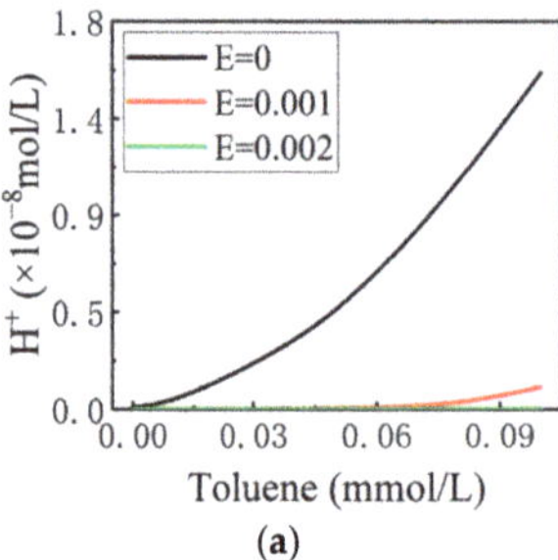
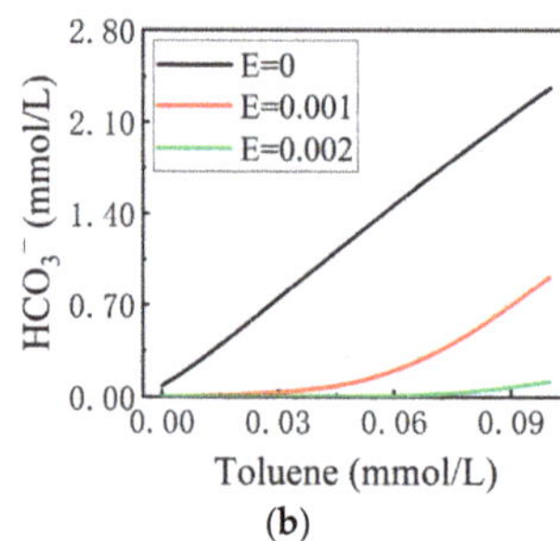
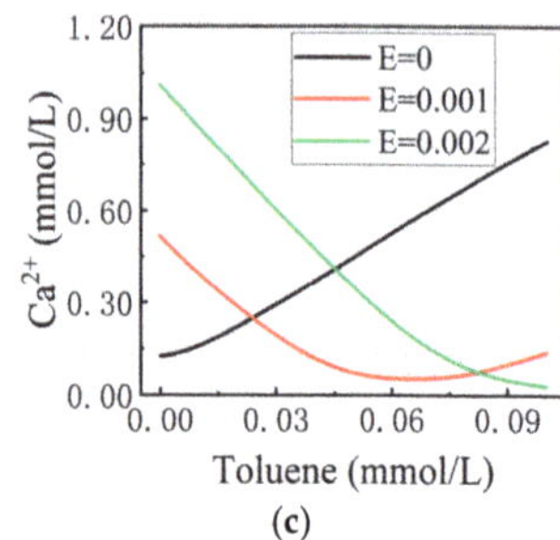

(a)   (b)   (c)

Figure 5. Variation in ion concentration with the milligram equivalent of other ions. (**a**) Variation in H^+ with toluene concentration; (**b**) variation in HCO_3^- with toluene concentration; (**c**) variation in Ca^{2+} with toluene concentration.

Based on the ion concentration in groundwater, Figure 5 shows the influence of complete mineralization on calcium carbonate when E is 0 mol/L, 0.001 mol/L, and 0.002 mol/L. According to the given information, it appears that the trend of ion concentration change is the same in all three cases, as the reaction mechanism is identical. As can be seen from the figure, the smaller the value of E is, the greater the concentrations of H^+ and HCO_3^- in the corresponding solution at the initial concentrations.

The amplitude of variation in the H^+ and HCO_3^- concentrations increases as the value of E decreases. The change in Ca^{2+} concentration is slightly different from that of the above two ions. E affects the initial concentration of calcium ion as follows: the larger the value of E is, the smaller the initial concentration. Moreover, E has an impact on the change trend of the Ca^{2+} concentration. When E is 0, the Ca^{2+} concentration always increases; when E is 0.002, the Ca^{2+} concentration always shows a downward trend; when E is 0.001, the Ca^{2+} concentration shows a trend of first decreasing and then increasing, and there is an inflection point near a toluene concentration of 7×10^{-5} mol/L. The variation in Ca^{2+} concentration is complex, possibly due to the complex relationship between Ca^{2+} and H^+ in the model.

3.3. Fitting Results

The analysis of the experimental results related to toluene mineralization presented in Figure 6 indicates that nitrate concentrations decreased with the increasing amount of toluene degraded. The figure shows that toluene was reduced by 0.02 mmol/L on the first day, suggesting that after domestication, the microorganisms quickly adapted to the experimental environment. The adaptation period of microorganisms in this study was much shorter than that in other studies. Toluene decreased gradually with time. It was degraded by more than 80% on the fourth day of the experiment and was approximately fully degraded within eight days.

The figure shows that the concentration of NO_3^- decreased gradually with time and was degraded by more than 80% on the fourth day of the experiment. The formation of NO_2^- was observed as a temporary phenomenon. According to A. PEÑA-CALVA, A. OLMOS-DICHAR, a similar phenomenon was observed and was called a respiratory pattern [46].

The CO_2 resulting from the complete degradation of toluene reacts directly with $CaCO_3$ and water in the following reactions:

$$CaCO_3 + CO_2 + H_2O = Ca(HCO_3)_2 \tag{23}$$

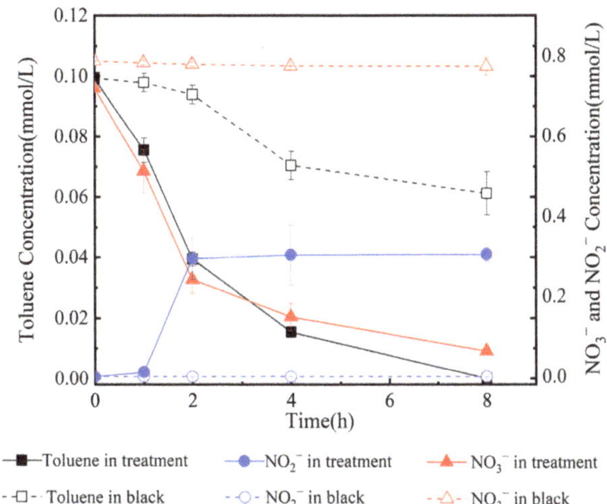

Figure 6. Variations in toluene, nitrate, and nitrite concentrations with toluene degradation. Values represent mean values and the $\pm$ range from three independent samples.

If the above reaction occurs, the magnitude of the change in HCO_3^- : Ca^{2+} should be 2:1. A graph of the amount of change in the concentration of HCO_3^- : Ca^{2+} during the experiment is shown in Figure 7. The slope of the plot is 5.29, and the ratio of the change in the HCO_3^- concentration to the change in the Ca^{2+} concentration is much greater than 2:1. This image also shows that mineralization promotes the increase in HCO_3^- concentration.

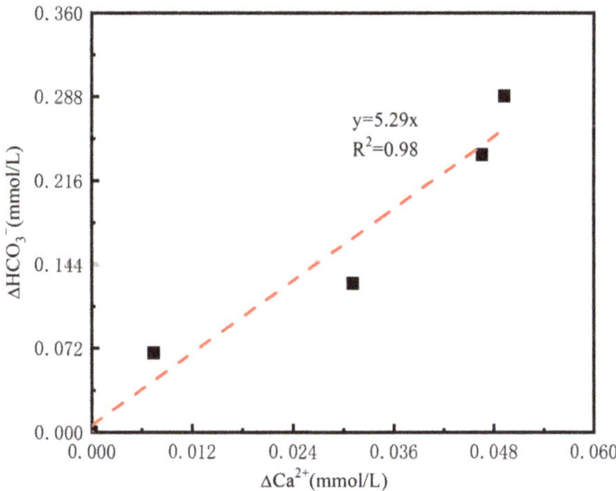

Figure 7. Ca^{2+} vs. HCO_3^- production for experiments in which calcite was the solid phase.

To determine the degradation process of toluene and nitrate in the system, the theoretical and practical correspondence between toluene and nitrate was calculated using the following formula. The consumption of NO_3^- was equal to the concentration of the experimental group NO_3^- plus the reduction of the experimental group NO_3^-. The consumption of NO_3^- and the residual amount of NO_3^- were measured, and the reduction of NO_3^- was calculated. Actual toluene consumption versus theoretical consumption is shown in Figure 8. The slope between the theoretical and actual consumption of toluene is

0.93, which indicates that the degradation roughly followed Equations (12) and (13), which would satisfy the presupposition of the model.

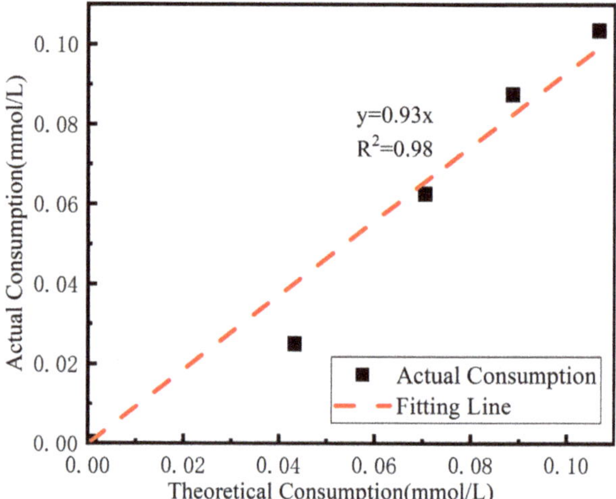

Figure 8. Actual toluene consumption versus theoretical consumption.

Based on the results shown in Figures 6–8 and the physics models introduced earlier, it is necessary to fit the models' equations to the experimental data.

The parameters of the model were set as follows:

The ratio for both nitrate and nitrite at each monitoring time was calculated. Because the proportion of toluene consumed by microbial growth compared to total toluene content (f) has little influence, the median of 6.50% was substituted into the model calculation. According to the experimental design, V was 280 mL, Va was 250 mL, and T was 20 °C.

The SSE, MSE, and RMS error of the Ca^{2+} concentration of the theoretical model and the experimental model were calculated and were 3.15×10^{-9}, 6.31×10^{-10}, and 2.51×10^{-5}, respectively. Initially, the Ca^{2+} concentration was lower than the model-derived value, but it gradually increased and eventually became greater than the model-derived Ca^{2+} concentration (Figure 9a).

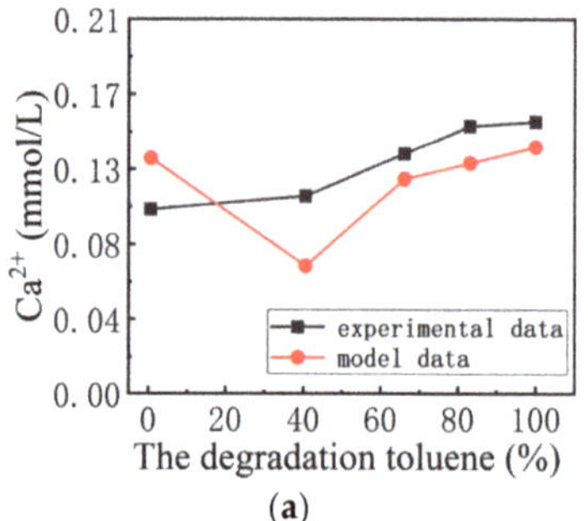
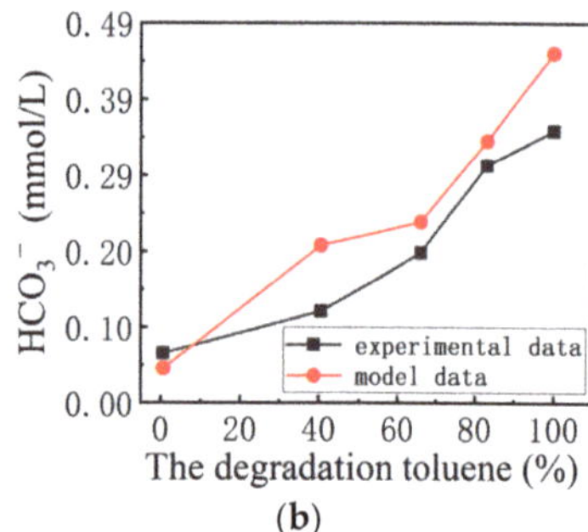

Figure 9. The fitting diagram of experimental data and model data. (**a**) The fitting diagram of Ca^{2+}; (**b**) the fitting diagram of HCO_3^{-}.

The SSE, MSE, and RMS error of the HCO_3^{-} concentration of the theoretical model and the experimental model were calculated to be 2.04×10^{-8}, 4.08×10^{-8}, and 6.38×10^{-5}, respectively. As can be seen from Figure 9b, among the five points, except for the second point, the gap is large, and the other points are well fitted.

After the above rough analysis, the fitting situation of the ions was studied.

Temperature affects the dissolution of calcium carbonate in groundwater. At the first point, the calcium carbonate may not be completely dissolved.

Toluene is converted to metabolic intermediates, such as benzylsuccinate and benzene carboxylic acid, which affect the pH value in the aqueous carbonate system [47,48]. The experimental data of the Ca^{2+} concentration showed slightly higher values than the model data. This may be because the intermediate product is acidic, which affects the ionization of carbon dioxide and promotes the dissolution of calcium carbonate.

The first two points may have belonged to the stage when microorganisms were adapting to their environment. The second to third points marked the period of rapid conversion of nitrate to nitrite, when the concentration of nitrate and nitrite changed rapidly, so there may have been a great difference in monitoring concentration. In addition, this stage also represented the exponential period of microbial growth and reproduction, and the toluene consumption of microorganisms may have been much higher than 5%, which may be the reason for the significant difference between the third point and the model. In the later period, microorganisms were in the decline period, and microorganisms had almost stopped consuming toluene, so the fitting results of the latter two points were slightly poor.

To ensure the survival of microorganisms, small amounts of Mg^{2+} and phosphate plasma were added to the experiment as necessary nutrients. The influence of various ions in the system also caused a difference between the model and the experiment.

3.4. Implications

The derivation of the model is based on part of the life cycle, which refers to the whole process of producing nitrogen, carbon dioxide, and water from the reaction of the reactants toluene and nitrate [49]. The effects of toluene degradation on the calcium carbonate system include many qualitatively complex reactions, and the mathematical model still has some limitations. Therefore, numerical simulation can be used to guide and apply the actual practice in the future, and the model can be better generalized [50]. Thus, it can serve to calculate the amount of degradation of petroleum pollutants by calculating the change in inorganic ion concentrations in the field, avoiding a series of problems associated with field organic tests, such as delayed test results, dangerous test drugs, and secondary pollution of the test solution. At the same time, with the help of the model, the degradation of toluene in a fixed period can be quickly determined, which is of great significance for improving the distribution efficiency of regional groundwater organic pollution.

In the model, NO_3^- is converted to N_2 by the action of microorganisms. According to the research, fungi are capable of this reaction. They possess enzymes involved in denitrification, including nitrate reductase (Nar) and nitrite reductase (Nir). Similarly, bacteria such as nitrite-reducing bacteria and facultative denitrifying bacteria also possess these enzymes and can carry out the reaction in the model [42,51].

In general, denitrifying bacteria have diverse living conditions. The pH range for these microorganisms is generally between 6.5 and 9, and they can thrive in environments with temperatures ranging from 10 to 40 degrees Celsius. For bacteria, the reaction occurs under anaerobic conditions. For fungi, a small amount of oxygen can promote the reaction, but excessive oxygen inhibits the reaction [52]. The environment of most groundwater systems is suitable for the growth of microorganisms. At present, the living environment of microorganisms is not considered in the derivation of the model. In the future, the existing research will be expanded by a series of experiments with different initial conditions, including toluene concentration, intermediate organic acids, and initial pH.

4. Conclusions

Based on hydrogeochemical and organic biomineralization theory, this study obtained conceptual and mathematical models of the effect of toluene mineralization on the dissolution and precipitation of $CaCO_3$.

It was found that several factors may influence the dissolution and precipitation of $CaCO_3$. The first is whether the nitrate is thoroughly reduced to N_2. When nitrate is thoroughly reduced, $CaCO_3$ will precipitate. On the contrary, calcium carbonate dissolution and precipitation depends on the degree of nitrate reduction. The second influencing factor is the proportion of toluene consumed by microbial growth compared to total toluene content (f). For a complete reaction, the greater the consumption of microorganisms is, the smaller the amount of toluene consumed in the reaction and the smaller the precipitation of calcium carbonate. The third factor is temperature (T). The Ca^{2+} concentration in the system is not a monotonic function that varies with temperature. The milligram equivalent of other ions (E) may also affect the dissolution and precipitation of $CaCO_3$ in a nonmonotonic function style.

In summary, toluene mineralization impacts the dissolution and precipitation of $CaCO_3$, which changes with the change in relevant parameters in the mathematical model. The established model provides a tool for evaluating biodegraded hydrocarbons by measuring the NO_3^-, NO_3^-, and Ca^{2+} concentration variations in groundwater.

Author Contributions: Conceptualization, M.Z. and L.S.; methodology, Z.N.; writing—original draft preparation, S.G. and X.W.; writing—review and editing, Y.Z. and S.Y.; validation, Y.X.; formal analysis, C.G. All authors have read and agreed to the published version of the manuscript.

Funding: This research was funded by the CAGS Research Fund, grant number SK202207; the Central Leading Local Science and Technology Development Fund Project, grant number 236Z4204G; the National Natural Science Foundation of China, grant number 42007171; the Sinopec Science Department Project, grant number 322082; the Hebei Natural Science Foundation, grant number D2022504009; and the Collaborative Innovation Center for sustainable utilization of water resources and optimization of industrial structure in Hebei Province, grant number XTZX202102.

Institutional Review Board Statement: Not applicable.

Informed Consent Statement: Not applicable.

Data Availability Statement: The data presented in this study are available on request from the corresponding author.

Conflicts of Interest: Author Shuai Yang, Yan Xie, Xinzhe Wang was employed by the company SINOPEC Research Institute of Safety Engineering Co., Ltd. The remaining authors declare that the research was conducted in the absence of any commercial or financial relationships that could be construed as a potential conflict of interest.

References

1. Zhao, B.; Huang, F.; Zhang, C.; Huang, G.; Xue, Q.; Liu, F. Pollution characteristics of aromatic hydrocarbons in the groundwater of China. *J. Contam. Hydrol.* **2020**, *233*, 103676. [CrossRef] [PubMed]
2. Wang, L.; Wu, W.-M.; Bolan, N.S.; Tsang, D.C.W.; Li, Y.; Qin, M.; Hou, D. Environmental fate, toxicity and risk management strategies of nanoplastics in the environment: Current status and future perspectives. *J. Hazard. Mater.* **2021**, *401*, 123415. [CrossRef] [PubMed]
3. Busygina, T.; Rykova, V. Scientometric analysis and mapping of documentary array on the issue "Oil and petroleum products in soil and groundwater". *Environ. Sci. Pollut. Res.* **2020**, *27*, 23490–23502. [CrossRef] [PubMed]
4. Hu, S.; Xiao, C.; Liang, X.; Cao, Y.; Wang, X.; Li, M. The influence of oil shale in situ mining on groundwater environment: A water-rock interaction study. *Chemosphere* **2019**, *228*, 384–389. [CrossRef] [PubMed]
5. Singh, S.; Gupta, V.K. Biodegradation and bioremediation of pollutants: Perspectives strategies and applications. *Int. J. Pharmacol. Biol. Sci.* **2016**, *10*, 53.
6. Wu, Z.; Liu, G.; Ji, Y.; Li, P.; Yu, X.; Qiao, W.; Wang, B.; Shi, K.; Liu, W.; Liang, B.; et al. Electron acceptors determine the BTEX degradation capacity of anaerobic microbiota via regulating the microbial community. *Environ. Res.* **2022**, *215*, 114420. [CrossRef] [PubMed]
7. Huang, H.; Jiang, Y.; Zhao, J.; Li, S.; Schulz, S.; Deng, L. BTEX biodegradation is linked to bacterial community assembly patterns in contaminated groundwater ecosystem. *J. Hazard. Mater.* **2021**, *419*, 126205. [CrossRef]
8. Davidson, C.J.; Hannigan, J.H.; Bowen, S.E. Effects of inhaled combined Benzene, Toluene, Ethylbenzene, and Xylenes (BTEX): Toward an environmental exposure model. *Environ. Toxicol. Pharmacol.* **2021**, *81*, 103518. [CrossRef]

9. Cheng, L.; Wei, W.; Guo, A.; Zhang, C.; Sha, K.; Wang, R.; Wang, K.; Cheng, S. Health risk assessment of hazardous VOCs and its associations with exposure duration and protection measures for coking industry workers. *J. Clean. Prod.* **2022**, *379*, 134919. [CrossRef]

10. Alenezi, R.A. BTEX and nitrogen oxides fume assessments in fuel dispense stations and residential areas. *J. Eng. Res.* **2022**, *10*, 50–59. [CrossRef]

11. Alenezi, R.A.; Aldaihan, N. Impact of fuel dispensing stations in the vicinity residential homes on the indoor and outdoor air quality. *Int. J. Environ. Sci. Technol.* **2019**, *16*, 2783–2796. [CrossRef]

12. Péquin, B.; Cai, Q.; Lee, K.; Greer, C.W. Natural attenuation of oil in marine environments: A review. *Mar. Pollut. Bull.* **2022**, *176*, 113464. [CrossRef] [PubMed]

13. Haider, F.U.; Ejaz, M.; Cheema, S.A.; Khan, M.I.; Zhao, B.; Liqun, C.; Salim, M.A.; Naveed, M.; Khan, N.; Nunez-Delgado, A. Phytotoxicity of petroleum hydrocarbons: Sources, impacts and remediation strategies. *Environ. Res.* **2021**, *197*, 111031. [CrossRef] [PubMed]

14. Kharangate-Lad, A.; D'Souza, N.C. Current Approaches in Bioremediation of Toxic Contaminants by Application of Microbial Cells; Biosurfactants and Bioemulsifiers of Microbial Origin. In *Rhizobiont in Bioremediation of Hazardous Waste*; Springer: Singapore, 2021; pp. 217–263. [CrossRef]

15. Zhang, M.; Guo, C.; Shi, C.; Ning, Z.; Chen, Z. A quantitative redox zonation model for developing natural attenuation-based remediation strategy in hydrocarbon-contaminated aquifers. *J. Clean. Prod.* **2021**, *290*, 125743. [CrossRef]

16. Budania, R.; Dangayach, S. A comprehensive review on permeable reactive barrier for the remediation of groundwater contamination. *J. Environ. Manag.* **2023**, *332*, 117343. [CrossRef] [PubMed]

17. Weelink, S.A.B.; van Eekert, M.H.A.; Stams, A.J.M. Degradation of BTEX by anaerobic bacteria: Physiology and application. *Rev. Environ. Sci. Bio/Technol.* **2010**, *9*, 359–385. [CrossRef]

18. Dordević, D.; Jančíková, S.; Vítězová, M.; Kushkevych, I. Hydrogen sulfide toxicity in the gut environment: Meta-analysis of sulfate-reducing and lactic acid bacteria in inflammatory processes. *J. Adv. Res.* **2021**, *27*, 55–69. [CrossRef]

19. Chen, X.; Sheng, Y.; Wang, G.; Guo, L.; Zhang, H.; Zhang, F.; Yang, T.; Huang, D.; Han, X.; Zhou, L. Microbial compositional and functional traits of BTEX and salinity co-contaminated shallow groundwater by produced water. *Water Res.* **2022**, *215*, 118277. [CrossRef]

20. Sohrabi, T.; Shakiba, M.; Mirzaei, F.; Pourbabaee, A. BTEX biodegradation using *Bacillus* sp. in a synthetic hypoxic aquatic environment: Optimization by Taguchi-based design of experiments. *Int. J. Environ. Sci. Technol.* **2022**, *19*, 5571–5578. [CrossRef]

21. Han, K.; Hong, U.; Park, S.; Kwon, S.; Kim, Y. In situ field method for evaluating biodegradation potential of BTEX by indigenous heterotrophic denitrifying microorganisms in a BTEX-contaminated fractured-rock aquifer. *Environ. Technol.* **2021**, *42*, 1326–1335. [CrossRef]

22. Sra, K.S.; Ponsin, V.; Kolhatkar, R.; Hunkeler, D.; Thomson, N.R.; Madsen, E.L.; Buscheck, T. Sulfate Land Application Enhances Biodegradation in a Petroleum Hydrocarbon Smear Zone. *Groundw. Monit. Remediat.* **2023**, *43*, 44–59. [CrossRef]

23. Zhu, H.H.; Chao, J.I.A.; Xu, Y.L.; Yu, Z.M.; Yu, W.J. Study on numerical simulation of organic pollutant transport in groundwater northwest of Laixi. *J. Groundw. Sci. Eng.* **2018**, *6*, 293–305. [CrossRef]

24. Yang, L.-Z.; Liu, C.-H. Study on the characteristics and causes of carbon tetrachloride pollution of karst water in eastern suburbs of Jinan. *J. Groundw. Sci. Eng.* **2015**, *3*, 331–341. [CrossRef]

25. Jiang, H.; Chen, D.; Zheng, D.; Xiao, Z. Anaerobic mineralization of toluene by enriched soil-free consortia with solid-phase humin as a terminal electron acceptor. *Environ. Pollut.* **2023**, *317*, 120794. [CrossRef] [PubMed]

26. He, S.; Li, P.; Su, F.; Wang, D.; Ren, X. Identification and apportionment of shallow groundwater nitrate pollution in Weining Plain, northwest China, using hydrochemical indices, nitrate stable isotopes, and the new Bayesian stable isotope mixing model (MixSIAR). *Environ. Pollut.* **2022**, *298*, 118852. [CrossRef] [PubMed]

27. Lahjouj, A.; Hmaidi, A.E.; Bouhafa, K. Spatial and statistical assessment of nitrate contamination in groundwater: Case of Sais Basin, Morocco. *J. Groundw. Sci. Eng.* **2020**, *8*, 143–157.

28. Liang, C.; Li, C.; Zhu, Y.; Du, X.; Yao, C.; Ma, Y.; Zhao, J. Recent advances of photocatalytic degradation for BTEX: Materials, operation, and mechanism. *Chem. Eng. J.* **2023**, *455*, 140461. [CrossRef]

29. Zainab, R.; Hasnain, M.; Ali, F.; Dias, D.A.; El-Keblawy, A.; Abideen, Z. Exploring the bioremediation capability of petroleum-contaminated soils for enhanced environmental sustainability and minimization of ecotoxicological concerns. *Environ. Sci. Pollut. Res.* **2023**, *30*, 104933–104957. [CrossRef]

30. De Muynck, W.; De Belie, N.; Verstraete, W. Microbial carbonate precipitation in construction materials: A review. *Ecol. Eng.* **2010**, *36*, 118–136. [CrossRef]

31. Fedorov, K.; Sun, X.; Boczkaj, G. Combination of hydrodynamic cavitation and SR-AOPs for simultaneous degradation of BTEX in water. *Chem. Eng. J.* **2021**, *417*, 128081. [CrossRef]

32. Spence, M.J.; Bottrell, S.H.; Thornton, S.F.; Richnow, H.H.; Spence, K.H. Hydrochemical and isotopic effects associated with petroleum fuel biodegradation pathways in a chalk aquifer. *J. Contam. Hydrol.* **2005**, *79*, 67–88. [CrossRef] [PubMed]

33. Ferguson, B.; Agrawal, V.; Sharma, S.; Hakala, J.A.; Xiong, W. Effects of Carbonate Minerals on Shale-Hydraulic Fracturing Fluid Interactions in the Marcellus Shale. *Front. Earth Sci.* **2021**, *9*, 695978. [CrossRef]

34. Cai, P.; Ning, Z.H.; Zhang, N.; Zhang, M.; Guo, C.; Niu, M.; Shi, J. Insights into Biodegradation Related Metabolism in an Abnormally Low Dissolved Inorganic Carbon (DIC) Petroleum-Contaminated Aquifer by Metagenomics Analysis. *Microorganisms* **2019**, *7*, 412. [CrossRef] [PubMed]
35. Laczi, K.; Erdeiné Kis, Á.; Szilágyi, Á.; Bounedjoum, N.; Bodor, A.; Vincze, G.E.; Kovács, T.; Rákhely, G.; Perei, K. New frontiers of anaerobic hydrocarbon biodegradation in the multi-omics era. *Front. Microbiol.* **2020**, *11*, 590049. [CrossRef] [PubMed]
36. Yu, B.; Yuan, Z.; Yu, Z.; Xue-song, F. BTEX in the environment: An update on sources, fate, distribution, pretreatment, analysis, and removal techniques. *Chem. Eng. J.* **2022**, *435*, 134825. [CrossRef]
37. Jiang, W.; Sheng, Y.; Wang, G.; Shi, Z.; Liu, F.; Zhang, J.; Chen, D. Cl, Br, B, Li, and noble gases isotopes to study the origin and evolution of deep groundwater in sedimentary basins: A review. *Environ. Chem. Lett.* **2022**, *20*, 1497–1528. [CrossRef]
38. Min, Z. Mechanism and Simulation of Response on GroundwaterChemical Compositions to Hydrodynamic Conditions. *Min. Sci. Technol.* **2009**, *19*, 435–440.
39. *ISO 10304-1:2007/Cor 1:2010*; Water Quality—Determination Of Dissolved Anions By Liquid Chromatography Of Ions—Part 1: Determination Of Bromide, Chloride, Fluoride, Nitrate, Nitrite, Phosphate And Sulfate. ISO Copyright Office: Geneva, Switzerland, 2010.
40. Burland, S.M.; Edwards, E.A. Anaerobic benzene biodegradation linked to nitrate reduction. *Appl. Environ. Microbiol.* **1999**, *65*, 529–533. [CrossRef]
41. Durand, S.; Guillier, M. Transcriptional and Post-transcriptional control of the nitrate respiration in bacteria. *Front. Mol. Biosci.* **2021**, *8*, 667758. [CrossRef]
42. Song, T.; Zhang, X.; Li, J.; Wu, X.; Feng, H.; Dong, W. A review of research progress of heterotrophic nitrification and aerobic denitrification microorganisms (HNADMs). *Sci. Total Environ.* **2021**, *801*, 149319. [CrossRef]
43. Mitchell, M.J.; Jensen, O.E.; Cliffe, K.A.; Maroto-Valer, M.M. A model of carbon dioxide dissolution and mineral carbonation kinetics. *Proc. R. Soc. A Math. Phys. Eng. Sci.* **2009**, *466*, 1265–1290. [CrossRef]
44. Cordova, C.E. Plows, Plagues, and Petroleum: How Humans Took Control of Climate. *Holocene* **2010**, *20*, 653. [CrossRef]
45. Niu, J.; Liu, Q.; Lv, J.; Peng, B. Review on microbial enhanced oil recovery: Mechanisms, modeling and field trials. *J. Pet. Sci. Eng.* **2020**, *192*, 107350. [CrossRef]
46. Peña-Calva, A.; Olmos-Dichara, A.; Viniegra-González, G.; Cuervo-López, F.M.; Gómez, J. Denitrification in presence of benzene, toluene, and m-xylene. *Appl. Biochem. Biotechnol.* **2004**, *119*, 195–208. [CrossRef] [PubMed]
47. Ahmed, S.; Kumari, K.; Singh, D. Different strategies and bio-removal mechanisms of petroleum hydrocarbons from contaminated sites. *Arab. Gulf J. Sci. Res.* **2023**. ahead of print. [CrossRef]
48. Muccee, F.; Ejaz, S.; Riaz, N. Toluene degradation via a unique metabolic route in indigenous bacterial species. *Arch. Microbiol.* **2019**, *201*, 1369–1383. [CrossRef] [PubMed]
49. Tsui, T.-H.; van Loosdrecht, M.C.; Dai, Y.; Tong, Y.W. Machine learning and circular bioeconomy: Building new resource efficiency from diverse waste streams. *Bioresour. Technol.* **2022**, *369*, 128445. [CrossRef]
50. Tsui, T.-H.; Zhang, L.; Zhang, J.; Dai, Y.; Tong, Y.W. Engineering interface between bioenergy recovery and biogas desulfurization: Sustainability interplays of biochar application. *Renew. Sustain. Energy Rev.* **2022**, *157*, 112053. [CrossRef]
51. Cao, S.; Du, R.; Zhou, Y. Coupling anammox with heterotrophic denitrification for enhanced nitrogen removal: A review. *Crit. Rev. Environ. Sci. Technol.* **2021**, *51*, 2260–2293. [CrossRef]
52. Rassamee, V.; Sattayatewa, C.; Pagilla, K.; Chandran, K. Effect of oxic and anoxic conditions on nitrous oxide emissions from nitrification and denitrification processes. *Biotechnol. Bioeng.* **2011**, *108*, 2036–2045. [CrossRef]

applied sciences

Article

Experimental Investigation about Oil Recovery by Using Low-Salinity Nanofluids Solutions in Sandstone Reservoirs

Nannan Liu [1], Shanazar Yagmyrov [1], Hengchen Qi [1] and Lin Sun [2,*]

1   School of Petroleum and Natural Gas Engineering, Changzhou University, Changzhou 213164, China; liunan2020@cczu.edu.cn (N.L.); shanazaryagmyrov@gmail.com (S.Y.); q450753020@163.com (H.Q.)
2   Institute of Hydrogeology and Environmental Geology, Chinese Academy of Geological Sciences, Shijiazhuang 050061, China
*   Correspondence: lin_sun166@163.com

Abstract: Production of crude oil from matured oil reservoirs has major issues due to decreased oil recovery with water channeling; however, the low-salinity water flooding technique is more commonly used to maximize recovery of the remaining oil. In this study, we demonstrated a new hybridization technique of combining low-salinity water and nanofluids; this was achieved by using experiments such as contact angle measurement with water of different salinity levels and nanofluid concentrations, core displacement, and NMR (nuclear magnetic resonance) between low-/high-permeability rock. The trial results demonstrated that the test with KCl-1+NF outperformed those with other compositions by changing the original contact angle from 112.50° to 53.3° and increasing formation production up to 15 cc. In addition, we saw that when 2 PV of KCl-1+NF was injected at a rate of 5 mL/min, the middle pores' water saturation dropped quickly to 73% and then steadily stabilized in the middle and late stages. Regarding the novel application of the hybridization technique, the insights presented in this paper serve as a helpful resource for future studies in this field.

Keywords: low-salinity water; silica nanofluids; EOR; wettability alteration; NMR

Citation: Liu, N.; Yagmyrov, S.; Qi, H.; Sun, L. Experimental Investigation about Oil Recovery by Using Low-Salinity Nanofluids Solutions in Sandstone Reservoirs. *Appl. Sci.* **2024**, *14*, 23. https://doi.org/10.3390/app14010023

Academic Editor: Matt Oehlschlaeger

Received: 15 November 2023
Revised: 17 December 2023
Accepted: 18 December 2023
Published: 19 December 2023

1. Introduction

According to forecasts that global energy consumption will go up by roughly 50% over the next 20 years, the world's energy demand is likely to continue rising in the future. The oil and gas industries are required to find a cost-effective strategy to recover remaining oil from reservoirs to fulfill the increasing global energy demand [1]. To extract the oil that primary or secondary recovery methods were unable to produce, EOR (enhanced oil recovery) techniques are utilized. The largest oil reserves in the world are found in sandstone reservoirs. A sandstone reservoir is a layer of sandstone that contains petroleum that may be extracted using existing technology. Sandstone reservoirs often contain fluid-filled pores, stable minerals, and supplemental minerals. Grains of sand must range in size from 1/8 to 2 mm for a rock to be referred to as sandstone [2–7]. Compared to oil-wet reservoirs, recovering oil from water-wet reservoirs is far simpler [8]. EOR is attracting increasing interest, since oil is in greater demand than ever before, hence the need to increase oil production [9].

Nanofluids, which are nanotechnology-based fluids containing nanoparticles, have emerged as a promising solution for enhanced oil recovery. The advancement of nanotechnology has led to the introduction of nanofluids as an economical, effective, and environmentally friendly replacement for existing chemicals [10]. Nanofluids might interact with the rock surface by chemical and physical mechanisms when they are injected into an oil reservoir. The rock's wettability can be altered from oil-wet to water-wet or intermediate-wet by the nanoparticles adhering to the surface and forming a thin layer. Better oil recovery is encouraged by this change in wettability because it strengthens the capillary forces between the fluid and the rock. A nanoparticle is one with a diameter of 1–100 nanometers (nm) [11]. A technique called "nanofluid flooding" is used in EOR operations to increase the efficiency of oil recovery by dispersing nanoparticles in a fluid and injecting them into oil reservoirs [12]. When creating a nanofluid, selecting the right nanoparticles is essential. For example, silica nanoparticles have demonstrated encouraging outcomes in modifying wettability and enhancing oil recovery. Because of their large surface area and ability to absorb the rock surface, these nanoparticles can facilitate oil displacement by lowering the interfacial tension between water and oil. Through several mechanisms, such as decreased interfacial tension and altered wettability, nanofluid flooding can enhance oil recovery. During the reduction of interfacial tension, the tension that exists between the oil and the injected nanofluid can be lessened by the presence of nanoparticles dispersed throughout the fluid. By lowering the interfacial tension, the oil can be more efficiently displaced by the nanofluid, increasing oil recovery. When oil and water come into contact, the nanoparticles group together to produce a stable layer that helps release trapped oil. The fluid's contact angle with the reservoir rock surface can be altered by nanoparticles. Wetness conditions may shift from oil-wet to water-wet or intermediate-wet as a result of this shift [13].

For decades, it was considered that injecting anything other than saline water into a reservoir could permanently destroy it, but we noticed that by reducing the salinity of injected water, trapped oil molecules can be more easily released from the surface of rock [14]. At the moment, the primary problem associated with enhanced oil recovery is wettability. The goal of the suggested EOR technique was to lower the oil's affinity and increase oil recovery by turning the reservoir from an oil-wet state to a water-wet. Wettability is important because the amount of oil recovered from water-wet reservoirs is higher than that from oil-wet reservoirs. Low-salinity water (LSW) flooding is a novel approach that has been effectively applied in several industries. It can convert an oil-wet sandstone reservoir to a water-wet one by changing the wetting characteristics of the reservoir rock. One of the most crucial factors to take into account is the ionic concentration and composition of low-salinity water. In our research, we demonstrated that oil molecules are bonded to clay particles on the surface of rock with the help of divalent cations. The electrical force of highly ionically concentrated saline water will force divalent cations to the clay surface [15]. If the salinity of water is reduced, divalent cations can expand, and mono-valent ions can access and replace their places [16]. LSW flooding is a new technology that has been successfully employed in various areas and affects the reservoir rock's wetting properties, thus increasing oil recovery; it can turn an oil-wet sandstone reservoir into a water-wet one [17]. The ionic composition of injected water is one of the main parameters to be considered. A study revealed deionized water has more potential to be used as an EOR technique than regular seawater. By injecting the right concentration of LSW, we can change the wettability of the reservoir [18].

To address the limitations of low-salinity water flooding, we propose a hybrid technique that combines low-salinity water flooding with nanofluids. This approach in EOR can improve the disadvantages of LSW flooding such as water channeling and improve sweep efficiency during flooding operations [19]. The nanofluids can modify fluid–rock interactions, while LSW can alter the wettability and promote spontaneous imbibition, leading to higher oil recovery [20]. LSW flooding can change the wettability of the rock surface by making it more favorable for displacement [21]. Nanofluids, on the other hand, can further enhance wettability alteration by providing additional surface activity through nanoparticle adsorption and interaction with the rock surface [22]. This combined effect can lead to improved oil mobilization and displacement [23].

In this study, we used sandstone rocks as the research object, and the treated thin slices were immersed in formation crude oil to form an oil film. The wettability angle of the thin slices' surface in both LSW solution and pure water at different times was measured. The formation of an oil film on the sandstone rock surface and the mechanism of wettability alternation were explained. We concluded that LSW-based NF flooding has many advantages for use as an EOR technique. Below, we present all the needed data and the method of our investigation.

2. Experiments

2.1. Experimental Materials

We used Berea sandstone core samples provided by Shengli Oilfield (Dongying, China) during these experiments. The specifications of core samples are shown in Table 1.

Table 1. Core sample properties.

Formation	Average Permeability (mD)	Average Porosity (%)	Dimensions	
			Diameter (mm)	Thickness (mm)
Sandstone	560	26	25	2~4

X-ray diffractometer (XRD) and core analyses were provided by Shengli Oilfield (Dongying, China) specialists; details are summarized in Table 2.

Table 2. XRD mineral composition test results.

SiO_2	Al_2O_3	FeO/Fe_2O_3	TiO_2	CaO	MgO	Alkalis	H_2O	N/D	Total
85.38	6.47	1.23	0.75	1.35	0.20	2.33	1.53	0.76	100

The density of light crude oil is 0.85 g/cm^3 with a viscosity of 65 cp; our core samples were acquired from a heavy oil reservoir (15 MPa, 80 °C) in Shengli Oilfield, Dongying, Shandong, China. Detailed specifications are listed in Table 3.

Table 3. Crude oil properties.

Gravity (API)	TAN (mg/gm)	Kinematic Viscosity (cSt)	Initial Oil Saturation (%)
20.6	0.8	325	57

The different salinity levels of KCl were compounded, and named KCl-1 (TDS = 500 mg/L) and KCl-2 (TDS = 6500 mg/L). Low-salinity formulations are listed in Table 4.

Table 4. Brine formulations.

Brine	K^+	Cl^-	TDS, mg/L
KCL-1	262.2	237.8	500
KCL-2	3408.6	3091.4	6500

The nanoparticles that we used are mono-dispersed commercial hydrophilic silica nanoparticles (SiO_2) of 99% purity. The surface area of SiO_2 is 180–600 m^2/g, and average particle size ranges from 20 nm to 30 nm.

2.2. Experimental Apparatus and Main Procedures

We divided our experiment into two main steps. The first step is to obtain the wettability measurements of rock core samples by using the contact angle method with different solutions including saline solutions, nanofluids, and a mixture solution of salt and nanofluids. The second step is to observe the effects of the hybridization on EOR, based on the core displacement results from nuclear magnetic resonance. The core displacement experiments provide data on breakthrough times, flow rates, and the displacement efficiency of various compositions. The fluid distribution and saturation changes that occurred within the sandstone cores during the displacement process could be seen and characterized thanks to the NMR analysis. This study verified the quick decrease in water saturation and the stabilization that followed, which showed that the oil had been successfully displaced. The details on each step can be found below.

2.2.1. Contact Angle Measurement

A commonly used method for determining the wettability characteristics of a solid surface by dripping a liquid droplet is contact angle measurement. It offers important insights regarding the interaction between liquid and solid at the interface, and is especially pertinent when researching phenomena associated with rock wetting. Rock wettability also can be measured by using the contact angle value. Angle (θ) created at the contact line of the liquid–fluid interface and the solid substrate is typically used to calculate the contact angle [24]. In applications involving oil recovery, the contact angle of a droplet of water or oil is measured for a rock surface that has already been treated with the substance anticipated to have an impact on the wettability [25–27]. The experimental steps were as follows and shown in Figure 1:

1. We polished the sandstone rock slices and washed and dried them with deionized water, immersed the slices in light crude oil, and then placed them inside a drying oven at 60 °C for 10 days. After taking out pieces from the oil, we rinsed them quickly with N-Heptane (C_7H_{16}). We then left them to air dry at room temperature for 1 day. We used contact angle measurement equipment to measure the wetting angle;
2. At room temperature, we prepared KCl LSW by adding 500 mg/6500 mg KCl to 1 L of deionized water. The solution is stirred for 1 h by using a magnetic stirrer to make sure that all KCl was fully dissolved in the deionized water. For the SiO_2 solution, we added 4 g of SiO_2 nanoparticles to 1 L of deionized water. The solution was stirred for 1 h using a magnetic stirrer, and after that, we used an ultrasonic cleanser to fully break up and dissolve all the formed SiO_2 clumps. Then, the rock slices were immersed in the solutions, and we then took pictures of the wetting angle at 6 h, 12 h, 24 h, 36 h, 48 h, 60 h, and 72 h timepoints;
3. Using contact angle measuring equipment, we measured the wetting angle for each solution at different times;
4. After obtaining the best result from the LSW and NF solutions, we mixed them to form a mixed hybrid solution. We placed the oil-wet rock slices in the solutions, and then took pictures of the wetting angle at 6 h, 12 h, 24 h, 36 h, 48 h, 60 h, 72 h timepoints;
5. To summarize, oil-wet rock slices were soaked in different solutions. By measuring the wetting angle and recording the difference in the change in the angles, we analyzed and compared them. Using, Excel v.2021 and Origin v.2022 we obtained post-processing results.

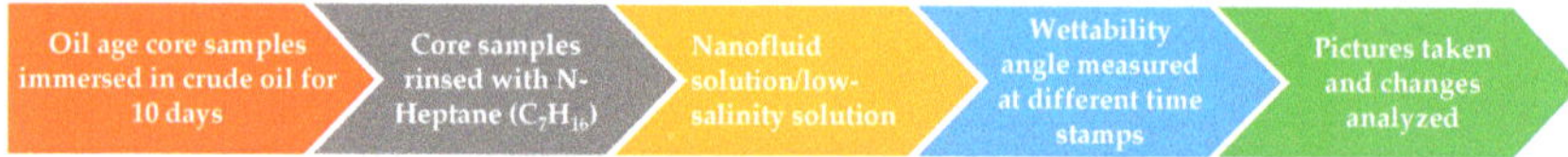

Figure 1. Wettability measurement experiment flow chart.

2.2.2. Core Displacement

A core displacement experiment is a laboratory method that we used to model and examine fluid flow and displacement processes in a porous medium. It is sometimes referred to as a displacement test or core flood test. To investigate the displacement efficiency, fluid flow behavior, and several factors influencing fluid–fluid and fluid–rock interactions, one fluid (LSW, NF, and a mixed hybrid solution) must be injected into a prepared core sample saturated with another fluid (light crude oil). The detailed core flooding experimental steps are as follows and shown in Figures 2 and 3:

6. Core Preparation: Cylindrical rock samples, known as core plugs, are extracted from reservoir rocks and prepared for experimentation. The cores are cleaned, dried, and saturated with light crude oil. Saturation is carried out by using kerosene. First, we vacuum-cleaned rock cores to be sure that any other foreign matter was not left in the rock pores. After that, we saturated them with kerosene at 5 MPa for 1 day.

7. Saturation and Injection: Core plugs were placed in a core holder, and the desired fluid (low-salinity water, nanofluid, and mixed hybrid solution) was injected into the core with a certain pressure and flow rate. The injection was performed using a displacement pump, also known as an ISCO pump.

8. Fluid Flow Monitoring: To evaluate fluid flow behavior, saturation profiles, and recovery efficiency during the injection process, injected fluid volumes and the pressure drop of the core were tracked. These data aid in assessing how well flooding with nanofluid and low-salinity water works to remove oil from the core.

9. Recovery and Analysis: After the displacement process, the core was extracted, and the residual oil saturation was measured. Additional analysis, such as fluid composition analysis or imaging techniques, can be performed to further evaluate the fluid–rock interactions and recovery mechanisms.

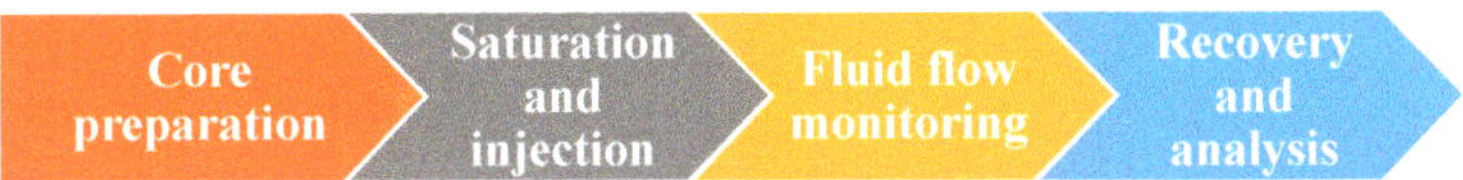

Figure 2. Core displacement experiment flow chart.

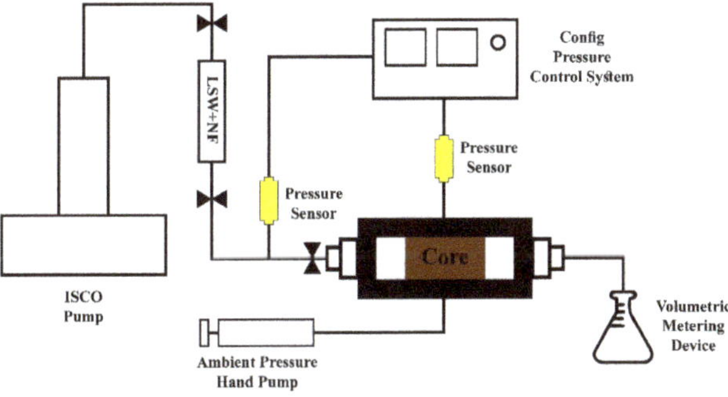

Figure 3. Core displacement experiment scheme.

2.2.3. NMR (Nuclear Magnetic Resonance)

Reservoir rocks and other porous media can be studied using NMR, an effective method for studying fluid behavior and interactions [28]. NMR provides insights into fluid distribution, pore size distribution, fluid-phase characterization, and wettability [29]. We used $MnCl_2$ solution for NMR spectroscopy instead of distilled water, because manganese chloride solution is a frequent paramagnetic shift reagent in NMR spectroscopy. Paramagnetic shift reagents are chemicals that interact with the NMR instrument's magnetic field, causing a change in the chemical shift of adjacent nuclei, resulting in a shift in the NMR [30]. The Mn_2^+ ion in manganese chloride has an unpaired electron, making it paramagnetic. When Mn_2^+ ions are given to a sample, they interact with the magnetic field of the NMR instrument, causing a paramagnetic shift in adjacent nuclei such as proton (1H) and carbon-13 (13C). This shift can help to improve the resolution and sensitivity of the NMR spectra by enhancing the NMR signals of these nuclei. Manganese chloride is a popular paramagnetic shift reagent in NMR spectroscopy because it is widely available, reasonably priced, and gives strong signal enhancement for a wide variety of nuclei [31]. The detailed experimental steps of NMR are as follows and shown in Figure 4:

10. Sample Preparation: Similar to the core displacement method, rock core samples were extracted and prepared for experimentation. The cores were cleaned, dried, and saturated with kerosene. For NMR, we used unsaturated dry rocks and rocks saturated with kerosene and deionized water, first to observe all rock information, and then to observe the difference they make.

11. NMR Measurements: The saturated cores were placed within an NMR instrument, and NMR measurements were performed. NMR signals were generated by the hydrogen nuclei (protons) present in the fluids within the rock pores.

12. Fluid Characterization: Information on fluid-phase characteristics, including fluid composition, saturation, and distribution, was provided by NMR. It helps in understanding how low-salinity water, nanofluids, and mixed hybrid solutions interact with the reservoir rock, and how they affect the wettability and displacement of oil.

13. Time-Lapse Monitoring: NMR can also be used for time-lapse monitoring to observe fluid displacement and flow processes within rock cores. This helps in tracking changes in fluid saturation, flow patterns, and recovery efficiency during low-salinity water, nanofluid, and mixed hybrid solution flooding experiments.

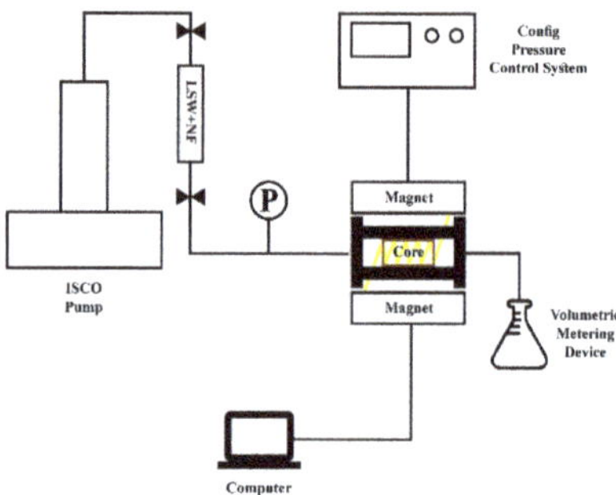

Figure 4. Schematic diagram of NMR experimental setup.

3. Results

3.1. Wettability Changes

The contact angle of the rock sample progressively rose, and its development rate was rapid, as seen in Figure 5. The wetting angle increment was mild and stabilized after the seventh day. As the wetting angle shifted from the original 22.5° to 118°, the rock sample's surface changed from water-wet to oil-wet.

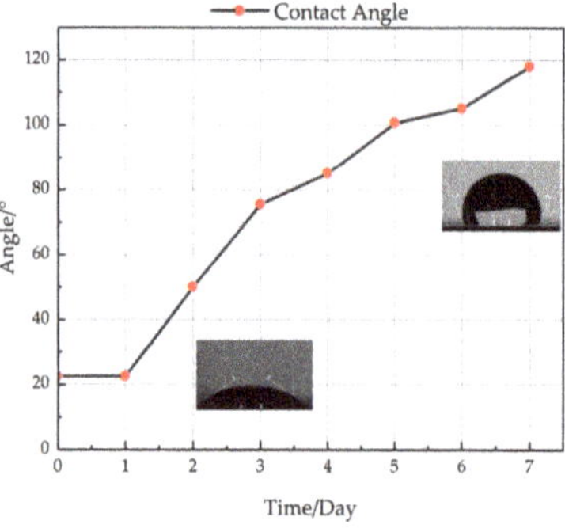

Figure 5. Progression of the static contact angle of rock sample over time (days).

After that, we treated rock samples with pure water, KCL-1, and KCL-1+NF solutions. As shown in Figure 6, it is possible to see how low salinity affects the change in wettability. Figure 6a shows the effect of pure water, Figure 6b the effect of KCL-2 solution, Figure 6c that of KCL-1 solution, and Figure 6d that of the mixed KCL-1 and NF solution. The experiment's findings showed that as opposed to pure water, KCl changed the contact angle by making it more water-wet. The combination solution including KCL-1 and NF showed the biggest alterations in wettability, followed by KCL-1 solution. The KCL-1+NF's contact angle showed a notable change in wettability, changing the initial angle from 112.50° to 53.3°.

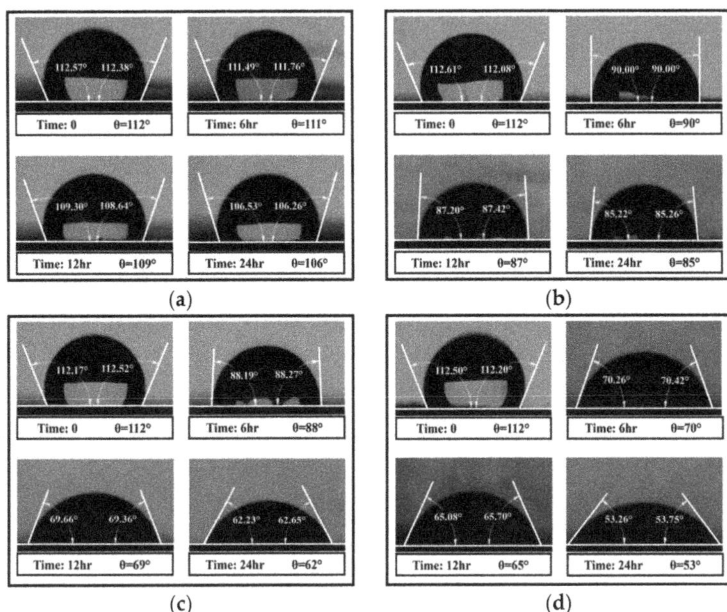

Figure 6. Contact angle variation for different solutions: (**a**) pure water; (**b**) KCl-2 solution; (**c**) KCl-1 solution; (**d**) KCl-1+NF solution.

3.2. Core Flooding

To observe the effects of brines during core flooding, we used pure water, KCl-1, KCl-2, and KCL-1+NF with SAS concentration set at 1/0 wt%. The KCl-1+NF injection pattern is depicted by the red curve in Figure 7. The injection pattern of KCl-2 is represented by the yellow curve. Finally, the pure water injection pattern is represented by the green curve. As demonstrated in Figure 7, in the case of KCl-1+NF injection, the residual oil recovery rate increases up to 0.5 PV directly after the injection. This is the key indicator of the favorable conditions generated by LSW's wettability-altering features. Starting from 0.5 PV to 2 PV, the recovery rate is not visible, passing the barrier when a minimum of 2 PV was produced. Our observation showed during this interval, the maximum recovery rate was achieved by the KCl-1+NF injection pattern. This is because of the significant wettability changes. Oil recovery can be increased in addition to the wettability adjustment by lowering the IFT between the displacing fluid and oil. This was primarily the function of the mixed composition's injected NF, which also allows for improved mobility control and sweep efficiency.

3.3. Fluid-Flow Characteristics Based on NMR

Matching MR images are displayed in Figure 8, showing the situation after the injection rate is changed from 1 mL/min to 5 mL/min. The displacement areas are visible in the high-permeability and low-permeability layers at 5 PV. As the injection continues, the high-permeability layer's displacement front moves clearly, and the low-permeability layer's displacement front's position barely changes between 10 and 15 PV. Conversely, we discovered that in both the high and low-permeability layers, the displacement front moves at the same pace.

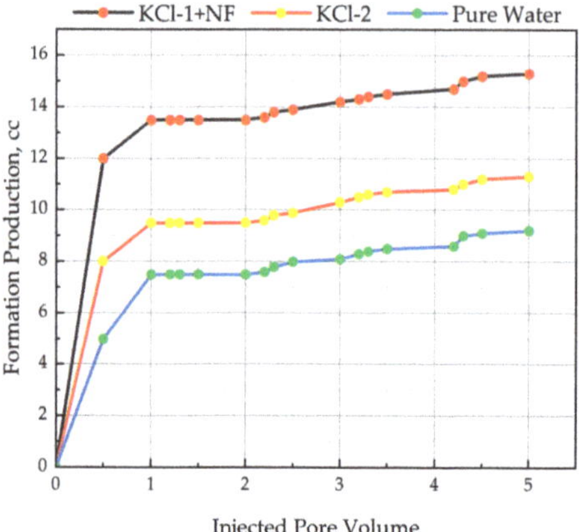

Figure 7. The total amount of water generated over time as a result of oil injection.

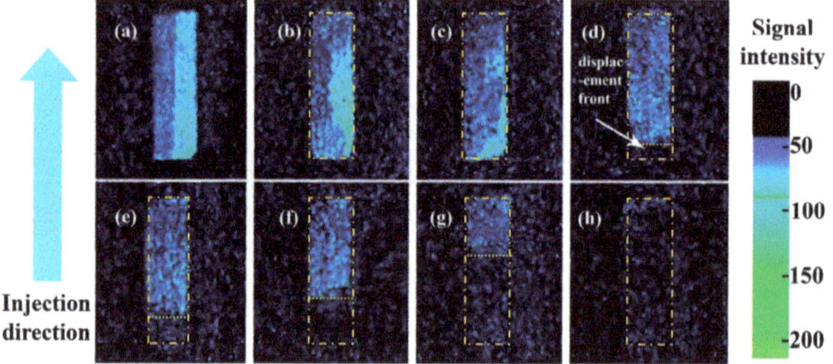

Figure 8. Images taken by MR: (**a**) 0.0 PV; (**b**) 0.6 PV; (**c**) 1.1 PV; (**d**) 4.0 PV; (**e**) 5.5 PV; (**f**) 7.5 PV; (**g**) 10.5 PV (**h**) 15.0 PV.

When 2 PV of KCl-1+NF is injected at a rate of 5 mL/min, Figure 9 shows how the water saturation in the middle pores quickly declines to 73%, and then gradually flattens in the middle and late stages. The water saturation in the medium pores decreases to 35.9% at 15 PV, which is a significantly higher percentage than what KCl-2 produced. Raising the injection rate reduces the impact of fluid displacement in the medium and small holes, as shown by the observation that the water saturation in small pores only decreases to 63.6%. This may be the result of a high fluid input rate decreasing jamming capacity. The fast injection rate quickly causes a shear-thinning effect, which drastically lowers the fluid's apparent viscosity. Increasing the injected solution sweep efficiency in the small and medium holes becomes more difficult when the blocking resistance in the large pores decreases.

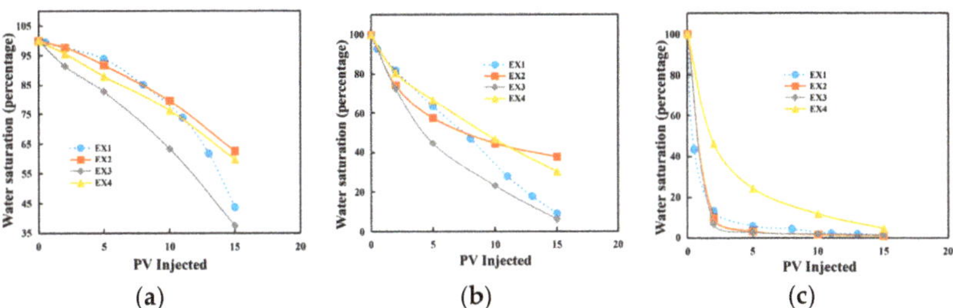

Figure 9. Water saturation during injection in various kinds of pores: (**a**) small (φ24.7%); (**b**) medium (φ25.5%); (**c**) large (φ27.3%).

4. Discussion

This study took into account the benefits of combining LSW with NF as well as the problems associated with NF flooding and LSW, such as sweep efficiency and the influence of wettability alteration. Then, to increase the effectiveness of low-salinity flooding, a novel hybrid EOR technique of combining LSW with NF to alter the wettability of the rock surface was presented. This technique is known as LSWNF (low-salinity water + nanofluid) flooding. Throughout this research, various experimental methods were employed to analyze the effects of different concentrations of low-salinity water and nanofluids on wettability alteration and enhanced oil recovery. The results demonstrated that the composition with KCl-1+NF outperformed other compositions in terms of wettability alteration and enhanced oil recovery. This finding indicates that a specific combination of LSW and NF can have a significant impact on improving oil recovery efficiency. The most significant findings and accomplishments from this research include the following.

First, nanofluids can alter the wettability of the reservoir rock further, enhancing the displacement of oil by water. The nanoparticles in the nanofluids interact with the rock surface, modifying the surface properties and improving the spreading and penetration of the injected fluid. This leads to improved oil recovery by reducing the residual oil saturation and increasing the sweep efficiency.

Second, the presence of nanoparticles in the nanofluids can also exhibit unique fluid behavior, such as reduced viscosity and increased thermal conductivity. These properties can enhance the fluid flow within the reservoir, allowing for better fluid mobility and improved displacement of oil. The reduced viscosity of the nanofluids can mitigate the challenges of a high mobility ratio between injected fluids and the reservoir oil, resulting in more efficient oil displacement.

Furthermore, the hybridization of nanofluids and low-salinity water can potentially address the issue of water channeling. The nanoparticles in the nanofluids can act as solid diverting agents, plugging the high-permeability zones and redirecting the injected fluids into the low-permeability areas. This helps in achieving a more uniform sweep of the reservoir, preventing the bypassing of oil and improving overall oil recovery efficiency.

Both NF and LSW have a positive impact on the wettability alteration of the rock surface. However, oil binds to the rock surface as KCl and SiO_2 concentrations rise. As a result, choosing compositions with lower LSW and NF concentrations is a smart practice.

The KCl-1+NF injection pattern from the core flooding trials produced the maximum residual oil recovery, which was followed by the KCl-2 and pure water injection patterns. The presence of NF has been shown to increase the amount of oil that can be collected, since it increases the mobility and sweep efficiency of the residual oil. Additionally, it was discovered that the KCl injection pattern produced the maximum recovery for the same interval. Its high wettability variations are the cause of this.

The magnetic resonance images and the T2 spectrum test results obtained from the NMR investigations may demonstrate the migration properties of LSW + NF in heterogeneous cores, and the displacement of the solution in different porosities. The results showed that due to the initial injection stage's low flow resistance, the saturation in big pores increased quickly. At 15.0 PV, the saturation increased slightly to 95.8%. At 15 PV, the medium pores' occupation climbed gradually to 91.5%.

5. Conclusions

In conclusion, a viable strategy for improved oil recovery from mature oil reservoirs is the hybridization of nanofluids with low-salinity water floods. The synergistic effects of combining these two approaches improve fluid movement, sweep efficiency, and wettability change. This study's findings add to our knowledge of the possible advantages of this hybrid strategy, and offer insightful information for future research as well as possible applications in the realm of oil reservoir production. The increasing need for energy worldwide makes it imperative to investigate new and economical ways to maximize oil recovery from current reserves. Combining low-salinity water floods with nanofluid hybridization is an achievable approach to ensure sustainable and effective use of oil resources, while also helping to fulfill the world's growing energy demand.

Author Contributions: Writing original draft, investigation, conceptualization, methodology, supervision, validation, funding acquisition N.L.; writing review and editing, software, investigation, data curation S.Y.; investigation, visualization, supervision H.Q.; writing review and editing, funding acquisition, investigation, supervision, visualization L.S. All authors have read and agreed to the published version of the manuscript.

Funding: This research was funded by the National Natural Science Foundation of China (Grant No. 42007171); the Natural Science Foundation of Hebei Province (Grant No. D2021504034); the National Science and Technology Major Project (Grant No. 2016ZX05031-002); and the Postgraduate Research & Practice Innovation Program of Jiangsu Province (Grant No. KYCX20_2577).

Institutional Review Board Statement: Not applicable.

Informed Consent Statement: Not applicable.

Data Availability Statement: The data presented in this study are available on request from the corresponding author. The data are not publicly available due to links' access being geoblocked outside of China.

Acknowledgments: The authors thank the anonymous reviewers for their constructive and valuable opinions.

Conflicts of Interest: The authors declare that they have no known competing financial interests or personal relationships that could have appeared to affect the work reported in this paper.

References

1. Afolabi, R.O.; Yusuf, E.O. Nanotechnology and global energy demand: Challenges and prospects for a paradigm shift in the oil and gas industry. *J. Pet. Explor. Prod. Technol.* **2019**, *9*, 1423–1441. [CrossRef]
2. Aghaeifar, A.; Dusseault, M.B.; Esfahani, M.R. A comprehensive review of nanofluid flooding in enhanced oil recovery: Effectiveness, applications, and challenges. *J. Pet. Sci. Eng.* **2019**, *177*, 30–49.
3. Mohammadi, S.; Ghazanfari, M.H. Polymer flooding in enhanced oil recovery: Recent advances and challenges. *J. Pet. Sci. Eng.* **2019**, *181*, 106244.
4. Caineng, Z.; Guangya, Z.; Shizhen, T.; Suyun, H.; Xiaodi, L.; Jianzhong, L.; Dazhong, D.; Rukai, Z.; Xuanjun, Y.; Lianhua, H.; et al. Geological features, major discoveries and unconventional petroleum geology in the global petroleum exploration. *Pet. Explor. Dev.* **2010**, *37*, 129–145. [CrossRef]
5. Wu, Y.; Li, M.; Wang, G.; Zhao, G. A Comprehensive Review of Surfactant-Based Enhanced Oil Recovery: Multiple Mechanisms, Challenges, and Future Perspectives. *Energy Fuels* **2020**, *34*, 12543–12566.
6. Schön, J.H. Rocks—Their classification and general properties. In *Developments in Petroleum Science*; Elsevier: Amsterdam, The Netherlands, 2015; Volume 65, pp. 1–19.
7. Zhang, S.; Gao, J.; Tang, S.; Wang, S. A review of low-salinity water flooding in enhanced oil recovery: Mechanisms, experimental studies, and field applications. *J. Pet. Sci. Eng.* **2020**, *189*, 107066.
8. Aminu, M.S.; Sepehrnoori, K.; Johns, R.T. Wettability alteration mechanisms and oil recovery improvement during low-salinity waterflooding: A review. *J. Pet. Sci. Eng.* **2019**, *173*, 1128–1146.

9. Alizadeh, A.H.; Piri, M.; Khishvand, M. A comprehensive review of low-salinity waterflooding mechanisms in carbonate reservoirs. *Fuel* **2019**, *255*, 115839.

10. Hussein, A.K. Applications of nanotechnology in renewable energies—A comprehensive overview and understanding. *Renew. Sustain. Energy Rev.* **2015**, *42*, 460–476. [CrossRef]

11. Li, X.; Huang, S.; Rong, H. A review of polymer flooding: Theories, development, and field applications. *J. Pet. Explor. Prod. Technol.* **2019**, *9*, 1775–1791.

12. Xu, Z.; Luo, P.; Li, M.; Li, X. A review of chemical flooding methods for enhanced oil recovery: From water-soluble polymers to surfactants. *J. Pet. Sci. Eng.* **2019**, *182*, 106281.

13. Alotaibi, M.B.; Alzaabi, A.M.; Alotaibi, S.N. A comprehensive review of EOR techniques for sandstone reservoirs. *J. Pet. Sci. Eng.* **2018**, *166*, 950–973.

14. Khishvand, M.; Kohshour, I.O.; Alizadeh, A.H.; Piri, M.; Prasad, S. A multi-scale experimental study of crude oil-brine-rock interactions and wettability alteration during low-salinity waterflooding. *Fuel* **2019**, *250*, 117–131. [CrossRef]

15. Xie, Q.; Liu, F.; Chen, Y.; Yang, H.; Saeedi, A.; Hossain, M.M. Effect of electrical double layer and ion exchange on low salinity EOR in a pH-controlled system. *J. Pet. Sci. Eng.* **2019**, *174*, 418–424. [CrossRef]

16. Li, X.; Li, G.; Wang, S. A review of low salinity water flooding: Mechanisms, experiments, and field applications. *J. Pet. Sci. Eng.* **2018**, *166*, 846–855.

17. Pu, W.; Zhang, H.; Huang, S. A review of alkaline flooding for enhanced oil recovery: Laboratory evaluation, pilot tests, and field applications. *J. Pet. Explor. Prod. Technol.* **2018**, *8*, 603–618.

18. Liu, N.; Ju, B.S.; Yang, Y.; Brantson, T.S.; Wang, J.; Tian, Y.P. Experimental study of different factors on dynamic characteristics of dispersed bubbles rising motion behavior in liquid-saturated porous media. *J. Petrol. Sci. Eng.* **2019**, *180*, 396–405. [CrossRef]

19. Liu, N.; Chen, X.L.; Ju, B.S.; Chen, X.L.; He, Y.F.; Yang, Y.; Brantson, T.S.; Tian, Y.P. Microbubbles generation by an orifice spraying method in a water-gas dispersion flooding system for enhanced oil recovery. *J. Petrol. Sci. Eng.* **2021**, *198*, 108196. [CrossRef]

20. Mahmoudi, A.H.; Pishvaie, M.R.; Ghazanfari, M.H. Hybrid enhanced oil recovery techniques: A review. *J. Pet. Sci. Eng.* **2019**, *183*, 106397.

21. Khishvand, M.; Alizadeh, A.H.; Oraki Kohshour, I.; Piri, M.; Prasad, R.S. In situ characterization of wettability alteration and displacement mechanisms governing recovery enhancement due to low-salinity waterflooding. *Water Resour. Res.* **2017**, *53*, 4427–4443. [CrossRef]

22. Eltoum, H.; Yang, Y.L.; Hou, J.R. The effect of nanoparticles on reservoir wettability alteration: A critical review. *Pet. Sci.* **2021**, *18*, 136–153. [CrossRef]

23. Rahman, R.U.; Nasrallah, R.A. Saline Water Flooding: A Promising Enhanced Oil Recovery Technique. *J. Pet. Sci. Eng.* **2019**, *176*, 724–742. [CrossRef]

24. Marmur, A. Soft contact: Measurement and interpretation of contact angles. *Soft Matter* **2006**, *2*, 12–17. [CrossRef] [PubMed]

25. Adejare, O.O.; Nasrallah, R.A.; Nasr-El-Din, H.A. A procedure for measuring contact angles when surfactants reduce the interfacial tension and cause oil droplets to spread. *SPE Reserv. Eval. Eng.* **2014**, *17*, 365–372. [CrossRef]

26. Liu, Y.; Lu, J.; Li, M. A review on nanofluid flooding for enhanced oil recovery: Mechanisms, applications, and challenges. *J. Mol. Liq.* **2020**, *306*, 112925. [CrossRef]

27. Zhang, X.; Babadagli, T. A review of thermal recovery technologies for heavy oil reservoirs. *J. Pet. Sci. Eng.* **2018**, *165*, 452–468.

28. Song, Y.Q.; Kausik, R. NMR application in unconventional shale reservoirs–A new porous media research frontier. *Prog. Nucl. Magn. Reson. Spectrosc.* **2019**, *112*, 17–33. [CrossRef]

29. Baban, A.; Keshavarz, A.; Amin, R.; Iglauer, S. Residual Trapping of CO_2 and Enhanced Oil Recovery in Oil-Wet Sandstone Core–A Three-Phase Pore-Scale Analysis Using NMR. *Fuel* **2023**, *332*, 126000. [CrossRef]

30. Carbajo, R.J.; Neira, J.L. *NMR for Chemists and Biologists*; Springer: Heidelberg, Germany, 2013; Volume 162.

31. Yonker, C.R.; Hoffmann, M.M. NMR Investigation of High-Pressure, High-Temperature Chemistry and Fluid Dynamics. In *Supercritical Fluid Technology in Materials Science and Engineering: Syntheses: Properties, and Applications*; CRC Press: Boca Raton, FL, USA, 2002; p. 59.

 *applied sciences*

Article

Analysis of Pathogen-Microbiota Indicator Responses in Surface Karst Springs under Various Conditions in a Rocky Desertification Area: A Case Study of the Xiaojiang Watershed in Yunnan

Weichao Sun [1,2,3], Xiuyan Wang [1,2], Zhuo Ning [1,2], Lin Sun [1,2,*] and Shuaiwei Wang [1,2,*]

[1] Institute of Hydrogeology and Environmental Geology, Chinese Academy of Geological Sciences, Shijiazhuang 050061, China; 3020190008@email.cugb.edu.cn (W.S.); wxiuyan9948@163.com (X.W.); ningzhuozhuo@163.com (Z.N.)

[2] Key Laboratory of Groundwater Remediation of Hebei Province & China Geological Survey, Shijiazhuang 050083, China

[3] School of Chinese Academy of Geological Sciences, China University of Geosciences (Beijing), Beijing 100086, China

* Correspondence: lin_sun166@163.com (L.S.); tairan_w@163.com (S.W.)

Abstract: The Xiaojiang watershed in Luxi, Yunnan, is a typical rocky desertification area, in which karst groundwater pollution is severe and water resources are scarce. This article takes the watershed as an example and investigates the response mechanisms of surface karst spring water quality to agricultural pollution in rocky desertification areas. Specifically, the study was conducted as follows: (I) A total of 108 water samples from 54 sources were collected during the dry and wet seasons for analysis. (i) Principal component and correlation analyses identified the main pollution indicators in the soil surface karst zone of the area, including total bacterial count, total coliforms, COD, pH, and redox potential. (ii) It was also discovered that surface soil, impacted by agricultural activities, directly contributes to groundwater pollution in the soil surface karst zone. (II) Local soil was used to prepare soil columns under various conditions for simulation. The findings indicate: (i) Temperature significantly affects the surface karst springs, with higher temperatures leading to more pronounced water quality responses, increased enrichment of pathogen-microbiota indicators, and degraded water quality. (ii) Soil porosity substantially influences the water quality of surface karst springs. Increased porosity results in looser soil, more oxidizing conditions in the storage matrix, reduced pathogen-microbiota development, and consequently, less water pollution. This study offers theoretical and technical references for evaluating, monitoring, and issuing early warnings for pathogenic bacteria-microbiota pollution in groundwater in rocky desertification areas.

Keywords: rocky desertification; surface karst spring; groundwater pollution; pathogenic microbiota; simulated in situ experiment

Citation: Sun, W.; Wang, X.; Ning, Z.; Sun, L.; Wang, S. Analysis of Pathogen-Microbiota Indicator Responses in Surface Karst Springs under Various Conditions in a Rocky Desertification Area: A Case Study of the Xiaojiang Watershed in Yunnan. *Appl. Sci.* **2024**, *14*, 1933. https://doi.org/10.3390/app14051933

Academic Editor: Fulvia Chiampo

Received: 31 December 2023
Revised: 3 February 2024
Accepted: 23 February 2024
Published: 27 February 2024

1. Introduction

Southwest China hosts the world's largest exposed karst region, where karst water resources, totaling 2039.67×10^4 m^3/year, comprise 23.39% of the nation's groundwater [1]. However, challenges such as high population density, uneven spatial and temporal distribution of these resources, intricate storage and burial dynamics, and significant pollution reduce per capita water availability [2,3]. Karst areas, characterized by rocky desertification, present a complex and diverse range of underground water pollution. This leads to a highly sensitive and vulnerable groundwater system, which is particularly susceptible to pathogen-microbiota pollution [4,5]. Research indicates that pathogen-microbiota indicators in groundwater are not merely detrimental to human health but also potentially harbor or transmit other harmful microbes. These indicators serve as proxies for microbial viruses posing threats to human health [6–9]. Consequently, investigating the response changes of pathogen-microbiota indicators in groundwater within rocky desertification areas holds substantial theoretical and practical significance.

The soil–water matrix environment plays a pivotal role in exploring the response mechanisms of groundwater to pathogen-microbiota pollution. The sequence of interactions, spanning from groundwater contamination through infiltration to microbial pollution, culminates in the response of pathogen toxicology indicators. This sequence is intimately linked with the evolutionary processes of microbes [4,5].

Soil, as a three-phase porous medium, exhibits substantial purification capabilities attributed to its unique structure and the presence of mineral ions. Within this porous matrix environment, the evolution and migration of groundwater microbes are influenced by the soil's self-purification capacities, which play a crucial role in retaining microbes, thereby substantially diminishing the microbial content that enters the groundwater [10,11]. The factors affecting microbial evolution in soil solid porous media can be classified into three categories: microbial factors, soil medium factors, and soil environmental factors [12,13].

Microbial factors encompass the microbe type, strain size, surface charge, hydrophobicity, and chemotactic nature of the community. The surface charge of microbes significantly influences their adsorption within the soil–water matrix environment [14]. Moreover, their migration and attenuation are associated with factors such as microbial type [8,15–17], surface hydrophobicity [18], individual size [19], community chemotaxis [12,20], and population heterogeneity [21,22].

Soil medium factors encompass soil particle size, pore structure, moisture content, mineral composition, and overall content. The dynamics of solute movement in the soil–water matrix environment adhere to mechanisms such as interception, leaching, and physicochemical filtration, as delineated in previous studies [23,24]. The ionic content of clay minerals demonstrates pronounced adsorption [25,26] and desorption [27,28] effects on microbes. Microbial mobility is enhanced in larger pores [29,30]. An increase in moisture content correlates with greater microbial retention [29,31], and smaller effective particle sizes are associated with higher microbial retention [32–34].

Soil environmental factors encompass pH, temperature, flow rate, and ionic strength. Within certain limits, elevated temperatures enhance the activity of microbial pathogens [35,36]. The migration capacity of microbes notably increases with the environmental pH [11,13]. Enhanced hydrodynamic conditions facilitate microbial migration capacity while diminishing retention effects [32,36]. An increased ion concentration amplifies microbial migration and penetration capabilities [37,38]. However, molecular diffusion effects are generally inconsequential in most microbial transport processes [39,40].

In summary, both domestic and international experts and scholars have extensively researched the response mechanisms, evolutionary processes, and patterns of pathogen-microbiota in the soil–water matrix environment. However, the response evolution of groundwater pathogen-microbiota indicators involves a complex interplay between the matrix environment and the microbes themselves. The mechanisms of response under varying conditions remain intricate, with many aspects still obscure, underscoring the need for in-depth exploration. Therefore, a water quality analysis of surface karst springs in a rocky desertification area was conducted through field sampling and laboratory tests in this paper, which provided theoretical and technical references for exploring the response mechanisms of pathogenic bacteria and microorganisms in groundwater in a rocky desertification area, which is of great scientific significance.

2. Materials and Methods

The Xiaojiang Watershed, situated in southeastern Yunnan Province (Figure 1), features a terrain that descends from the northeastern highlands to the southwestern lowlands, extending northeasterly over an area of approximately 1009.28 km^2. It has a population of 200,400. This watershed exemplifies a typical independent karst water system, encompassing comprehensive processes of recharge, flow, and discharge. The area is marked by thin surface soil, sparse vegetation, pronounced karst geological structures, extensive fractures, and complex hydrogeological formations [1,2,5]. Given that surface karst springs are a vital source of drinking water in the region, analyzing the response of pathogenic microbes to the water quality of these springs within the typical rocky desertification context of the Xiaojiang Watershed is of substantial practical importance.

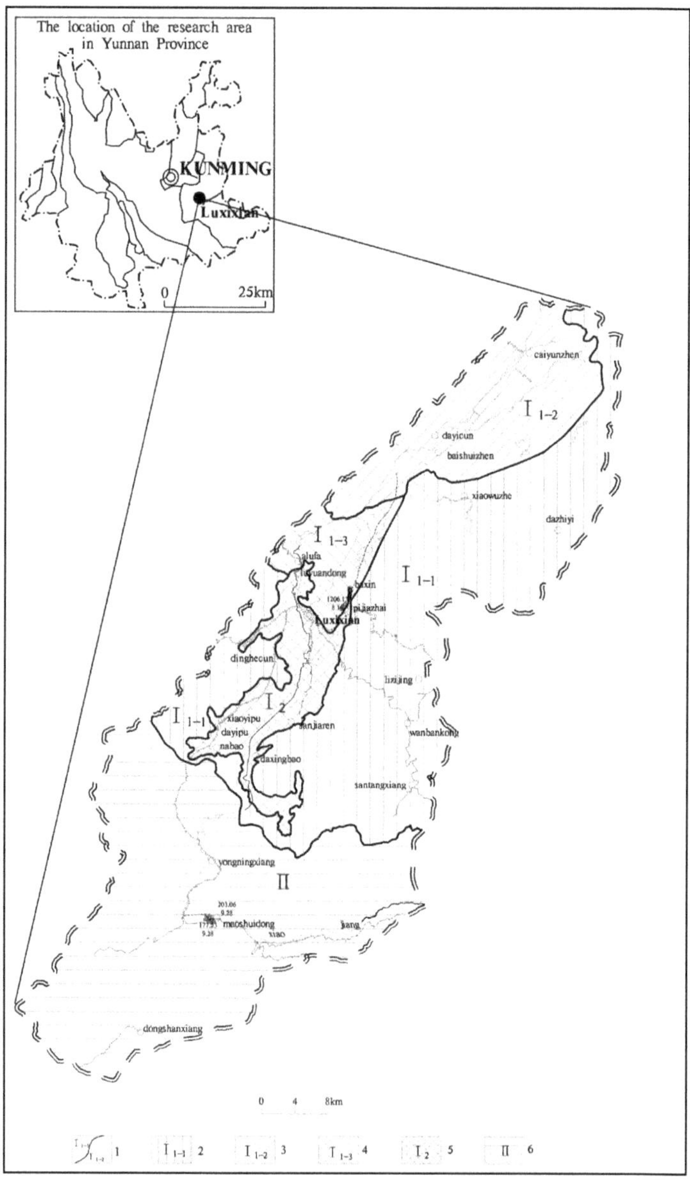

Figure 1. Hydrogeological zoning map: (1) hydrogeological zoning boundaries; (2) surrounding rocks outside karst basins in mountainous areas (I_{1-1}); (3) upstream karst basin, karst hilly plateau trough valley area (I_{1-2}); (4) karst peak cluster and depression areas around the basin bottom (I_{1-3}); (5) basin bottom covered karst discharge runoff area (I_2); (6) Xiaojiang karst valley (II).

The surface spring is located in Wanbankong Village, Santang Township, Luxi County. The village now has a population of more than 900 people, who mainly farm. The upper layer of the surface spring is a shallow surface karst aquifer of pure carbonate rock with a high degree of rock fissures. The physical and biological weathering and the karst process form a large number of karst voids, such as dissolved pores and network dissolved gaps. The atmospheric precipitation seeps into the surface karst water through the solution holes and gaps, and the karst runoff seeps into the surface to form the surface spring, where the aquifer and the bottom water-barrier layer are exposed [1–3,5].

2.1. Field Sampling

Following comprehensive field surveys, a total of 108 water samples were collected from 54 sources within the basin. This included 18 natural spring outlets, 13 rivers, 12 reservoirs, and 11 wells, during both the dry and wet seasons. The spatial distribution of these sampling points is depicted in Figure 2. The field sampling adhered strictly to the technical specifications presented in [41]. Subsequent analyses of these samples were carried out by the Testing Center of the Institute of Hydrogeology and Environmental Geology at the Chinese Academy of Geological Sciences, Ministry of Natural Resources. This analysis followed the Standard Examination Methods [42].

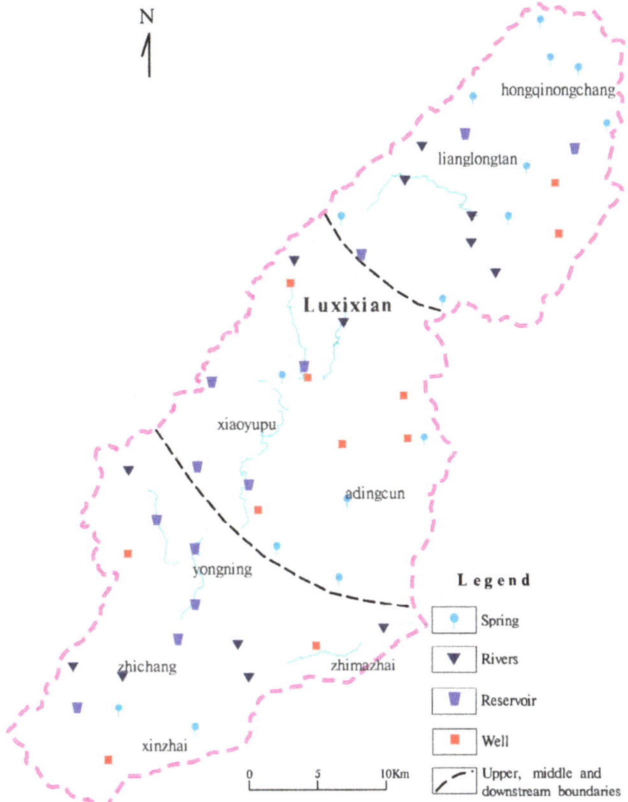

Figure 2. Research area watershed scope and sampling layout map.

2.2. Soil Column Experiment

Undisturbed soil samples were collected from various depths within the soil surface karst zone of the Xiaojiang Watershed. Indoor density was measured using the ring knife method, volume was measured using the wire method, moisture content was measured using the weighing method, and specific gravity was measured using the heavy bottle method to obtain the basic physical parameters of the undisturbed soil (Figure 3). To replicate the natural conditions of the soil surface karst zone soil layers in the Xiaojiang Watershed, taking local undisturbed soil, from the surface to the bedrock depth, we packed the undisturbed soil per 10 cm layers. After returning to the laboratory, the 60 cm soil column was filled according to the same physical parameters (density, porosity, moisture content, etc.) of the in situ soil (every 10 cm depth corresponds to the same layer). Additionally, outlets were strategically installed at 10 cm intervals along the side of the column to facilitate the study [43–45].

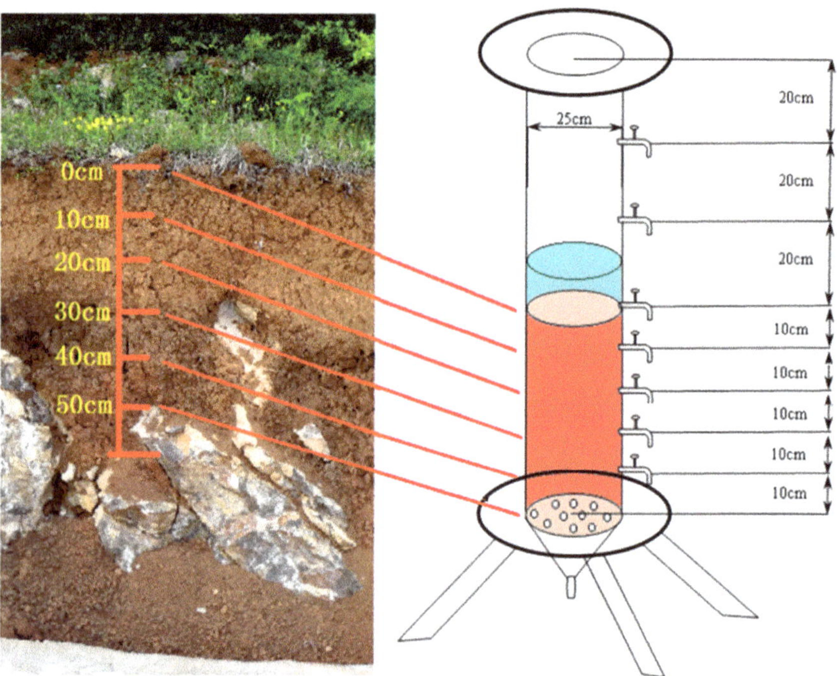

Figure 3. The indoor soil column experiment simulated in situ permeation.

2.3. Experimental Conditions

Informed by the annual temperature variation chart of the Xiaojiang Watershed and the distribution range of physical and mechanical indicators of the soil surface karst zone soil (Tables 1 and 2), temperature and porosity rate, as controllable conditions, were chosen as variables. Experimental setups, representing different levels of these factors, were established to conduct indoor simulated in situ soil water environment percolation process experiments.

Table 1. The monthly average temperature in the research area over many years.

Month	Jen	Feb	Mar	Apr	May	Jun	Jul	Aug	Sept	Oct	Nvo	Dec
$T_{ave}(°C)$	9.5	11	15	18	22	24	26	29	29	23	15	12

Table 2. The experimental condition table.

Experimental Numbers	Temperature (°C)	Porosity
1	10	0.5
2	20	0.5
3	30	0.5
4	10	0.55
5	20	0.55
6	30	0.55
7	10	0.6
8	20	0.6
9	30	0.6

According to weather data from 1958 to 2023, the average annual temperature in the Xiaojiang River Basin is 19.5 °C, with an annual temperature of $\geq$9.5 °C. The coldest month is January, with an average monthly temperature of 9.5 °C, while the hottest month is July, with an average monthly temperature of 29 °C. Therefore, the indoor test temperature conditions are set at 10 °C, 20 °C, and 30 °C.

2.4. Sampling and Analyses

Water samples were systematically collected from various depths (10 cm, 20 cm, 30 cm, 40 cm) of the soil column outlets at predetermined intervals: 0, 1, 3, 6, 10, 15, 21, 28, 35, 42, 49, and 56 days. In the case of orthogonal experiments, wastewater samples were specifically gathered at different depths on days 7 and 9. These samples were secured in 100 mL sterile glass bottles and subsequently stored at 4 °C. The water samples underwent testing for pH, redox potential, COD, total bacterial count, total coliform count, and other physical, chemical, and microbiological indicators of groundwater, in accordance with the Standard Examination Methods [42].

2.5. Statistical Evaluation

The water quality of the field water samples was analyzed completely. Data analysis was conducted using the Nemero index method (Equations (1) and (2)), and the water quality comprehensive grade of each water source was calculated (Table 3).

$$P_i = \frac{C_i}{S_i} \tag{1}$$

where P_i is the single-factor pollution index, C_i is the single-factor pollution measured value, and S_i is the single-factor pollution evaluation standard value.

$$P_s = \sqrt{\frac{\overline{P}^2 + P_{imax}^2}{2}} \tag{2}$$

where P_s is the composite pollution index, $\overline{P}$ is the single-factor index average, and P_{imax} is the maximum value of the single-factor index.

Table 3. Water quality comprehensive pollution grade classification standard.

P_s	$P_s \leq 0.7$	$0.7 < P_s \leq 1.0$	$0.7 < P_s \leq 1.0$	$0.7 < P_s \leq 1.0$
Grade	1	2	3	4

Then, the data analysis was conducted using Origin Pro 8 and SPSS 19 statistical software, and water quality indexes obtained from indoor soil column experiments were collected. Aiming at pathogen-microorganism (total bacterial community and total *Escherichia coli*), correlations, variances, significance, distribution characteristics, and consistency, tests of the data were calculated. A threshold of ≤ 0.05 was set for the consistency test to ensure statistical significance.

3. Results

3.1. Determination of Pollution Indicators

Utilizing pre-human-engineering-activity baseline values from the basin, pollution indicators for karst water quality at 54 water source locations were established based on test results (Table S1). These results were derived from seven water sources in the study area, which included seven population drinking water source areas (Baishuitang, Yanjinggou, Wuzhe Shuiku, Zuyuandi, Aobushan, Wulang Haizi), as well as one karst spring water source (Pijiazhai Daquan). The establishment of these baselines followed the guidelines of the report presented in [43] in conjunction with the Standard Examination Methods [42].

The Nemero index method was used to calculate the degree of water pollution and the water quality grade in the study area (Figure 4). The results are as follows:

Figure 4. Research area watershed scope and water quality.

(i) The value of P_i was calculated, and the contribution degree of the water quality grade was analyzed. The identified pollution indicators include total bacterial count, total coliforms, total hardness (CaCO$_3$), HCO^{3-}, oxygen demand (COD), pH, Ca^{2+}, Mg^{2+}, chroma, odor, visible objects, and turbidity, ordered by their individual factor contribution. Collectively, the first eight factors account for 89.82% of the cumulative pollution affecting the water quality in the area.

Notably, total hardness (CaCO$_3$), HCO^{3-}, oxygen demand (COD), pH, Ca^{2+}, Mg^{2+}, chroma, odor, visible objects, and turbidity align with the chemical characteristics of carbonate rocks in rocky desertification karst areas [43]. In addition, ORP as a comprehensive water quality indicator should also be considered. Consequently, the laboratory experiments prioritized pathogen-microbes (total bacterial count, total coliforms), COD, and pH as the target characteristic pollution indicators.

(ii) Under the influence of long-term agricultural activity, the thin surface soil layer not only fails to purify surface water but also acts as a repository for pollution. Field surveys revealed that the primary sources of karst groundwater pollution in the area are domestic sewage, livestock breeding, and agricultural cultivation, which significantly pollute both water resources and soil layers.

3.2. Temperature Variations and Karst Groundwater Quality Indicator Responses

3.2.1. Changes in pH of Groundwater at Various Depths under Different Temperature Conditions

Indoor experiments at different temperatures measured the pH values of groundwater at various depths, as shown in Table 4 and Figure 5.

Table 4. The pH values of groundwater under different temperature and depth conditions.

Depth/(m)	Temperature		
	10 °C	20 °C	30 °C
0.1	6.18	6.30	6.32 (*max*)
0.2	5.99	6.12	6.25
0.3	5.94	6.05	6.10
0.4	5.88 (*min*)	5.98	6.03
0.5	5.92	6.02	6.08
Average value E	5.98	6.09	6.16
Range R	0.3	0.32	0.29

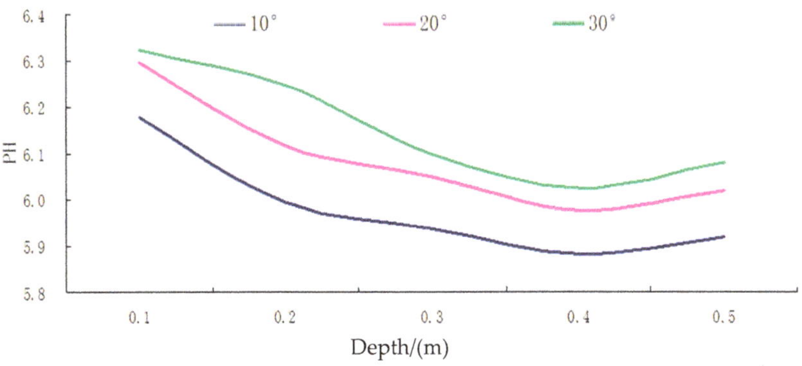

Figure 5. The pH curve of different depth of groundwater under different temperature conditions.

(i) As shown in the results, under varying temperature conditions, the pH value of groundwater in the soil surface karst zone exhibited fluctuations within the range of 5.88–6.32, typically demonstrating acidic characteristics. The variation range was relatively narrow at 0.44, indicating modest fluctuation levels.

(ii) As demonstrated in Figure 5, across the entire soil surface karst zone, the overall trend shows that the groundwater pH tends to decrease with increasing soil depth. However, this rate of change gradually diminishes, reaching an extreme value at a depth of 0.4 m. This is followed by a slight upward fluctuation between 0.4 m and 0.5 m, although the extent of this increase is minimal.

(iii) The pH value of karst groundwater tends to increase with higher temperatures, exhibiting smaller fluctuations and a tendency towards stability. Conversely, in winter, with the lowest temperatures, groundwater pH reaches its minimum value.

3.2.2. Changes in ORP (Oxidation-Reduction Potential) of Groundwater at Various Depths under Different Temperature Conditions

In the experiments conducted under varying temperature conditions, the oxidation-reduction potential (ORP) values of groundwater were measured at different depths. The results of these measurements are detailed in Table 5 and Figure 6.

Table 5. The ORP values of different depths of groundwater under different temperature conditions.

Depth/(m)	Temperature		
	10 °C	**20 °C**	**30 °C**
0.1	43.11	15.67 (*min*)	24.33
0.2	49.00	26.33	34.33
0.3	47.33	33.67	40.83
0.4	56.00 (*max*)	37.33	45.83
0.5	47.23	31.28	39.66
Average value E	48.53	28.86	37.00
Range R	12.89	21.67	22.67

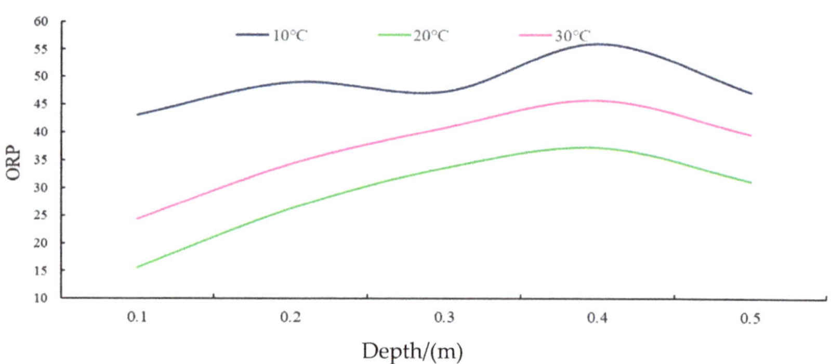

Figure 6. The ORP curve for different depths of groundwater under different temperature conditions.

(i) As shown in the results, under varying temperature conditions, the ORP (oxidation-reduction potential) values in the groundwater of the soil surface karst zone exhibited a range of fluctuation between 15.67 and 56.00. This range suggests an overall low potential and strong reduction characteristic, with a variation extent of 40.33, indicating considerable and unstable fluctuations.

(ii) As demonstrated in Figure 6, the general trend in the ORP values of groundwater across the entire soil surface karst zone is an increase with greater soil depth, signifying a decrease in reduction and an increase in oxidation. Nevertheless, this rate of change gradually decreases, peaking at a depth of 0.4 m. Beyond this point, a minor decreasing fluctuation trend is observed between 0.4 and 0.5 m depth, though the extent of this decline is minimal.

(iii) Under different temperature conditions, lower temperatures correlate with higher karst groundwater ORP, displaying minimal fluctuations and a tendency towards stability.

Conversely, with the highest temperatures, groundwater ORP shows the most significant fluctuations and is least stable.

3.2.3. Changes in COD (Chemical Oxygen Demand) of Groundwater at Various Depths under Different Temperature Conditions

Indoor experiments at different temperatures measured the chemical oxygen demand (COD) values of groundwater at various depths, as shown in Table 6 and Figure 7.

Table 6. The COD values of different depths of groundwater under different temperature conditions.

Depth/(m)	Temperature		
	10 °C	20 °C	30 °C
0.1	7.87	10.64	14.05
0.2	7.74 (*min*)	9.84	16.47
0.3	10.21	12.20	17.21
0.4	12.34	14.58	21.32
0.5	14.39	19.86	25.15 (*max*)
Average value E	10.51	13.42	18.84
Range R	6.65	10.02	11.10

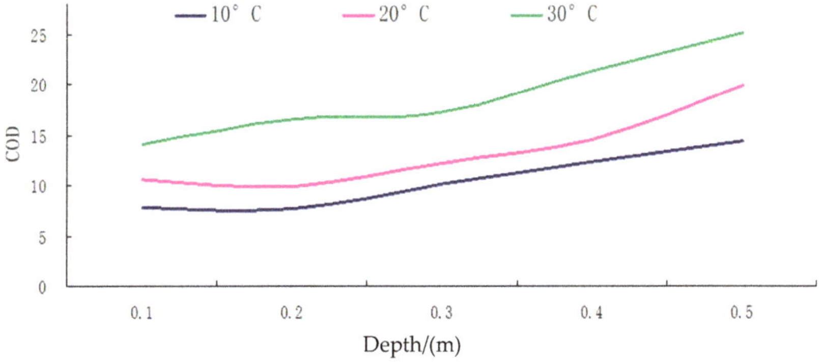

Figure 7. The COD curve for different depths of groundwater under different temperature conditions.

(i) As shown in the results, under various temperature conditions, the chemical oxygen demand (COD) values of groundwater in the soil surface karst zone demonstrated fluctuations between 7.74 and 25.15, with a range of 17.41. This significant fluctuation indicates that the groundwater in this karst surface zone generally exhibits high chemical oxygen demand and is substantially polluted.

(ii) As demonstrated in Figure 7, the overall trend of COD values throughout the soil surface karst zone tends to increase with increasing soil depth, displaying a generally linear positive correlation.

(iii) At consistent temperature conditions, with a temperature rise, the COD values in karst groundwater increase, exhibiting more significant fluctuations and instability. When temperatures reach their lowest, groundwater COD values are at their minimum, with the smallest fluctuations, indicating greater stability.

3.2.4. Changes in TBC (Oxidation-Reduction Potential) of Groundwater at Various Depths under Different Temperature Conditions

Indoor experiments under different temperature conditions measured the total bacterial count (TBC) values of groundwater at various depths, as shown in the Table 7 and Figure 8.

Table 7. The TBC of different depths of groundwater under different temperature conditions.

Depth/(m)	Temperature		
	10 °C	20 °C	30 °C
0.1	38,102	41,825	290,667
0.2	46,483	51,089	293,333
0.3	38,900	47,019	340,000
0.4	42,792	45,521	278,667
0.5	26,900 (*min*)	28,900	386,667 (*max*)
Average value E	38,635	42,871	317,867
Range R	19,583	22,189	108,000

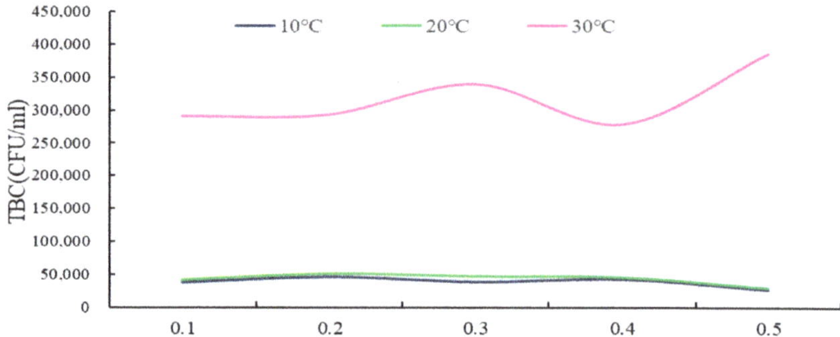

Figure 8. The TBC curve for different depths of groundwater under different temperature conditions.

(i) As shown in the results, under varying temperature conditions, the total bacterial count (TBC) value range of groundwater in the soil surface karst zone was observed to fluctuate between 26,900 and 386,667, with a substantial range of 359,767, indicating significant fluctuations. This suggests that the groundwater in this karst surface zone is subject to severe TBC exceedance, indicative of serious microbial pollution.

(ii) As demonstrated in Figure 9, across the entire soil surface karst zone soil layer, at lower temperatures, there is a general trend of decreasing groundwater TBC values with increasing soil depth, although the range of this change is quite minimal. Conversely, at higher temperatures, the overall trend is for groundwater TBC values to disperse more with increasing soil depth.

(iii) Under different temperature scenarios, with the summer temperature rise, karst groundwater TBC values increase, displaying larger fluctuations and more instability. When temperatures are at their lowest, groundwater TBC values reach their minimum and exhibit the smallest fluctuations, indicating greater stability.

3.2.5. Changes in TEC (Total *Escherichia coli*) of Groundwater at Various Depths under Different Temperature Conditions

Indoor experiments under different temperature conditions measured the total *Escherichia coli* (TEC) values of groundwater at various depths, as indicated in Table 8 and Figure 9.

(i) As shown from the results, under diverse temperature conditions, the total *E. coli* count (TEC) values in the groundwater of the soil surface karst zone exhibited fluctuations between 60 and 1600, with a range of 1540. This substantial fluctuation indicates that the groundwater in this karst surface zone experiences severe TEC exceedance, signifying serious microbial pollution.

(ii) As demonstrated in Figure 9, throughout the entire soil surface karst zone soil layer, at lower temperatures, the general trend shows a decrease in groundwater TEC values with increasing soil depth, although the range of this change is relatively minor. At higher temperatures, groundwater TEC values initially decrease and then tend to stabilize as soil depth increases.

(iii) Under different temperature scenarios, with the summer temperature rise, karst groundwater TEC values increase, demonstrating larger fluctuations and more instability. When temperatures are at their lowest, groundwater TEC values reach their minimum, with the least fluctuation, indicating greater stability.

Table 8. The TEC of different depths of groundwater under different temperature conditions.

Depth/(m)	Temperature		
	10 °C	20 °C	30 °C
0.1	222	470	1600 (*max*)
0.2	186	467	1178
0.3	196	432	1068
0.4	200	457	1137
0.5	60 (*min*)	139	1158
Average value E	173	393	1228
Range R	162	331	532

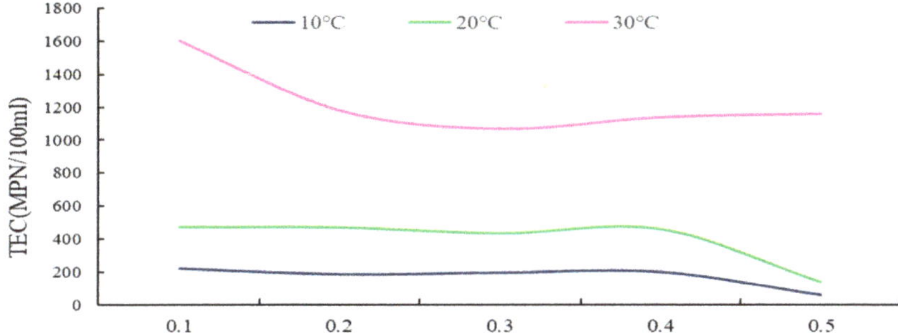

Figure 9. The TEC curve for different depths of groundwater under different temperature conditions.

3.3. Response of Karst Groundwater Quality Indicators to Changes in Porosity

3.3.1. Changes in pH of Groundwater at Various Depths under Different Porosity Conditions

Indoor experiments under different porosity conditions of undisturbed soil tested the pH values of groundwater at various depths, as indicated in Table 9 and Figure 10.

(i) As shown in the results, in the soil surface karst zone under various porosity conditions, the pH value range of groundwater exhibited fluctuations between 5.91 and 6.29, typically acidic, with a fluctuation range of 0.38, indicating relatively minor fluctuations.

(ii) Within the soil layer of the surface karst zone, with smaller porosity, karst groundwater pH values tend to increase, exhibiting greater fluctuations and more instability.

(iii) As demonstrated in Figure 10, throughout the entire soil surface karst zone, the general trend of groundwater pH values showed a tendency to decrease with increasing soil depth. The change was approximately linear in the 0.1−0.3 m depth range, reaching extreme values at a depth of 0.3 m under varying porosity rates. This was followed by a slight upward fluctuation trend with increasing soil depth, although the magnitude of this rise was quite minimal.

Table 9. The pH of different depths of groundwater under different porosity conditions.

Depth/(m)	Porosity		
	0.60	0.55	0.50
0.1	6.16	6.28	6.29 (*max*)
0.2	6.02	6.10	6.15
0.3	5.91	6.02	6.03
0.4	5.91 (*min*)	5.95	5.96
0.5	5.93	5.97	6.01
Average value E	5.99	6.06	6.09
Range R	0.25	0.33	0.34

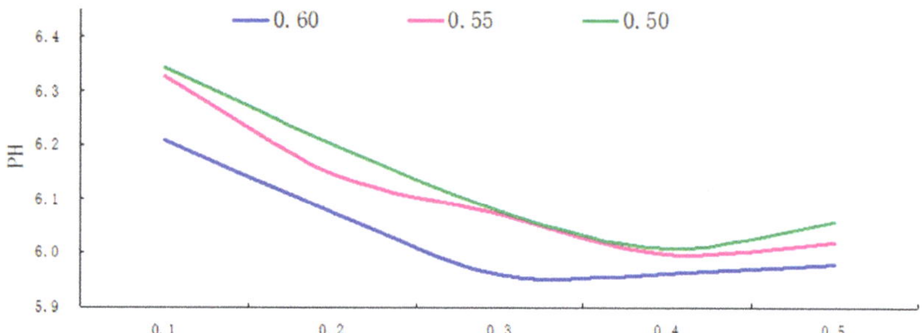

Figure 10. The PH curve for different depths of groundwater under different porosity conditions.

3.3.2. Changes in ORP (Oxidation-Reduction Potential) of Groundwater at Various Depths under Different Porosity Conditions

Indoor experiments under different soil porosity conditions tested the oxidation-reduction potential (ORP) values of groundwater at various depths, as indicated in Table 10 and Figure 11.

(i) As shown from the results, in the soil surface karst zone with varying porosity rates, the oxidation-reduction potential (ORP) values of groundwater fluctuated between 22.50 and 49.72, indicative of overall low potential and strong oxidizing conditions. The variation range of 27.22 suggests substantial and unstable fluctuations.

(ii) Within the soil layer of the soil surface karst zone, the smaller the porosity results in minimized groundwater ORP values and secondary stability, exhibiting greater fluctuations and more instability.

(iii) As demonstrated in Figure 11, across the entire soil surface karst zone, the general trend is for groundwater ORP values to increase with increasing soil depth, although the rate of change gradually diminishes. An extreme value is observed at a depth of 0.4 m, followed by a minor downward fluctuation trend between 0.4 and 0.5 m depth; however, the extent of this downward fluctuation is minimal.

Table 10. The ORP values of different depths of groundwater under different porosity conditions.

Depth/(m)	Porosity		
	0.60	0.55	0.50
0.1	35.89	24.72	22.50 (*min*)
0.2	42.17	36.83	30.67
0.3	48.67	38.50	34.67
0.4	49.72 (*max*)	46.17	43.28
0.5	45.62	37.78	35.89
Average value E	44.41	36.80	33.40
Range R	13.83	21.44	20.78

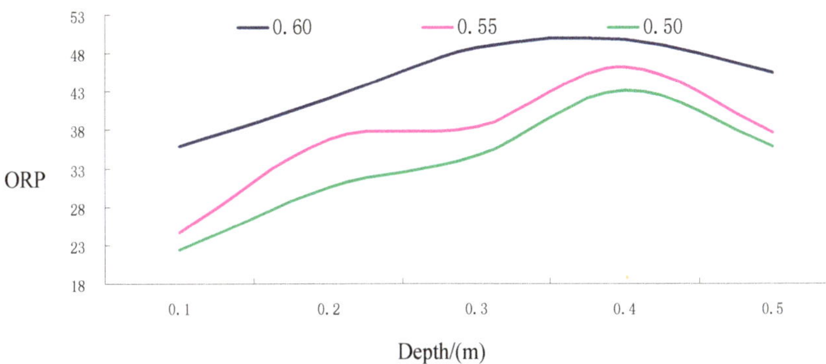

Figure 11. The ORP curve for different depths of groundwater under different porosity conditions.

3.3.3. Changes in COD (Chemical Oxygen Demand) of Groundwater at Various Depths under Different Porosity Conditions

Indoor experiments under different soil porosity conditions tested the chemical oxygen demand (COD) values of groundwater at various depths, as indicated in Table 11 and Figure 12.

Table 11. The COD values of different depths of groundwater under different porosity conditions.

Depth/(m)	Porosity		
	0.60	0.55	0.50
0.1	8.76	10.76	13.02
0.2	7.62	8.25	12.65
0.3	7.19 (*min*)	10.28	18.83
0.4	10.44	11.65	21.49
0.5	16.24	24.35	28.78 (*max*)
Average value E	10.05	13.06	18.95
Range R	9.05	16.10	16.13

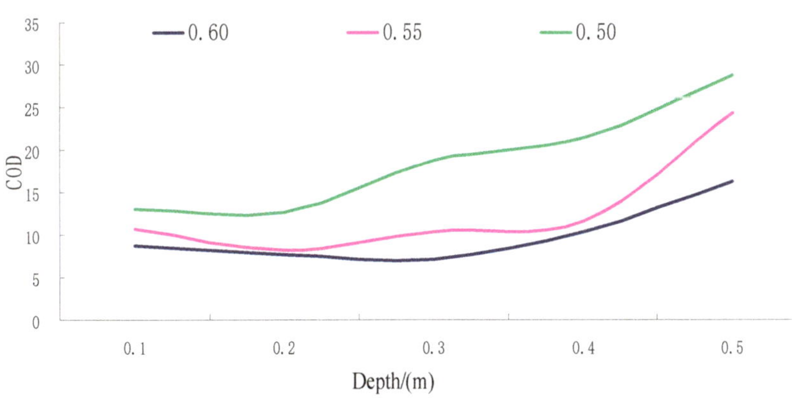

Figure 12. The COD curve for different depths of groundwater under different porosity conditions.

(i) As shown from the results, in the soil surface karst zone with different porosity rates, the range of chemical oxygen demand (COD) values in groundwater fluctuated between 7.19 and 28.78, with a variance of 21.59. This significant fluctuation suggests that the groundwater in this karst surface zone generally has a high COD, indicative of substantial pollution.

(ii) In the soil surface karst zone soil layer, with smaller porosity, the fluctuations in groundwater COD values are also reduced, indicating increased stability.

(iii) As demonstrated in Figure 12, throughout the entire soil surface karst zone, the overall trend shows that groundwater COD values tend to increase with increasing soil depth, exhibiting a non-linear positive correlation.

3.3.4. Changes in TBC (Total Bacterial Count) of Groundwater at Various Depths under Different Porosity Conditions

Indoor experiments under different soil porosity conditions tested the total bacterial count (TBC) of groundwater at various depths, as indicated in Table 12 and Figure 13.

Table 12. The TBC values of different depths of groundwater under different porosity conditions.

Depth/(m)	Porosity		
	0.60	0.55	0.50
0.1	109,275 (*min*)	118,925	152,060
0.2	114,750	124,133	162,689
0.3	113,667	127,973	184,280
0.4	116,592	121,809	178,578
0.5	114,633	131,467	186,367 (*max*)
Average value E	113,783	124,861	172,795
Range R	7317	12,542	23,677

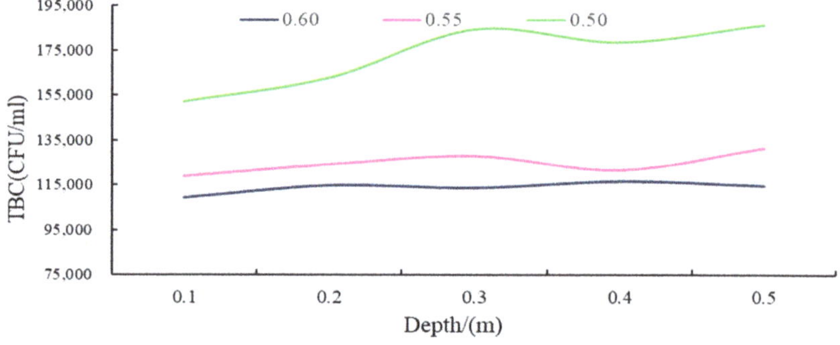

Figure 13. The TBC curve for different depths of groundwater under different porosity conditions.

(i) As shown in the results, in the soil surface karst zone, under varying porosity rates, the Total Bacterial Count (TBC) values of groundwater fluctuated between 109,275 and 186,367 CFU/mL, with a range of 77,092. These significant fluctuations suggest that the groundwater in this karst surface zone generally experiences severe TBC exceedance, indicative of serious microbial pollution.

(ii) Within the soil layer of the soil surface karst zone, with smaller porosity, the groundwater TBC values increase, with the average being approximately 1.5 times higher than in looser soil layers, and fluctuations are larger and more unstable.

(iii) As demonstrated in Figure 13, across the entire soil surface karst zone, as soil becomes denser or the soil particle size decreases, the general trend is for groundwater TBC values to increase with increasing soil depth, typically exhibiting non-linear changes. In contrast, when the soil layer is looser or the soil particle size is larger, changes in groundwater TBC values with increasing depth are relatively smaller.

3.3.5. Changes in TEC (Total *Escherichia coli*) of Groundwater at Various Depths under Different Porosity Conditions

Indoor experiments under different soil porosity conditions tested the total *Escherichia coli* (TEC) values of groundwater at various depths, as indicated in Table 13 and Figure 14.

Table 13. The TEC values of different depths of groundwater under different porosity conditions.

Depth/(m)	Porosity		
	0.60	0.55	0.50
0.1	480	540	839
0.2	418	445	968
0.3	357 (*min*)	415	998
0.4	432	451	1020
0.5	366	386	1100 (*max*)
Average value E	411	447	985
Range R	123	154	261

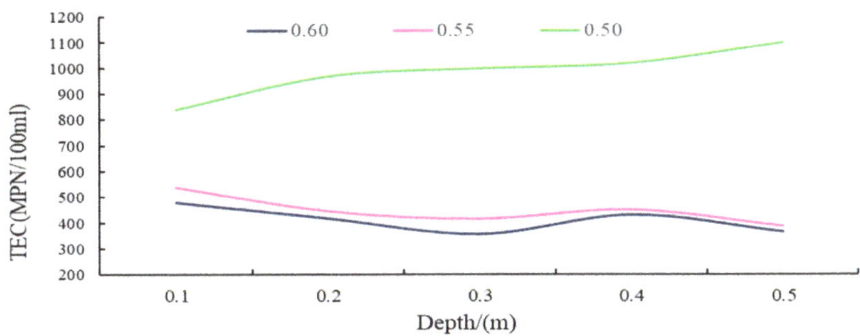

Figure 14. The TEC curve for different depths of groundwater under different porosity conditions.

(i) As shown in the results, in the soil surface karst zone, under varying porosity rates, the total E. coli count (TEC) values in groundwater fluctuated between 357 and 1100 MPN/100 mL, with a variance of 743. This significant fluctuation suggests that the groundwater in this karst surface zone typically experiences severe TEC exceedance, indicative of serious microbial pollution.

(ii) As demonstrated in Figure 14, throughout the entire soil surface karst zone, as soil becomes denser or the particle size decreases, the general trend is for groundwater TEC values to increase with increasing soil depth, displaying predominantly non-linear changes. Conversely, in conditions where the soil layer is looser or the soil particle size is larger, the changes in groundwater TEC values with increasing depth are relatively smaller, and the fluctuations are less pronounced.

(iii) Within the soil layer of the soil surface karst zone, with smaller porosity, the groundwater TEC values increase, with the average being approximately 2.4 times higher than in looser soil layers, and fluctuations are larger and more unstable.

4. Discussion and Conclusions

4.1. Water Quality and Pollution Status

In the study area, exposed bedrock, developed cracks, thin surface rooted soil zone, sparse vegetation, uneven annual rainfall, and strong monsoon effect lead to uneven distribution of groundwater resources in the inland basin in terms of morphology and burial, contributing to severe rocky desertification, water scarcity, and notable pollution. The primary pollution sources in the basin include dispersed domestic waste, farmyard manure application in fields, and unregulated livestock farming, leading to a high concentration of total bacterial count, total coliforms, and other microbes in the soil. The research focused on identifying primary pollution indicators in the soil surface karst zone of the study area, including total bacterial count, total coliforms, COD, pH, and oxidation-reduction potential. It was observed that under long-term agricultural activity, the thin surface soil layer serves as a direct source of pollution for groundwater in the soil surface karst zone. The uneven distribution of water resources, combined with intense pollution and the local residents' limited awareness of water resource conservation, has resulted in significant ecological degradation.

4.2. The Temperature and Porosity Effects on Groundwater Quality

The results of the present study have demonstrated that temperature was the most significant factor influencing groundwater pathogen-microbiota indicators. Mesophilic microbes, such as bacteria and coliforms, exhibit optimal growth within specific temperature ranges. Coliforms, which are commonly found in the intestines of animals and humans, demonstrate optimal growth at temperatures ranging from 30 °C to 40 °C, with the most rapid reproduction occurring at 37 °C. The quantity, metabolism, and activity are subject to change when the temperatures are not conducive to their survival [44–48]. This is consistent with our finding that TBC and TEC values are lower below 30 °C. In summer, with high temperatures, the average groundwater TEC values are approximately seven times higher than in other seasons. This suggests that during summer, the microbial TBC and TEC exceedance in soil surface karst zone groundwater is severe, rendering it unsuitable as a source of potable water for human use.

Soil consists of particles of varying sizes, and its particle size distribution, coupled with porosity, constitutes one of its critical physical properties. These aspects significantly influence the evolution of groundwater microbes. Research by scholars such as Bradford and Harter, using saturated soil columns with varied particle sizes, has demonstrated that as particle size diminishes, microbial concentration at the column outlet decreases, while retention at the inlet increases [32,33]. Furthermore, David's experiments on agricultural grassland soil revealed that particles of different sizes possess distinct adsorption surface energies. For instance, only 2% of *Escherichia coli* were retained in soil particles ≥ 31 μm, compared to 65% in particles <2 μm [49]. In studies involving Bacillus subtilis spores in soil, Minyoung found that doubling the soil particle size could increase the migration speed of spore microbes by 82%, likely due to the impact of larger pores in sandy soil on both adsorption capacity and hydrophobic interactions [50]. This is consistent with our finding that in denser soil or with smaller soil particle sizes, the groundwater TBC and TEC values increase. This phenomenon underscores that in rocky desertification areas under agricultural conditions, more frequent tilling and looser soil quality correlate with higher groundwater TBC levels, with the smallest and most stable fluctuations.

4.3. Implication

The results highlight that temperature exerts the most significant influence on the quality of karst groundwater. Elevated temperatures lead to increased pH, COD, and microbial indicator (total bacterial count and total coliforms) values in groundwater, along with wider ranges in ORP values. When the temperature exceeds 25 °C, the total bacterial count markedly increases with rising temperatures, while below 25 °C, there is minimal response to temperature variations. Total coliform concentration gradually increases with temperature, showing a more rapid rise above 15 °C. At the same temperature, depth shows little impact on water quality changes in surface karst springs. When porosity exceeds 0.55, the hydrodynamic cycle is accelerated, and soil oxygen content increases, resulting in higher ORP values and lower pH, indicative of a pronounced oxidative environment in the soil–water matrix. This suggests that the surface charge of soil mineral particles increases, thereby enhancing the adsorption capacity of the soil mineral surface for microbial indicators in karst groundwater, leading to more enriched pathogen-microbiota indicators. On the contrary, when porosity is less than 0.55, the surface soil layer exerts the opposite effect on karst groundwater microbes, reducing the concentration of pollution indicators. This indicates that denser soil surface karst zone soil or smaller soil particle sizes possess some capacity to purify karst groundwater quality.

Supplementary Materials: The following supporting information can be downloaded at: https://www.mdpi.com/article/10.3390/app14051933/s1, Table S1: Concentration values and evaluation results of water pollution.

Author Contributions: Conceptualization, L.S.; data curation, W.S. and Z.N.; funding acquisition, X.W.; writing—original draft, W.S.; writing—review and editing, S.W. All authors have read and agreed to the published version of the manuscript.

Funding: This research was supported by Hebei Natural Science Foundation: D2022504009; National Natural Science Foundation of China (No. 42330714); National Key R&D Projects of China (No. 2016 YFC0502502); Basal Research Fund of Iheg (CAGS: SK202314, SK202319).

Data Availability Statement: The original contributions presented in this study are included in the article. Further inquiries can be directed to the corresponding author.

Conflicts of Interest: The authors declare that the research was conducted in the absence of any commercial or financial relationships that could be construed as a potential conflict of interest.

References

1. Congxi, C. Geology of karst rocky mountainous areas in southwestern China-Ecological Environment and Governance. *Geol. China* **1999**, *4*, 11–13.
2. Zhang, L.; Yamane, T.; Satoh, E.; Amagasaki, K.; Kawataki, T.; Asahara, T.; Furuya, K.; Nukui, H.; Naganuma, H. Establishment and partial characterization of five malignant glioma cell lines. *Neuropathology* **2005**, *25*, 136–143. [CrossRef]
3. Jiang, Y.J.; Yuan, D.X.; Zhang, C. Impact of Land Use Change on Soil Properties in a Typical Karst Agricultural Region: A Case Study of Xiaojiang Watershed, Yunnan. *Acta Geogr. Sin.* **2005**, *50*, 911–918. [CrossRef]
4. Ning, Z.; Wang, S.; Guo, C. The impact of environmental factors on the transport and survival of pathogens in agricultural soils from karst areas of Yunnan province, China: Laboratory column simulated leaching experiments. *Front. Microbiol.* **2023**, *14*, 1143900. [CrossRef] [PubMed]
5. Wang, S. *Microbial-Toxicological Combined Response Mechanism and Simulation of Groundwater Polluted by Farming in Desertification Areas*; Chinese Academy of Geological Sciences: Beijing, China, 2019.
6. Charudattan, R. Biological Control of Water Hyacinth by Using Pathogens: Opportunities, Challenges, and Recent Developments//Biological and integrated control of water hyacinth: Eichhornia crassipes. In Proceedings of the Second Meeting of the Global Working Group for the Biological and Integrated Control of Water Hyacinth, Beijing, China, 9–12 October 2000; Australian Centre for International Agricultural Research (ACIAR): Sydney, Australia, 2001.
7. Pang, R.-l.; Wang, S.-y.; Wang, R.-p. Study on the Enrichment and Migration Characteristics of Heavy Metals in Soil-Grapevine System. *J. Ecol. Rural. Environ.* **2019**, *35*, 515–521. [CrossRef]
8. Yuan, S.; Tan, Z.; Huang, Q. Migration and transformation mechanism of nitrogen in the biomass–biochar–plant transport process. *Renew. Sustain. Energy Rev.* **2018**, *85*, 1–13. [CrossRef]
9. Mohammadi, K.H.; Karim, G.H.; Razavilar, V. Study on the growth and survival of Escherichia coli O157:H7 during the manufacture and storage of Iranian white cheese in brine. *Iran. J. Vet. Res.* **2009**, *10*, 346–351. [CrossRef]
10. Andrew, F.B. Numerical simulation of chemical migration in physically and chemically heterogeneous porous media. *Water Resour. Res.* **1993**, *29*, 3709–3726. [CrossRef]
11. Li, Y.; Hou, Q.; Wang, S. Assembly of abundant and rare maize root-associated bacterial communities under film mulch. *Appl. Soil Ecol.* **2023**, *182*, 1–11. [CrossRef]
12. Jiang, W.; Meng, L.; Liu, F. Distribution, source investigation, and risk assessment of topsoil heavy metals in areas with intensive anthropogenic activities using the positive matrix factorization (PMF) model coupled with self-organizing map (SOM). *Environ. Geochem. Health* **2023**, *45*, 6353–6370. [CrossRef]
13. Wang, X.; Sun, L.; Wang, S. Development and application of multi-field coupled high-pressure triaxial apparatus for soil. *Groundw. Sci. Eng. Engl. Version* **2023**, *11*, 308–316. [CrossRef]
14. Bitton, G. Adsorption of viruses onto surfaces in soil and water. *Water Res.* **1975**, *9*, 473–484. [CrossRef]
15. Sun, L.; Wang, S.-W.; Guo, C.-J.; Shi, C.; Su, W.-C. Using pore-solid fractal dimension to estimate residual LNAPLs saturation in sandy aquifers: A column experiment. *J. Groundw. Sci. Eng.* **2022**, *10*, 87–98. [CrossRef]
16. Mallén, G.; Maloszewski, P.; Flynn, R. Determination of bacterial and viral transport parameters in a gravel aquifer assuming linear kinetic sorption and desorption. *J. Hydrol.* **2005**, *306*, 21–36. [CrossRef]
17. Sinton, L.W.; Braithwaite, R.R.; Hall, C.H. Tracing the Movement of Irrigated Effluent into an Alluvial Gravel Aquifer. *Water Air Soil Pollut.* **2005**, *166*, 287–301. [CrossRef]
18. Hermansson, M.; Kjelleberg, S.; Norkrans, B. The hydrophobicity of bacteria? An important factor in their initial adhesion at the air-water interface. *Arch. Microbiol.* **1981**, *128*, 267. [CrossRef]

19. Jiang, W.; Liu, H.; Sheng, Y. Distribution, Source Apportionment, and Health Risk Assessment of Heavy Metals in Groundwater in a Multi-mineral Resource Area, North China. *Water Qual. Expo. Health* **2022**, *14*, 807–827. [CrossRef]

20. Yang, C.; Liu, Y.; Liu, S.Q. Mechanism of microbial clogging effect on single-well push-pull test. *Saf. Environ. Eng.* **2023**, *30*, 199–204.

21. Foppen, J.W.; Herwerden, M.V.; Schijven, J. Measuring and modelling straining of Escherichia coli in saturated porous media. *J. Contam. Hydrol.* **2007**, *93*, 236–254. [CrossRef]

22. Haznedaroglu, B.Z.; Bolster, C.H.; Walker, S.L. The role of starvation on Escherichia coli adhesion and transport in saturated porous media. *Water Res.* **2008**, *42*, 1547–1554. [CrossRef]

23. McDowell-Boyer, L.M.; Hunt, J.R.; Sitar, N. Reply to "Comments on 'Particle transport through porous media' by Laura M. McDowell-Boyer, James R. Hunt, and Nicholas Sitar". *Water Resour. Res.* **1987**, *23*, 1699. [CrossRef]

24. McDowell-Boyer, L.M.; Hunt, J.R.; Sitar, N. Particle transports through porous media. *Water Resour. Res.* **1986**, *22*, 1901–1921. [CrossRef]

25. Reneau, R.B.; Hagedorn, C.; Degen, M.J. Fate and Transport of Biological and Inorganic Contaminants from On-Site Disposal of Domestic Wastewater. *J. Environ. Qual.* **1989**, *18*, 135–144. [CrossRef]

26. Yee, N.; Fein, J.B.; Daughney, C.J. Experimental study of the pH, ionic strength, and reversibility behavior of bacteria–mineral adsorption. *Geochim. Cosmochim. Acta* **2000**, *64*, 609–617. [CrossRef]

27. Wellings, F.M.; Lewis, A.L.; Mountain, C.W. Demonstration of Virus in Groundwater after Effluent Discharge onto Soil. *Appl. Microbiol.* **1975**, *29*, 751. [CrossRef]

28. Sander, M.; Pignatello, J.J. On the Reversibility of Sorption to Black Carbon: Distinguishing True Hysteresis from Artificial Hysteresis Caused by Dilution of a Competing Adsorbate. *Environ. Sci. Technol.* **2007**, *41*, 843–849. [CrossRef] [PubMed]

29. Stewart, B.A. The Influence of Macropores on the Transport of Dissolved and Suspended Matter Through Soil. In *Advances in Soil Science: Volume 3*; Springer: New York, NY, USA, 1985; pp. 95–120. [CrossRef]

30. Huysman, F.; Verstraete, W. Water-facilitated transport of bacteria in unsaturated soil columns: Influence of inoculation and irrigation methods. *Soil Biol. Biochem.* **1993**, *25*, 91–97. [CrossRef]

31. Jewett, D.G.; Logan, B.E.; Arnold, R.G. Transport of Pseudomonas fluorescens strain P17 through quartz sand columns as a function of water content. *J. Contam. Hydrol.* **1999**, *36*, 73–89. [CrossRef]

32. Bradford, S.A.; Simunek, J.; Bettahar, M. Modeling colloid attachment, straining, and exclusion in saturated porous media. *Environ. Sci. Technol.* **2003**, *37*, 2242–2250. [CrossRef]

33. Bradford, S.A.; Bettahar, M. Straining, attachment, and detachment of cryptosporidium oocysts in saturated porous media. *J. Environ. Qual.* **2005**, *34*, 469–478. [CrossRef]

34. Shahmohammadi-Kalalagh, S.; Beyrami, H.; Taran, F. Bromide Transport through Soil Columns in the Presence of Pumice. *Iran. J. Chem. Chem. Eng.* **2022**, *41*, 1305–1312. [CrossRef]

35. Jewett, D.G.; Hilbert, T.A.; Logan, B.E. Bacterial transport in laboratory columns and filters: Influence of ionic strength and pH on collision efficiency. *Water Res.* **1995**, *29*, 1673–1680. [CrossRef]

36. Gerba, C.P. Persistence of Viruses in Desert Soils Amended with Anaerobically Digested Sewage Sludge. *Appl. Environ. Microbiol.* **1992**, *58*, 636–641. [CrossRef]

37. Fontes, D.E.; Mills, A.L.; Hornberger, G.M. Physical and chemical factors influencing transport of microorganisms through porous media. *Appl. Environ. Microbiol.* **1991**, *57*, 2473–2481. [CrossRef] [PubMed]

38. Gannon, J.; Tan, Y.; Baveye, P.; Alexander, M. Effect of sodium chloride on transport of bacteria in a saturated aquifer material. *Appl. Environ. Microbiol.* **1991**, *57*, 2497–2501. [CrossRef] [PubMed]

39. Unc, A.; Goss, M.J.; Cook, S.; Li, X.; Atwill, E.R.; Harter, T. Analysis of matrix effects critical to microbial transport in organic waste-affected soils across laboratory and field scales. *Water Resour. Res.* **2012**, *48*, 154–167. [CrossRef]

40. Kato, S.; Jenkins, M.B.; Ghiorse, W.C. Chemical and physical factors affecting the excystation of *Cryptosporidium parvum oocysts*. *J. Parasitol.* **2001**, *87*, 575–581. [CrossRef] [PubMed]

41. *DD2008-01*; Technical Code for Groundwater Pollution Investigation and Evaluation. Ministry of Environment and Climate Change Strategy: Victoria, BC, Canada, 2021.

42. *GB/T 5750-2023*; Standard Test Method for Drinking Water. Code of China: Beijing, China, 2023.

43. Wang, Y. *The Yunnan Typical Regional Karst Groundwater Investigation and Geological Environmental Remediation Demonstration Report*; China Geological Survey: Beijing, China, 2003.

44. Conner, D.E. Growth and survival of Escherichia coli O157:H7 under acidic conditions. *Appl. Environ. Microbiol.* **1995**, *61*, 382–385. [CrossRef] [PubMed]

45. Vanbleu, E.; Vanderleyden, J. *Molecular Genetics of Rhizosphere and Plant-Root Colonization*; Springer: Amsterdam, The Netherlands, 2007. [CrossRef]

46. Sinton, L.W.; Mackenzie, M.L.; Karki, N. Transport of Escherichia coli and F-RNA bacteriophages in a 5 m column of saturated pea gravel. *J. Contam. Hydrol.* **2010**, *117*, 71–81. [CrossRef]

47. Zheng, Y.; Zhang, Z.; Chen, Y. Adsorption and desorption characteristics and mechanism of Cd in reclaimed soil under the influence of dissolved organic carbon. *J. Int. J. Coal Sci. Technol.* **2022**, *9*, 225–235.

48. Li, Z.Y. Assessment of future climate change impacts on water-heat-salt migration in unsaturated frozen soil using CoupModel. *Front. Environ. Sci. Eng.* **2021**, *15*, 145–161.

49. Haack, S.K.; Metge, D.W.; Fogarty, L.R. Effects on Groundwater Microbial Communities of an Engineered 30-Day in Situ Exposure to the Antibiotic Sulfamethoxazole. *Environ. Sci. Technol.* **2012**, *46*, 7478. [CrossRef] [PubMed]
50. Kim, M.; Choi, C.Y.; Gerba, C.P. Development and evaluation of a decision-supporting model for identifying the source location of microbial intrusions in real gravity sewer systems. *Water Res. J. Int. Water Assoc.* **2013**, *47*, 4630–4638. [CrossRef] [PubMed]

Article

A Comparative Study on Soil-Crop Selenium Characteristics in High-Incidence Areas of Keshan Disease in Chinese Loess and Black Soil

Jun Zhao [1,2], Zhu Rao [1,*], Siwen Liu [1], Lei Wang [2], Peng Wang [2], Tao Yang [2] and Jin Bai [2]

[1] National Research Center for Geoanalysis, Chinese Academy of Geological Science, Beijing 100037, China; 18092106153@163.com (J.Z.); siwenzliu@126.com (S.L.)
[2] Xi'an Geological Survey Center, China Geological Survey, Xi'an 710119, China; tleiwang@163.com (L.W.); nanqi0606@163.com (P.W.); yangtao008@163.com (T.Y.); 13991873977@139.com (J.B.)
* Correspondence: raozhu@126.com

Abstract: With the gradual emphasis on health by people, the research on the pathogenesis of endemic diseases has become increasingly in-depth. Through analyzing the environmental selenium characteristics and conducting a comparative study in typical areas of Chinese loess and black soil in this paper, it is concluded that the environmental selenium in the two regions has different characteristics. The soil in the loess area has the characteristics of high alkalinity, low selenium, and relatively high selenium availability, and the crops are selenium-deficient, while the soil in the black soil area has the characteristics of high organic matter, low selenium availability, and relatively high selenium in crops. The research concluded that the environmental occurrence mechanism of Keshan disease in the loess area and the black soil area is different. Keshan disease can be induced in both low-selenium and sufficient-selenium environments, and environmental selenium should be one of the inducing factors of Keshan disease. This research provides a reference for predicting the areas where Keshan disease occurs and for disease prevention, and it can also serve in the prevention and control of endemic diseases.

Keywords: China; loess area; black soil area; Keshan disease; environmental selenium

Citation: Zhao, J.; Rao, Z.; Liu, S.; Wang, L.; Wang, P.; Yang, T.; Bai, J. A Comparative Study on Soil-Crop Selenium Characteristics in High-Incidence Areas of Keshan Disease in Chinese Loess and Black Soil. *Appl. Sci.* **2024**, *14*, 5703. https://doi.org/10.3390/app14135703

Academic Editor: Rafael López Núñez

Received: 30 March 2024
Revised: 7 June 2024
Accepted: 17 June 2024
Published: 29 June 2024

1. Introduction

1.1. Preface

Keshan disease is an unidentified endemic cardiomyopathy [1]. It was discovered in Keshan County, Heilongjiang Province, in 1935, and was therefore named Keshan disease. By the end of 2015, Keshan disease was distributed in 16 provinces [2], municipalities, and autonomous regions in China, involving 328 counties, 2953 townships, and a population of about 60 million [3]. Geographically, it formed a broad band extending from the northeast to the southwest in China [4].

Keshan disease has a concentrated high-incidence season in both northern and southern disease areas. It is more common in winter in the northeast region and in summer in the southwest region. Nowadays, it occurs sporadically throughout the year. There was a high-incidence period from 1955 to 1978. After 1978, the incidence rate showed a downward trend. In 1959, the incidence rate of acute and subacute Keshan disease in the disease area reached 60.18/100,000, and 40.42/100,000 in 1970, with a fatality rate of more than 40% [1]. During the Keshan disease surveillance from 2000 to 2004, a total of 2096 cases of potential and chronic Keshan disease were detected nationwide in five years, including 113 new cases of potential Keshan disease and 29 cases of chronic Keshan disease, with an average annual incidence rate of 3.1‰ [5].

The main pathological changes in this disease are myocardial parenchymal degeneration, necrosis, and fibrosis, resulting in the failure of cardiac systolic and diastolic function.

It is an independent endemic cardiomyopathy and a serious local public health problem in China [6,7].

Due to its regional high incidence, studying the environmental mechanism of the onset of the disease to prevent its occurrence has always been a hot topic in scientific research.

This research conducts a comparison of the environmental selenium in the high-incidence areas of Keshan disease in Chinese loess and black soil, analyzes the differences between the two kinds of environmental selenium, and proposes that Keshan disease can occur in both high-selenium and low-selenium environments; in the loess area where selenium is generally deficient, it puts forward in-depth planting suggestions to maximize the selenium content of crops to ensure the selenium intake of the population; and this research provides a basis for the prediction and prevention of Keshan disease.

1.2. Previous Research Results on the Etiology of Keshan Disease

The etiology of Keshan disease is currently not clear. There are more than 10 theories about the etiology. Current research on the etiology of Keshan disease focuses on two aspects: biogeochemical etiology and biological etiology.

(1)   Biogeochemical etiology

A. Environmental element deficiency etiology

In the 1960s, the Keshan Disease Research Laboratory of Xi'an Medical University proposed the hypothesis of soil and water etiology and found that the pathological changes in white muscle disease caused by selenium deficiency in animals in Keshan disease areas were similar to those of Keshan disease, so it was suggested that selenium deficiency was the cause of Keshan disease [8]. However, selenium deficiency cannot explain all the epidemiological characteristics of Keshan disease. Keshan disease does not occur in areas with low selenium, and there is also Keshan disease in areas with sufficient selenium [9]. Moreover, selenium deficiency does not change correspondingly with the seasonal and annual high incidence of Keshan disease. There is no significant difference in selenium and blood selenium between children with and without the disease in the disease area, so selenium deficiency is not the only factor [10]. At the same time, epidemiological investigations found that manganese can promote the excretion of selenium, and abundant manganese can further aggravate myocardial damage, suggesting that abundant manganese is related to the onset of Keshan disease [11].

B. Lack of protein amino acids and vitamins

Experiments conducted by Yu Weihan and others found that the total weight gain, food utilization rate, and glutathione peroxidase activity in the whole blood of rats fed with low selenium and low protein were lower than those of rats fed with low selenium and high protein, and the detection rate of myocardial lesions was higher [12]. The supplementation of protein has a protective effect on the myocardium of rats. Through a comparative analysis of the amino acid content in the main grain in the disease area and the free amino acid content and antioxidant capacity in the plasma of the residents, it was found that the sulfur amino acid level in the population in the disease area was at a critical deficiency state, which may lead to a decrease in the antioxidant capacity of the body, thereby inducing the onset of Keshan disease [13].

Studies have shown that the total amount of vitamin E in the plasma of the population in the disease area is significantly lower than that in the non-disease area, and the content of alpha-tocopherol, which has the strongest antioxidant effect in vitamin E, is generally lower than that in the non-disease area. Studies have found that the content of polyunsaturated fatty acids in grain in the disease area is significantly increased, which can lead to a relative lack of vitamin E, thereby reducing the antioxidant capacity of the body and inducing the onset of Keshan disease [1].

(2)   Biological etiology

Previous studies have suggested that factors such as simple selenium deficiency cannot fully explain the epidemic pattern of Keshan disease, but some characteristics are consistent with the characteristics of biological factor infection, such as the isolation of multiple viruses from the myocardial tissue of Keshan disease, including Coxsackie B group virus [14], chlorogenic acid, T-2 toxin [15], and fusarium toxins [16]. It is believed that viruses are also factors that cause Keshan disease.

2. Materials and Methods

To conduct this study, soil and crop samples were collected in Xunyi County in the high-incidence area of Keshan disease in the Loess Plateau of China and Keshan County in the high-incidence area of black soil, and their selenium characteristics were analyzed and compared, in order to analyze the contribution of environmental selenium to Keshan disease and provide a reference for the prevention of endemic diseases.

2.1. Research in Xunyi County, Shaanxi Province

(1)   Overview of Xunyi County

Xunyi County, under the jurisdiction of Xianyang City, Shaanxi Province, is located in the north of Xianyang City, with a total area of 1811 square kilometers. The exposed strata mainly include the Quaternary, Tertiary, Triassic, and Cretaceous systems. There are nine main types of soil in Xunyi County, among which loess is the most widely distributed, followed by cinnamon soil and black loess. The soil samples collected in this study mainly include black loess in Zhitian Town, black loess in Tuqiao Town, and yellow loess in the southeast of Zhangbaosi Town.

(2)   The situation of Keshan disease in Xunyi

Shaanxi Province is a province with relatively severe Keshan disease. Xunyi County is located in the southern part of the Northern Shaanxi Plateau, and all 14 townships in the county are Keshan disease areas. In the mid-1960s, it was one of the historically severe Keshan disease areas (counties) in Shaanxi Province. In 1965, the incidence rate of acute and subacute Keshan disease in Xunyi County reached 86.8/100,000. Since 1990, Xunyi County has been listed as one of the first national Keshan disease surveillance sites in China [17]. There are 14 disease-stricken townships in Xunyi County, with a population of 265,448. A survey in 2005 showed that there were still 14 new cases of natural chronic Keshan disease, with an annual incidence rate of 0.53/10,000.

Houyiyang Village and Changshetou Village in Houzhang Township, and Dongcao Village in Tuqiao Town, Xunyi County, serve as national surveillance points for Keshan disease in Xunyi County. In 2003, there were still 50 potential cases and 3 chronic cases [18].

This time, Changshetou Village, Zaochi Village, and Dongcao Village in Xunyi County were selected as high-incidence study areas to conduct soil selenium research.

(3)   Sample collection and testing

Soil sampling was carried out across the county in August 2017, with a sampling area of approximately 1811 km^2. The sampling depth was 0–20 cm. Sample testing was conducted by the Xinjiang Geological Laboratory. Elements and oxides such as Al_2O_3, Cr, Ga, K_2O, Nb, P, Pb, V, Rb, SiO_2, Ti, Y, Cl, Zr, Br, Cu, Ba, CaO, Co, TFe_2O_3, MgO, Mn, Na_2O, Ni, Sr, and Zn were analyzed using X-ray fluorescence spectroscopy (XRF). Elements such as Ag, B, Sn, and Mo were analyzed using emission spectroscopy (AES). Elements such as Cd, Mo, U, Th, TL, Bi, Ge, W, Ga, Rb, Nb, Pb, Cr, Y, Cu, and Ce were analyzed using plasma mass spectrometry (ICP-MS). Elements such as As, Sb, Hg, and Se were analyzed using atomic fluorescence spectroscopy (AFS). Elements such as Be, CaO, Ce, Co, TFe_2O_3, La, Li, MgO, Mn, Na_2O, Ni, Sc, Sr, Zn, Al_2O_3, Ti, Ba, K_2O, and V were analyzed using inductively coupled plasma emission spectroscopy (ICP-OES). The actual number of samples analyzed in the whole county was 460, and 25 sets of wheat and its rhizosphere soil and 20 soil profile samples were collected in the high-incidence area.

(4)   Characteristics of selenium in the soil of the whole county

① 1:250,000 soil selenium characteristics

Based on the soil data of the whole county, the element characteristic table related to Keshan disease in Xunyi County is calculated, as shown in Table 1.

Table 1. Characteristics of soils in Xunyi County.

	Sample Size	Mean	Median	Maximum	Minimum
As (10^{-6})	427	13.07	13.35	15.80	6.65
CaO (10^{-2})	430	6.58	6.60	14.50	1.70
Cd (10^{-9})	429	174.33	170.00	230.00	118.84
Corg (10^{-2})	410	0.98	0.82	3.84	0.36
Cr (10^{-6})	427	69.04	69.10	84.20	48.90
Cu (10^{-6})	422	24.60	24.67	32.77	16.20
Hg (10^{-9})	405	28.77	27.13	229.00	12.20
MgO (10^{-2})	413	2.31	2.29	3.19	1.73
Mn (10^{-6})	423	652.9	650.0	887.37	432.0
Mo (10^{-6})	427	0.77	0.77	0.96	0.58
Pb (10^{-6})	431	22.41	22.40	26.60	18.20
PH	428	8.23	8.26	8.56	7.28
Se (10^{-6})	420	0.13	0.13	0.30	0.07
Zn (10^{-6})	422	68.51	68.25	96.20	43.50

The average selenium content in the soils of Xunyi County is 0.13×10^{-6}, while the national average is 0.29×10^{-6} [19]. Compared with the national level, Xunyi County is a selenium-deficient area, with an enrichment coefficient of 0.45. The analysis results of the total selenium content in the surface soils of Shaanxi Province show that the range is $0.018–17.618 \times 10^{-6}$, with an average of 0.118×10^{-6} [20]. Compared with the soils of Shaanxi Province, it is relatively stable. The selenium content in the surface soils of the Guanzhong area of Shaanxi Province is $0.034–2.628 \times 10^{-6}$, with an average content of 0.174×10^{-6} [21]. The enrichment coefficient is 0.75, and it still belongs to a low-selenium county in the loess area.

Blazina et al. (2014) divided China into four regions based on soil selenium content. Regions with a selenium content ≤ 0.1 mg/kg are selenium-deficient areas, those with selenium content of 0.1–0.2 mg/kg are low-selenium areas, those with a selenium content of 0.2–0.4 mg/kg are areas with normal selenium content, and those with a selenium content > 0.4 mg/kg are high-selenium areas [22].

All the soils in the whole Xunyi County area are alkaline and have less organic matter, and there is a general lack of selenium.

② Selenium characteristics of different soil types

According to the soil-type map of Xunyi County, the main soil types in this area are loessial soil, cinnamon soil, and black loam, and the elemental characteristics of the three soil types are shown in Table 2.

Table 2. Element characteristics of main soils in Xunyi.

Element	As (10^{-6})	CaO (10^{-2})	Cd (10^{-9})	Corg (10^{-2})	Cr (10^{-6})	Cu (10^{-6})	Hg (10^{-9})
Cinnamon soil	12.77	6.75	172.3	1.18	69.16	24.36	29.21
Black loam	13.93	6.37	189.2	0.74	70.78	25.78	33.74
Loessial soil	13.21	6.82	174.0	0.87	69.25	24.52	27.96

Element	Mn (10^{-6})	Mo (10^{-6})	Pb (10^{-6})	PH	Se (10^{-6})	Zn (10^{-6})	MgO (10^{-2})
Cinnamon soil	650.8	0.75	22.23	8.23	0.13	67.73	2.33
Black loam	670.3	0.79	22.94	8.2	0.14	69.47	2.24
Loessial soil	647.3	0.77	22.31	8.24	0.13	68.23	2.3

From the table, it can be seen that Se has similar contents in the three soils and is all in the low-selenium zone. It is known that Keshan disease mainly occurs in black loam. Although Se is relatively higher than that in loessial soil and cinnamon soil, it is still selenium-deficient.

③ Correlation of selenium elements

Using SPSS software (IBM spss statistics20) to calculate the correlation of elements and draw diagrams of the correlation between each element and selenium, as shown in Figure 1.

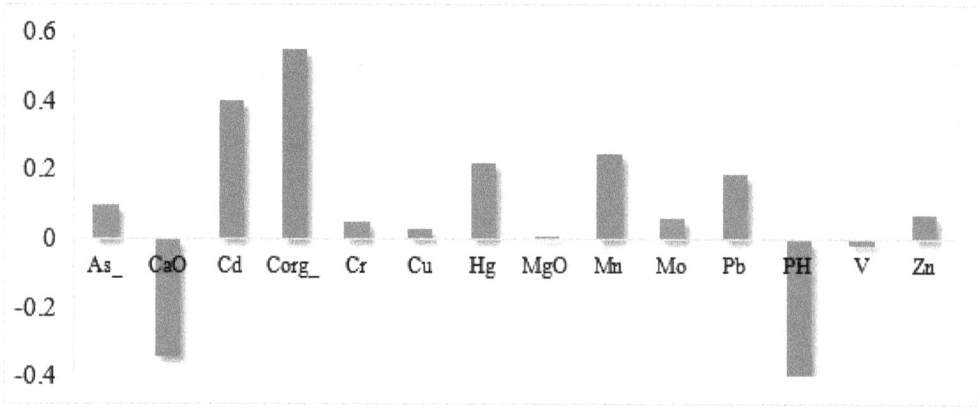

Figure 1. Correlation of each element with selenium.

It can be seen from the figure that the distribution of selenium in the soil of this area is positively correlated with Corg, Cd, Mn, Hg, Pb, etc. The most correlated is organic carbon. Due to the low level of soil organic matter, the effectiveness of Se is positively correlated with organic matter [23]. At the same time, Cd, Hg, Pb, etc., have an antagonistic effect on selenium, which can also affect the effectiveness of selenium. According to previous studies, the pH value range that is most suitable for selenium absorption is about 6.7–7.9 [24]. The soil in this area is 7.28–8.56, in the middle of the most suitable pH range. The closer the soil pH is to 7.28, the higher the total amount of Se, and the more the soil tends to be neutral, which further affects the absorption of selenium.

(5)   Characteristics of selenium in the high-incidence area of Keshan disease.

In order to further study the selenium content in the soil and crops in high-incidence areas, soil samples and crops were collected in high-incidence areas such as Dongcao Village, Tuqiao Town, Zhichian Town, and Changshetou Village, for a total of 25 sets of wheat and root soil, 2 soil profiles, etc.

① Characteristics of selenium in soil and crops in the high-incidence area.

The characteristics of wheat samples, the main crop in the area, are shown in the following Table 3.

Table 3. Element content in root soil in high-incidence areas.

	Pb (10^{-6})	Zn (10^{-6})	Cr (10^{-6})	Mo (10^{-6})	Ge (10^{-6})	Se (10^{-6})
	25.41	80.71	68.64	0.89	1.05	0.12
High-incidence area soil	Corg (10^{-2})	pH	Mn (10^{-6})	Ca (10^{-2})	Fe (10^{-2})	
	1.7	8.44	0.07	3.68	3.32	

It can be seen that the mean value of selenium in the soil of the high-incidence area is 0.12×10^{-6}, slightly lower than the average value of soil in the whole county, which is

0.13×10^{-6}. Compared with the Guanzhong area [25], the mean value of selenium in the soil of the Guanzhong Plain is 0.171×10^{-6}, and the selenium in the soil of the high-incidence area is significantly lower, with an enrichment coefficient of about 0.702.,as shown in Table 4.

Table 4. Characteristics of wheat in the high-incidence area.

	Pb (10^{-6})	Zn (10^{-6})	Cr (10^{-6})	Mo (10^{-6})
	0.04	20.72	0.4	0.89
Wheat in the high-incidence area	Se (10^{-6})	Mn (10^{-6})	Ca (10^{-2})	Fe (10^{-2})
	0.014	47.8	426.83	36.86

The mean value of wheat in the high-incidence area is 0.014×10^{-6}. Referring to the relevant selenium content standards for selenium-rich and selenium-containing soil in Shaanxi Province, it is the selenium-containing standard. Compared with the Guanzhong area [26], the mean value of selenium in the wheat of the Guanzhong Plain is 0.056×10^{-6}, and the enrichment coefficient is 0.25.

② Selenium availability in root zone soil

The existing forms of selenium in the soil determine the ability of crops to absorb this element [27]. Existing studies have shown that the forms of selenium in the soil can be roughly divided into water-soluble, ion-exchangeable, carbonate, humic acid, iron-manganese oxide-bound, organic matter-bound, and residual states [28,29]. Based on bioavailability considerations, water-soluble and ionic states are classified as available states; carbonate-bound and iron-manganese oxide-bound Se are the direct source of available Se and can be classified as potentially available states; while humic acid-bound, organic-bound, and residual-state Se can be classified as non-available Se.

There are significant differences in the valence content of selenium in the soil of this area, with most being in the residual and strong organic forms, as shown in Figure 2.

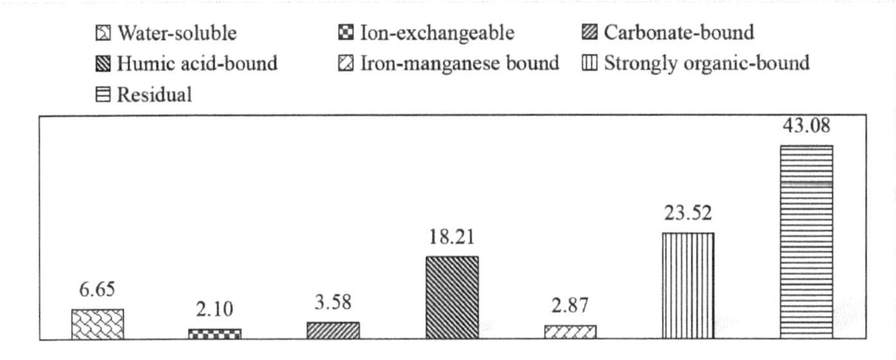

Figure 2. Proportion of selenium valence in soil.

It can be seen from Figure 3 that the forms of selenium in the area are in the order of residual > strong organic > humic acid > water-soluble > carbonate > iron-manganese binding > ion-exchangeable. By statistically analyzing the content of available Se, potentially available Se, and non-available Se in the area, it is found that Se in the area is mainly in the non-available state, accounting for 84.81% of the total Se, the available Se is only 8.75%, and the potentially available Se accounts for 6.45% of the total Se. Compared with the soil selenium in the tillage layer of soil in Zhejiang, Qinghai, and Jilin, the proportion of available Se is the same, which is higher than that of red soil and black soil [30–32].

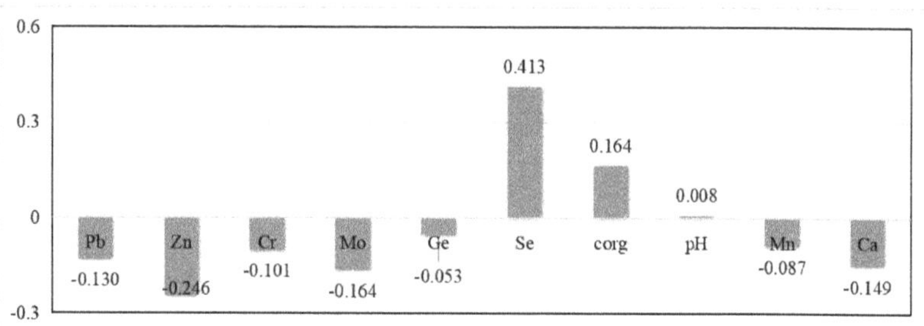

Figure 3. Correlation diagram of selenium effectiveness and various elements.

③ Relationship between available selenium in soil and content of soil elements.

Based on the characteristics of root zone soil elements and the availability of soil selenium, the SPSS software (IBM spss statistics20) was used to calculate the correlation between availability and each element, as shown in the histogram of the correlation between availability and each element.

It can be seen from the figure that the effectiveness of soil selenium is weakly positively correlated with the total amount of Se and organic matter. According to the research, when the overall content of soil organic matter is low, within a certain range, a higher content of soil organic matter will increase the content of organic selenium. Along with the mineralization of organic selenium, the small molecule organic selenium and inorganic selenium that are easily dissolved in the soil will increase, thereby improving the biological effectiveness of selenium [33]; the organic matter in the loess area is generally low, which makes the selenium effectiveness positively correlated with the organic matter, but when the organic matter rises to a certain extent, it can show a decrease in the effectiveness of Se. At the same time, the available selenium in the soil is negatively correlated with elements such as Zn, Mn, Mo, and Pb, reflecting that the lower the content of these metal elements in the soil, the higher the effectiveness of selenium.

④ Enrichment coefficient of wheat selenium

In order to reflect the enrichment of Se in the soil-agricultural crops, the enrichment factor (EF) is expressed by EF = wheat Se/root soil Se. The maximum value of EF in this area is 0.22, the minimum value is 0.1, and the average value is 0.14.

Based on the measurement results, the selenium coefficient of wheat and root soil was calculated. The correlation diagram of the enrichment coefficient of wheat and the various elements of the root soil is shown in Figure 4.

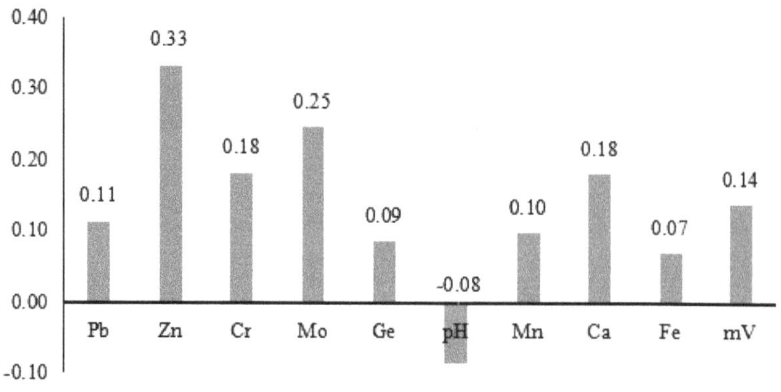

Figure 4. Correlation between wheat Se enrichment coefficient and root soil elements.

It can be seen from the figure that most elements are positively correlated with the wheat selenium enrichment coefficient, among which Mo and Zn have the greatest influence coefficient. The acidity and alkalinity are negatively correlated with the wheat selenium enrichment coefficient. It reflects that in this area, the wheat-producing areas have a high Mo and Zn, and the pH tends to be neutral. The higher the enrichment coefficient, the higher the selenium content in wheat.

⑤ Vertical profile selenium characteristics

Samples were collected from the profile PM01 in Dongcao Village in the high-incidence area, with a profile depth of 2.0 m and sampling every 20 cm; Samples were collected from the profile PM02 in Zhitian Town Zaochi Village, with a profile depth of 2.2 m and sampling every 20 cm. The results of the selenium and available selenium profile are shown in Figures 5 and 6.

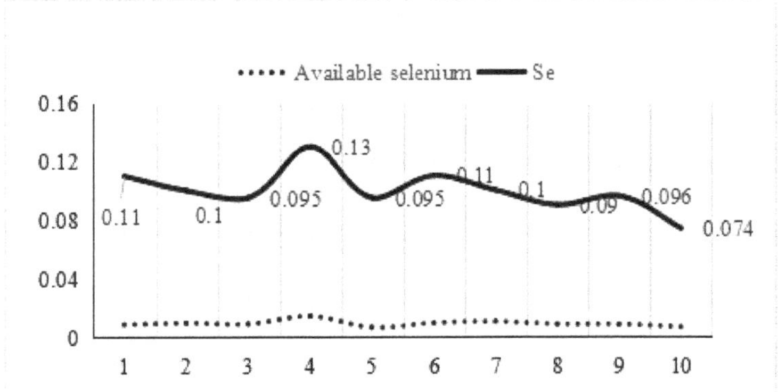

Figure 5. Depth variation diagram of soil selenium and available selenium in PM01 (10^{-6}).

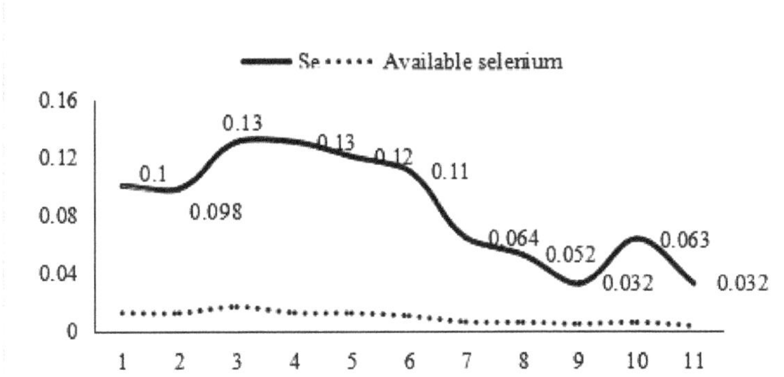

Figure 6. Depth variation diagram of soil selenium and available selenium in PM02 (10^{-6}).

From the curves of the change in selenium and available selenium with depth in the two profiles, it can be concluded that the overall selenium in the soil in this area is relatively low. From the surface to 2 m underground, the selenium shows a certain change trend: the total amount of soil selenium and available selenium in PM01 both reach a peak at 0.8 m; the soil selenium in PM02 changes with the depth, and the soil selenium first increases and then decreases, reaching a peak at 0.6–1.2 m; and then the soil selenium value is always lower than the surface selenium with the increase in depth, and the lowest value is at 1.8 m. The above indicates that the variety of the root activity layer of the crops in this area at 0.8 m can achieve the maximum absorption of soil selenium.

2.2. Comparison of Environmental Selenium Characteristics in Keshan County, Heilongjiang Province

Keshan County is under the jurisdiction of Qiqihar City, Heilongjiang Province. It is located in the western part of Heilongjiang Province, with a total area of 3320 square kilometers. It is the earliest place where Keshan disease was discovered. The geological background of the county is mainly Quaternary black soil. The high-incidence area of Keshan disease in the county is Guorong Village. According to the survey, there were a total of 286 people in Guorong Village in the winter of 1935, including 205 men and 81 women. In just two months, 36 people died of Keshan disease [9].

(1)   Comparison of soil selenium values in the entire area

The county has conducted a 1:250,000 land quality survey. The survey results are compared with the characteristics of elements such as As, Cao, Cd, Corg, Cr, Cu, Hg, Mgo, Mn, Mo, Pb, Ph, Se, and Zn in Xunyi County, as shown in Table 5.

Table 5. Comparison of soil element characteristics between Xunyi County and Keshan County.

	As (10^{-6})	Cd (10^{-9})	Corg (10^{-2})	Cr (10^{-6})	Cu (10^{-6})	MgO (10^{-2})
Xunyi County	13.07	0.174	0.98	69.04	24.6	2.31
Keshan County	10.65	0.09	2.28	66.09	20.77	1.281
ratio	1.23	1.85	0.43	1.04	1.18	1.8
	Mo (10^{-6})	Pb (10^{-9})	PH	Se (10^{-6})	Zn (10^{-6})	Mn (10^{-6})
Xunyi County	0.77	22.41	8.23	0.13	68.51	652.93
Keshan County	0.64	23.17	6.45	0.25	58.65	683.7
ratio	1.21	0.97	1.27	0.52	1.17	0.95

Compared with Keshan County, the soil in Xunyi County has higher contents of As, Cd, Cr, Cu, MgO, Mo, Pb, Ph, and Zn, indicating that the soil in Xunyi County is more alkaline, with higher contents of Cd and Cr within the safe range; Xunyi County has lower Mn and Pb contents, and the mean value of selenium in the soil of this county is only 0.52 times that of Keshan County, as shown in Table 6.

Table 6. Comparison of soil selenium in the high-incidence areas of Keshan disease in the two counties.

	Pb (10^{-6})	Zn (10^{-6})	Cr (10^{-6})	Mo (10^{-6})	Se (10^{-6})	Corg (10^{-2})	pH	Mn (10^{-2})
Xunyi County	25.41	80.71	68.64	0.89	0.12	1.7	8.44	0.07
Keshan County	25	63.35	66.18	0.63	0.33	3.1	5.8	0.08
ratio	1.02	1.27	1.04	1.42	0.36	0.55	1.46	0.92

In the high-incidence areas of Keshan disease in the two counties, Xunyi County has higher Ph and Mo contents and lower organic matter and Se.

(2)   Comparative analysis of the correlation of soil selenium elements

Based on the characteristics of root soil elements and the availability of soil selenium, the SPSS software (IBM spss statistics20) was used to draw a column chart of the correlation between the calculated soil selenium in Keshan County and various elements, as shown in Figure 7.

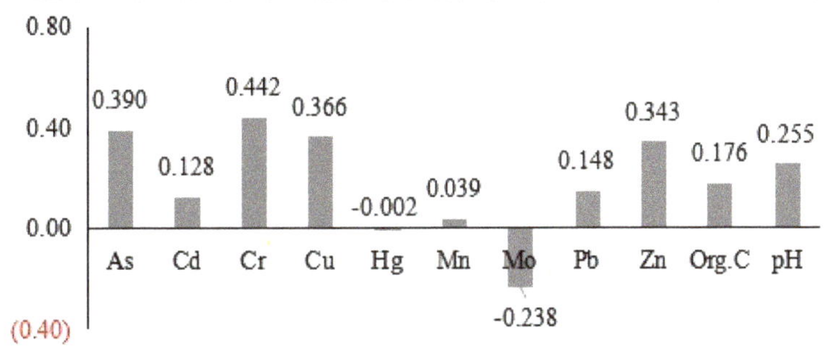

Figure 7. Correlation diagram of selenium in the surface soil of Keshan County with other elements.

It can be seen from the figure that when the selenium content in the soil of Keshan County increases, As, Cr, Cu, Zn, Ph, Org.C, etc., also increase, while Mo decreases. The organic matter in the black soil area is higher, which can reduce the availability of selenium. In the area where the selenium content in the soil of Xunyi County increases, Corg, Cd, Mn, Hg, Pb, etc., also increase. Among them, the organic matter can increase the availability of selenium due to its lower content.

(3)   Comparison of the availability of soil selenium.

According to previous data, in Keshan County, Se in the unavailable state occupies 90.37% of the total Se, and the available Se is only 6.03%; in Xunyi County, the unavailable Se occupies 84.81% of the total Se, and the available Se is only 8.75%.

Compared with Keshan County, the soil pH in Xunyi County is 8.4, while that in Keshan County is 5.8; the organic matter in Xunyi County is 1.7, while that in Keshan County is 3.1; the soil in Xunyi County has the characteristics of low organic matter, being alkaline, and high selenium availability.

(4)   Element contents in crops

Samples of the main crops and root soil in the area were collected near Gongrong Village, Xicheng Town, Keshan County, a high-incidence area. Samples of soybeans, rice, and corn were collected. See the element table of crops in Gongrong Village, Keshan County (Table 7) for details.

Table 7. Mean selenium content in crops in the high-incidence area of Keshan disease in Keshan County.

Crop	Zn (10^{-6})	Ge (10^{-6})	Se (10^{-6})	Cd (10^9)	Cr (10^{-6})	Pb (10^{-6})
maize	15.07	1.10	0.02	2.94	0.20	0.08
rice	16.47	2.12	0.038	3.70	0.10	0.05
soybean	38.25	1.16	0.04	28.12	0.21	0.08

(5)   Comparison of the variation in selenium with depth.

According to previous research [9], the selenium in the soil of Keshan County decreases continuously with the increase in depth, and the lowest value appears around 2 m. After 2 m, there is a phenomenon of increasing content, which may be related to the deep geological body, and the Se on the surface is relatively high. In contrast, in Xunyi County, the selenium in the soil increases gradually from the surface to 0.8 m, and the maximum value appears at 0.8 m, then decreases continuously,,as shown in Figure 8.

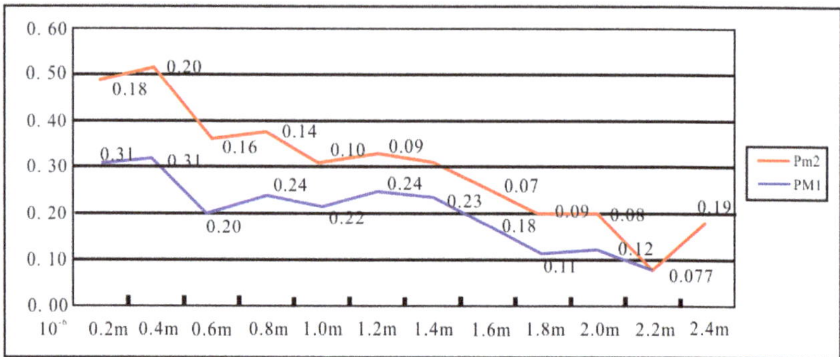

Figure 8. Curve diagram of the change in Se in Keshan County's soil with depth.

3. Discussion

The results of this study show that there are selenium-deficient and selenium-sufficient environments in the Keshan disease high-incidence areas, which is similar to the previous studies of scholars such as Wang Xiuhong, Fan Zhongxue, and Lei Yanxia [34–36], but also with some differences. The similarity lies in the presence of selenium deficiency in the Keshan disease areas, while the difference may be due to the existence of selenium-sufficient environments in the Keshan disease areas. This finding is of great significance for the prediction and prevention of Keshan disease, further enriching the research on the mechanism of Keshan disease. However, this study also has certain limitations. At the same time, we note that the distribution of elements in the soil is related to industrial activities, traffic emissions, and agricultural activities [37]. Therefore, future studies should further consider the secondary enrichment of environmental selenium to deeply explore the characteristics of environmental selenium and the prevention and treatment of this disease.

4. Conclusions

With the improvement of people's living standards, people are paying more and more attention to health. Regarding Keshan disease, as an endemic disease that still exists at present, its induction mechanism has always been unknown, and the research on the environmental element selenium has been ongoing. This time, the environmental selenium research is carried out in the high-incidence areas of Keshan disease in Chinese loess and black soil, and the following preliminary conclusions are drawn:

(1) As a typical area of loess, Shaanxi Xunyi has alkaline soil and low organic matter, and the soil selenium is in a state of selenium deficiency and low selenium. The selenium in the soil and wheat in the high-incidence area are both poor, indicating that Keshan disease is related to low selenium, which is consistent with the research results of previous scholars. The proposal of deep cultivation in the loess area is put forward, which can maximize the selenium content of crops and has a positive effect on the prevention of Keshan disease.

(2) Compared with the loess area, Keshan County in Heilongjiang, as a typical area of black soil, has the characteristics of a high soil organic matter, low selenium availability, and relatively high selenium in crops. Therefore, a selenium-sufficient environment can also induce Keshan disease, and selenium should be one of the factors inducing Keshan disease, which is a new discovery of this research.

(3) This research provides a reference for predicting the geographical area of Keshan disease onset and disease prevention, and it can also serve in local disease control and prevention.

This research was supported by the National Key R&D Program of China (Key Special Project for Marine Environmental Security and Sustainable Development of Coral Reefs 2021-01)

Author Contributions: Conceptualization, J.Z., Z.R., S.L., L.W. and P.W.; methodology, J.Z., Z.R., S.L., L.W. and P.W.; validation, J.Z., Z.R., S.L., L.W. and P.W.; formal analysis, J.Z., Z.R., S.L., L.W. and P.W.; investigation J.B.; resources, J.Z., Z.R., S.L., L.W. and P.W.; data curation, T.Y.; writing—original draft preparation, J.Z., Z.R., S.L., L.W. and P.W.; writing—review and editing, J.Z., Z.R., S.L., L.W. and P.W.; visualization, J.Z., Z.R., S.L., L.W. and P.W.; supervision J.Z., Z.R., S.L., L.W. and P.W.; project administration, J.Z., Z.R., S.L., L.W. and P.W. All authors have read and agreed to the published version of the manuscript.

Funding: This research was funded by the National Key R&D Program of China (Key Special Project for Marine Environmental Security and Sustainable Development of Coral Reefs 2021-01).

Data Availability Statement: The original contributions presented in the study are included in the article, further inquiries can be directed to the corresponding author.

Acknowledgments: Thanks to the technical and administrative support provided by the Xi'an Geological Survey Center.

Conflicts of Interest: The authors declare no conflicts of interest.

References

1. Zhu, Y. Research progress on the etiology of Keshan disease. *Foreign Med. Sci. Sect. Geogr.* **2009**, *30*, 496–498.
2. Health and Family Planning Commission. *China Health and Family Planning Statistical Yearbook*; Peking Union Medical College Press: Beijing, China, 2016; p. 272.
3. Yan, C. Current status of Keshan disease and progress in etiology. *Adv. Cardiovasc. Dis.* **2017**, *38*, 226–229.
4. Yang, J.Y.; Wang, T.; Wu, C.J.; Liu, C.B. Selenium level surveillance for the year 2007 of Keshan diasease in endemic areas and analysis on sueveillance results between 2003 and 2007. *Biol. Trace Elem. Res.* **2010**, *138*, 53–59. [CrossRef]
5. Wang, T.; Hou, J.; Li, Q. Summary analysis of the national Keshan disease surveillance for 5 years from 2000 to 2004. *Chin. J. Endem. Dis.* **2005**, *24*, 676–679.
6. Wang, T.; Hou, J.; Li, Q.; Zhang, L.; Li, X.; Gao, L.; Pei, J.; Deng, J.; Xu, B.; Dong, G. Summary analysis of the national Keshan disease surveillance in 2004. *Chin. J. Endem. Dis.* **2005**, *24*, 401–403.
7. Han, F.; Wang, S. Analysis of key issues in the control of Keshan disease, The latest medical information abstracts. *World Latest Med. Inf. Dig.* **2013**, *13*, 83.
8. Chen, J. An original discovery: Selenium deficiency and keshan disease (an endemic heart disease). *Asia Pac. J. Clin. Nutr.* **2012**, *21*, 320–326.
9. Zhao, J.; Zhang, Z.; Mu, H.; Wang, X.; Liang, X. Study on the characteristics of selenium in soil and crops in the high-incidence area of Keshan disease. *Northwestern Geol.* **2021**, *54*, 251–258.
10. Lei, C.; Niu, X.L.; Ma, X.K.; Wei, J. Is selenium deficiency really the cause of keshan disesase. *Environ. Geochem. Health* **2011**, *33*, 183–188. [CrossRef]
11. Zou, N.; Wan, H.; Liu, H.; Zhou, B. Relationship between selenium, vitamin E, zinc and myocardial injury in rats caused by low selenium and rich manganese. *Chin. J. Endem. Dis.* **2003**, *22*, 32–33.
12. Tl, B.; Han, F.; Yu, W. Protective effect of protein on myocardial injury in rats fed with grain from Keshan disease endemic area. *Chin. J. Endem. Dis.* **2002**, *21*, 6–9.
13. Qu, N.; Zhou, Y. Comparison of protein and amino acids in grain between Keshan disease endemic area and non-endemic area. *J. Hyg. Res.* **2000**, *4*, 251.
14. Cermelli, C.; Vinceti, M.; Scaltriti, E.; Bazzani, E.; Beretti, F.; Vivoli, G.; Portolani, M. Selenite inheibition of Coxsackievirus B5 replication:imliacations on the etiology of Keshan dissease. *J. Trace Elem. Med. Biol.* **2002**, *6*, 41–46. [CrossRef]
15. Bai, Y. Study on the DNA damage toxicity and oxidative stress mechanism of penicillium chlorophyllum toxin on HepG2 cells. *Full-Text Database Master's Theses China* **2012**, *19*, 58–61.
16. Yang, J. Causes, conditions and related factors of Keshan disease. *Chin. J. Endem. Dis.* **2012**, *31*, 499–505.
17. Yang, J.; He, X. Surveillance report of Keshan disease in Xunyi County, Shaanxi Province in 2005. *Chin. J. Endem. Dis.* **2006**, *25*, 680–682.
18. He, X.; Yang, J.; Chen, C.; Liu, H.; Deng, J.; Yan, X.; Liu, T.; He, L.; Luo, F. Surveillance report of Keshan disease in Xunyi County, Shaanxi Province in 2003. *Bull. Endem. Dis.* **2004**, *19*, 58–60.
19. China Environmental Monitoring Station. *Background Values of Elements in Chinese Soils*; China Environmental Science Press: Beijing, China, 1990.
20. Chen, D.; Ren, S. Selenium in soils of Shaanxi region. *Acta Pedol. Sin.* **1984**, *21*, 248–256.
21. Ren, R.; Wang, M.; Chen, J.; Chao, X.; Wang, H.; Xie, Y.; Meng, Q. Distribution characteristics and influencing factors of soil selenium in Guanzhong area of Shaanxi. *Miner. Explor.* **2018**, *9*, 1827–1833.
22. Blazina, T.; Sun, Y.; Voegelin, A.; Lenz, M.; Berg, M.; Winkel, L.H. Terrestrial selenium distribution in China is potentially linked to monsoonal climate. *Nat. Commun.* **2014**, *5*, 4717. [CrossRef]

23. Wang, S.; Liang, D.; Wei, W.; Wang, D. Relationship between soil properties and selenium forms based on path analysis. *Acta Pedol. Sin.* **2011**, *48*, 823–830.

24. Liu, S.; Yu, W. Research on selenium requirement of the Chinese people - the influence of other trace elements in the food of the residents in the Keshan disease area on the selenium requiremen. *J. Inst. Health* **1986**, 27–31.

25. Huang, J. *Spatial Distribution Characteristics of Selenium in Soil-Wheat in the Main Wheat Producing Areas of Shaanxi*; Northwest Agricultural University: Xianyang, China, 2018.

26. Wei, R.; Hou, Q.; Yang, Z.; Yin, G.; Zhong, C.; Deng, G.; Ma, Y. Analysis of selenium forms in root zone soil in Poyang Lake Basin, Jiangxi Province and its migration and enrichment rules. *Geophys. Geochem. Explor.* **2012**, *36*, 109–113.

27. Qin, H.; Zhu, J.; Li, S.; Lei, L.; Shang, L. Research progress on the analysis methods and forms of selenium in the environment. *Bull. Mineral. Petrol. Geochem.* **2008**, *27*, 180–187.

28. Qu, J.; Xu, B.; Gong, S. Determination of selenium forms in soils and sediments by continuous extraction technique. *Environ. Chem.* **1997**, *16*, 277–283.

29. Shen, Y.; Zhou, J. Occurrence state and migration and transformation of selenium in soil. *Geol. Anhui* **2011**, *21*, 186–191.

30. Fan, H.; Wen, H.; Ling, H.; Hu, R. Research status of selenium forms in the environment. *Earth Environ.* **2006**, *34*, 19–26.

31. Li, Y. Characteristics of selenium occurrence forms in selenium-rich soils in Zhejiang. *Geophys. Geochem. Explor.* **2007**, *31*, 95–99.

32. Song, X.; Wang, J.; Li, Z.; Ju, X.; Lai, Y. Analysis of selenium forms and valence states in selenium-rich soils in Qinghai Province. *Jiangsu Agric. Sci.* **2017**, *45*, 272–275.

33. Liang, D.; Peng, Q. Research progress on the transformation of selenium forms in soil and its effect on availability. *Adv. Biotechnol.* **2017**, *7*, 374–380.

34. Lei, Y.X.; Liu, Z.G.; Zhao, J.J.; Zhu, Y. Dynamic Observation on the Relationship between Keshan Disease and Trace Element Levels in the Internal and External Environments in Shaanxi Province. *J. Hyg. Res.* **2007**, *36*, 433–436.

35. Wang, X.H.; Xiang, Y.Z.; Qu, F.R.; Song, S.; Wang, L.; Guan, S. Study on the Relationship between Selenium Nutritional Levels in Internal and External Environments and the Incidence of Keshan Disease. *Chin. J. Control Endem. Dis.* **2005**, *26*, 351–353.

36. Fan, Z.; Li, P.; He, X. Low selenium in ecological environment and Keshan disease in Shaanxi province. *Micronutr. Health Res.* **2006**, *23*, 36–37.

37. Jiang, W.J.; Meng, L.S.; Liu, F.T.; Sheng, Y.Z.; Chen, S.M.; Yang, J.L.; Mao, H.R.; Zhang, J.; Zhang, Z.; Ning, H. Distribution, source investigation, and risk assessment of topsoil heavy metals in areas with intensive anthropogenic activities using the positive matrix factorization (PMF) model coupled with self-organizing map (SOM). *Environ. Geochem. Health* **2023**, *45*, 6353–6370. [CrossRef] [PubMed]

Article

Quantitative Assessment of Organic Mass Fluxes and Natural Attenuation Processes in a Petroleum-Contaminated Subsurface Environment

Yubo Xia [1], Bing Wang [2], Yuesuo Yang [3,4,*], Xinqiang Du [3] and Mingxing Yang [5]

1   Tianjin Center (North China Center for Geoscience Innovation), China Geological Survey, Tianjin 300170, China; sosodragon@163.com
2   Tianjin Geothermal Exploration and Development-Designing Institute, Tianjin 300250, China; wangbingdry@126.com
3   College of New Energy and Environment, Jilin University, Changchun 130021, China; duxq@jlu.edu.cn
4   Region Polluted Environment, Shenyang University, Shenyang 110044, China
5   School of Resource and Environment Engineering, Guizhou Institute of Technology, Guiyang 550003, China; yangmingxing@git.edu.cn
*   Correspondence: yuesuo@jlu.edu.cn or yangyuesuo@jlu.edu.cn; Tel.: +86-024-6226-7101

Abstract: We perceived a trend in the study and practice of petroleum-contaminate sites. Monitored natural attenuation (MNA) can reduce the contaminant concentrations in the soil and groundwater, and it is a method that can remediate the petroleum-contaminated site effectively. MNA is becoming a research focus. This study evaluated MNA using a series of lab-based bench-scale experiments and a large amount of monitoring data from field samplings. Based on the in-site total petroleum hydrocarbon (TPH) results, we used statistical methods, the Mann-Kendall test, and mass fluxes in order to evaluate the MNA of petroleum-contaminated sites in groundwater. The results showed that the TPH concentrations were decreasing, and the plume became smaller. The attenuation rate was from 0.00876 mg/d to 0.10095 mg/d; remediating the petroleum contamination site would cost 1.3 years to 10.6 years. The plume reached a quasi-steady state, and mass flux declined. The most essential process of MNA was biodegradation, and the second was sorption. During the monitoring period, 393 g of TPH was attenuated, including 355 g of TPH gradated by microbes. Biodegradation upstream of the plume was more serious. Iron(III) and manganese were the main electron acceptors utilized by microbes during the monitored period. MNA was in progress, and it can be an effective method to remediate the petroleum-contaminated site. Lab-based bench-scale experiments were performed with much monitoring data from the field samplings in order to understand the fate and transport mechanism of the petroleum contamination from the land surface to shallow groundwater according to site conditions.

Keywords: monitored natural attenuation; mass fluxes; petroleum-contaminated site; groundwater; Mann–Kendall test

Citation: Xia, Y.; Wang, B.; Yang, Y.; Du, X.; Yang, M. Quantitative Assessment of Organic Mass Fluxes and Natural Attenuation Processes in a Petroleum-Contaminated Subsurface Environment. *Appl. Sci.* **2023**, *13*, 12782. https://doi.org/10.3390/app132312782

Academic Editor: Dino Musmarra

Received: 18 October 2023
Revised: 26 November 2023
Accepted: 27 November 2023
Published: 28 November 2023

1. Introduction

The dependence on petroleum as an indispensable resource for our today's life is increasing. During petroleum exploitation, transporting, and processing, spillage and contamination of the groundwater and soil environment are hard to avoid. Petroleum hydrocarbons are very common contaminants in groundwater [1]. It is extremely important to clean up the contaminated soil and subsurface environment in a cost-effective manner [2,3]. MNA has become a preferred choice for the remediation of petroleum-contaminated groundwater for decades [4]. An increasing interest in implementing low-cost, environmentally friendly, and nonintrusive solutions such as MNA has been observed in recent years in many sites worldwide [5–8]. MNA involves low exposure risk for cleanup and requires less equipment and labor involvement than most other methods [7]. Monitoring for many years can be costly but may cost less than other methods [9]. In China, the extensive study of MNA at the laboratory scale is moving into the study of petroleum-contaminated sites due to the great real-world demand [10–12].

MNA can be used as the sole remedial solution [13], or when other methods will not work or are expected to take almost as long. Sometimes, MNA is used as a final cleanup step after other methods for polishing purposes [14,15]. Some of these naturally occurring physical, chemical, and biological processes can transform contaminants to less harmful forms or immobilize them in the subsurface, therefore reducing contaminant concentrations in groundwater [16–18]. Attenuation mechanisms encompass physical dilution, physicochemical sorption, ion exchange, chemical dissolution/precipitation or complexation, and microbial metabolic processes [19,20]. Many studies found that aerobic and anaerobic biodegradation are the major processes for the reduction of contaminant mass in the subsurface [21,22]. Biodegradation is generally considered the most important process for contaminated site remediation because it is destructive, unlike sorption, dilution, and volatilization [23]. Biodegradation would reduce the mass and toxicity of contaminants in the groundwater and soils. The biodegradation of aerobic microbes is more efficient than that of anaerobic microbes [24,25]. Once oxygen is completely used or becomes limiting, microbes utilize nitrate, followed by manganese, iron, and sulfate [26]. However, if the rate of biodegradation is not significant enough to contain the plume, MNA will not be sufficient to protect aquifers and downgradient receptors [27]. So, ensuring the attenuation effect is the first and foremost task by estimating the trend of a plume, monitoring the concentration, and calculating the remediation period [28–31]. Theories, such as the Mann–Kendall test, are often used to define the stability of a contaminant plume based on concentration trends at individual wells [32–34]. It is widely used in hydrology.

There are multiple methods for estimating biodegradation. First-order kinetics may be able to approximately reproduce the mass and dimensions of contaminant plumes that follow from a far more complex degradation model but are often regarded as imprecise or even conceptually incorrect [16,30,35]. One shortcoming, for example, is that their resolution is not fine enough to reflect plume fringes of the intrinsic biodegradation and their limiting factors of electron acceptors from the surrounding subsurface environment [36]. A substantial problem is the negligence of various hydrogeological variables and attenuation processes that may occur in subsurface environments (e.g., advection, dispersion, mixing). The resulting first-order rate constants thus describe bulk attenuation rather than local in situ biodegradation kinetics [37].

Another reactive model used for the in situ quantification of biodegradation is Michaelis–Menten (MM) kinetics, also known as Monod, no-growth kinetics, which presents the situations that the initial cell number is much greater than that could be produced using the substrate present at time zero [16,38]. The MM model originates from biochemistry and is generally applied for enzyme kinetics in the quasi-stationary state. It relates the maximum turnover rate of one specific enzyme to the present substrate concentration (single compound) in the medium. The MM model requires that the amount of enzyme remains constant during the experiment and is not particularly useful for in-site simulation.

Mass flux reflecting the contaminant mass loss through the control planes (CPs) perpendicular to the groundwater flow direction is often calculated to quantify the natural attenuation of petroleum hydrocarbons [39–41]. The 2D method and 3D mass balance approach for determining plume mass loss [42] have been used to estimate the natural attenuation. The 3D mass balance approach is more accurate but needs more monitoring information. Due to aquifer heterogeneity, the resulting plume delineation is strongly influenced by the positioning and number of monitoring wells [43–45]. In the application of the method, a few issues need to be addressed. Firstly, it is important to choose appropriate CPs and determine their lengths by the TPH concentration in groundwater, which can be an easy way but is actually sometimes not accurate. When enough monitoring wells are not available at the site, a pumping well is needed [41]. Secondly, the vertical distribution of TPH is to be delineated, and the mass quantity of plume compounds transformed between synoptic sampling events is estimated on the basis of dissolved phase concentrations, sorption characteristics, and groundwater flux [46,47]. Although the assumption of sorption ideality is often applied, laboratory and field experiments can often provide evidence of the solute behavior that deviates from ideality. Thirdly, it is hard to calculate the TPH fluxes at each CP at various times under the presence of microbial degradation [48]. Therefore, many assumptions are to be made generally during the assessment.

In this article, a study for the quantitative assessment of an organic contaminant plume at the typical petroleum contaminated site in NE China under a framework of the government core funding program (863) for clean and novel science and technology in the resources and environment section is reported. Extensive field surveys focusing on geology, hydrogeology, geochemistry, and fractions of the petroleum contaminants were carried out systematically in order to understand the contaminant hydrogeological processes and potential in situ remediation with enhanced monitored natural attenuation (e-MNA), air/bio sparging (AS/BS), and/or permeable reactive barriers (PRB). With the support of much monitoring data from the field samplings, a series of lab-based bench-scale experiments were conducted to understand the fate and transport mechanism of the petroleum contamination from land surface to shallow groundwater according to this site condition. The mathematical, statistical model, Mann–Kendall test, and mass flux were reported in this study. The major objectives of this study were to (1) evaluate the attenuation process of petroleum hydrocarbons in the subsurface, (2) assess the contributions of intrinsic biodegradation patterns and other processes by the mass flux approach, and (3) quantify the organic contaminant plume by using the Mann–Kendall model. The effectiveness of employing MNA as an in situ remedial strategy can then be achieved for this site.

2. Materials and Methods

2.1. Study Area

The study site is located in the east of Songnen fluvial plain, China. It is in the temperate continental monsoon climate, with a dry and windy spring, hot and rainy summer, cool fall, and cold winter. The average temperature over multiple years is 4.7 °C, but it has had the extreme climate of 37.5 °C in summer and −37.8 °C in winter in recent years. The average annual precipitation is 436 mm, and of that, over 70% occurs in summer (i.e., June–August).

The elevation of the ground level of the site is between 130.7 and 139.4 m above sea level. The aquifer layer is Holocene alluvium spread out over the valley of the first terrace. The surface layer of the aquifer is drab silt (Figure 1). The lower layer is fine sand, sand, and gravel, being 20–22 m thick. The groundwater table is from 3 to 4.5 m depth below ground level. The groundwater flow direction generally is from the southeast to northwest. The hydraulic gradient is from 0.002 to 0.008. According to slug tests, the hydraulic conductivity is between 0.5 and 12.9 m/d at this site.

There are four oil production wells around the site, and one of the faulty production wells in the upper gradient of the site has been leaking for some years, having also been fixed for some time. This faulty oil well has also produced a petroleum spill pit on the surface that was believed to be a main contamination source for this site (Figure 2). The most recent cut off was in April 2009, which should have cured the leak completely.

2.2. Sample Collection and Analysis

There are twelve boreholes and one potable well at the site for groundwater monitoring, all being located at the down gradient area of the petroleum spill (Figure 2). Perpendicular to the groundwater flow, three CPs were assigned for the plume assessment in this study, each of them containing two or three monitoring wells (Figure 2). All of the CPs were ensured by Formula (6) below, which ignored the influence of depth. The hydro-geochemical parameters were monitored from September 2009 to March 2010. These included six campaigns of TPH concentration sampled monthly, except for February 2010 from 13 monitoring wells. Other major ions including calcium, magnesium, sodium, potassium, chloride, and bicarbonate/carbonate were sampled in September 2009, January 2010, and March 2010 so that a good trend of hydro-geochemistry could be obtained. Electron acceptors and environment parameters including nitrate, nitrite, ammonium, total iron, iron(III), manganese, sulphate, sulfide, pH, oxidation reduction potential (ORP), conductivity, temperature, and salinity were also analyzed in September 2009, October 2009, November 2009, January 2010, and March 2010. The environment parameters were tested in situ during field sampling using Hanna, and then the sample for TPH, major ions, and electron acceptors were taken for further laboratory analysis. The TPH concentration before the pumping test was higher than after, and it showed that petroleum hydrocarbon concentration decreased with depth. To ensure the effect of MNA, the surface concentration of TPH was used in the estimation.

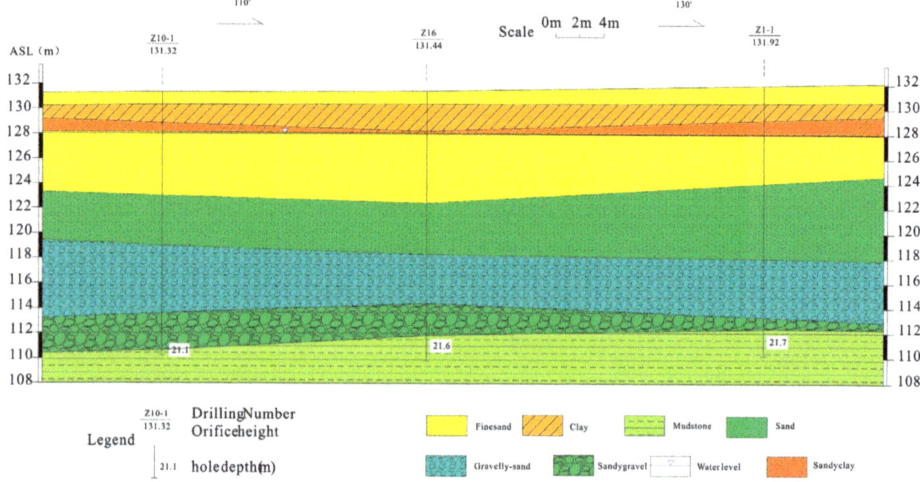

Figure 1. Geological cross-section A-A′ of the field site.

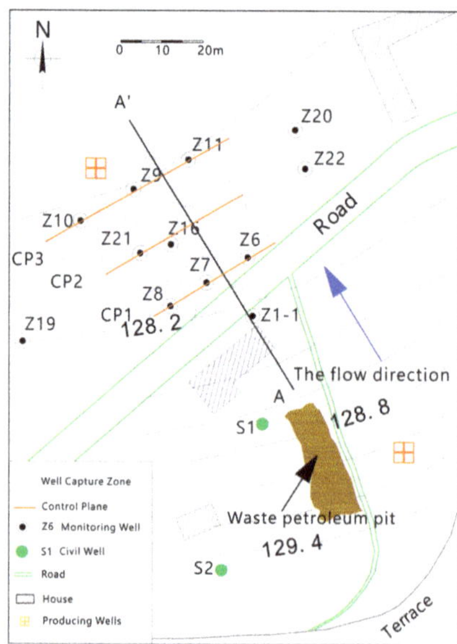

Figure 2. Site overview with the location of the employed wells and control planes.

The sample for TPH analysis was taken with amber glass bottles. For data quality control, samples for Fe/Mn were added with hydrochloric acid, and samples for sulfide were added with 1 mol/L sodium hydroxide and 1 mol/L zinc acetate as chemical stabilizers. The sampling bottles were attempted with no head space and sealed with aluminum foil paper, shipped avoiding sunlight, and kept at 4 °C for further lab analysis.

Total petroleum hydrocarbons were detected by infrared spectrophotometry (JDS-108U+). K^+ and Na^+ were determined by flame atomic absorption spectrophotometry; Ca^{2+}, Mg^{2+}, NO_3^-, Cl^-, and Fe(III) were determined by the titration method; and SO_4^{2-} was determined by the turbidity method. Analytical methods adopted the Quality Standards for Groundwater (GB/T 14848-2017) [49] of China. The other main instruments used to determine the groundwater properties included a multiparameter water quality analyzer (Hach-HQ40).

3. Results

3.1. Contamination and Attenuation Processes

3.1.1. Spatial Distribution of TPH

The plume of TPH was at a downgradient of the waste petroleum pit (WPP) that was based on the analysis of the flow direction and the distribution of the TPH concentration, in addition to the fact that the TPH concentration of Z10 and Z19 was excessively high. The TPH concentration of Z1-1 was 15.08 mg/L, which was highest in September 2009 because it is the nearest monitoring well to the WPP. The TPH concentration in the plume ranged from 6.21 mg/L to 15.08 mg/L. At the edge of the plume, the TPH concentration was 2.99 mg/L. The TPH concentration decreased, with distance ranging from 7.06 mg/L to 2.99 mg/L. The TPH concentration of Z10 was influenced by both downstream producing wells and upstream contamination, and it was the highest in October 2009. From November 2009 to March 2010, the TPH concentration of Z10 was higher than Z6, Z7, and Z8. The spatial distributing of TPH concentration was largely the same from September 2009 to March 2010. Contamination areas were at the downstream area of the WPP and Z10 (Figure 3).

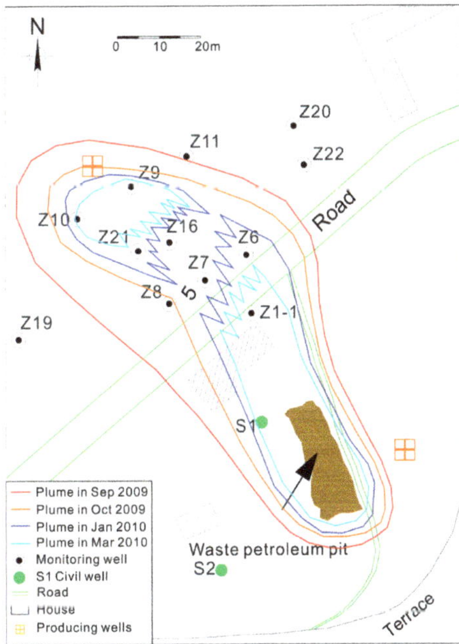

Figure 3. Distribution of TPH concentrations.

3.1.2. TPH along the Flowline

Figure 4 shows the TPH concentration along the flowline from the contaminating source. Abscissa represents the distance from the source to monitoring, and the origin point represents the edge of WPP (Figure 1, A-A'). The concentrations of TPH were reduced along the groundwater flow from source zones to downstream areas. Investigating the results revealed that the decreases of TPH concentrations were significant. The contamination, which was far away from the contamination sources, usually had a longer role of natural attenuation, including soil adsorption, dilution, mixture, and biodegradation. The trend was more obvious in September and October 2009.

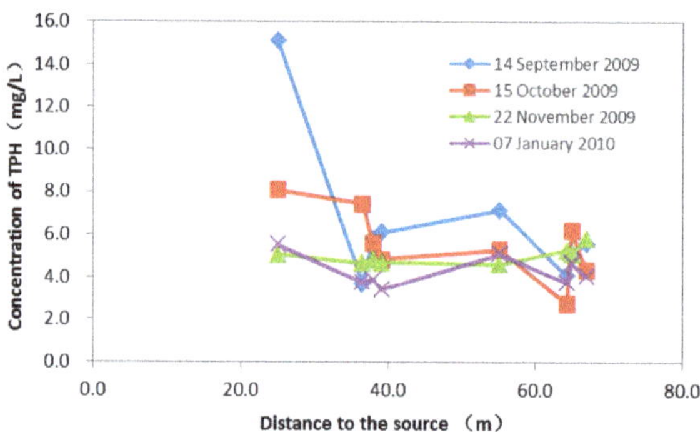

Figure 4. The TPH concentrations along the flowline from the contaminating source.

3.1.3. Temporal Variation of TPH

The distribution of TPH changed during the monitoring period. These changes mainly reflected that the TPH concentration was decreasing, and the plume was wilting. The highest concentration was 15.08 mg/L in September 2009, dropping to 4.77 mg/L in March 2010. Approximately 78% of TPH was attenuated. On the edge of the plume, the attenuation trend was smooth, and 45% of TPH was attenuated on average. The TPH concentrations increased from September 2009 to October 2009 in Z6, Z7, Z9, Z10, Z16, and Z21. This variation was due to the migration of contamination. The same trend occurred in Z11 from October 2009 to November 2009. In another period, the TPH concentration decreased. Overall, in almost all of the monitoring wells, the TPH concentration decreased (Figure 5), and 36.6% of TPH was attenuated on average in all of the monitoring wells. The final concentration of TPH was decreased to 6 mg/L.

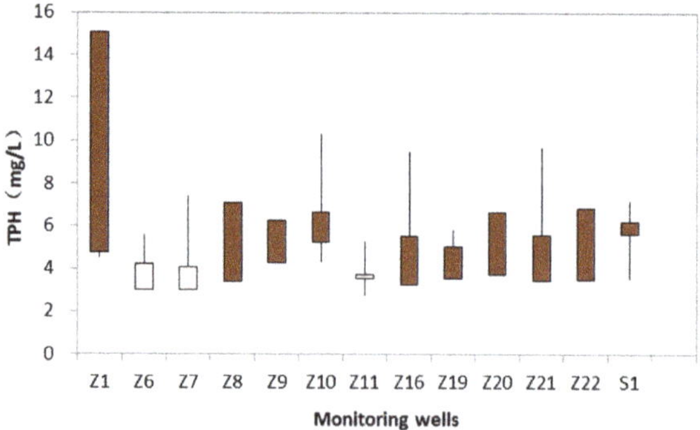

Figure 5. Temporal variation of the TPH concentrations.

3.1.4. Hydrogeochemistry and Biogeochemistry

Based on the investigation of groundwater level, the groundwater flow direction was from the southeast to the northwest (Figure 1). The main anion in the site was the bicarbonate ion, and the content was usually not less than 60%, except Z22. The distributing of bicarbonate ions associated with the TPH concentration, which was high at the downstream of WPP, reaching 2 mg/L, such as with Z6, Z7, Z8, and Z22. In peripheral monitoring wells (Z10, Z11, Z19, Z20, and S1), bicarbonate ion concentrations were lower than 1 mg/L (Figure 6). The evapotranspiration process reacted in the area where the bicarbonate ion concentration was high, and the cations were either sodium ions or sodium ions and calcium ions. Chloride ions were another major anion in the study area, and they were widely distributed downstream of the WPP. After the cleanup the source, the enrichment area of chloride ions became more significant. The main cations in the groundwater were sodium ions and calcium ions. The sodium ions were distributed in the west of the study area, and calcium ions were distributed in the east of the study area (Figure 6). The enrichment of calcium ions and bicarbonate was the result of the leaching process, and the enrichment of sodium ions and chloride ions was the result of the evapotranspiration process. The distribution of the hydrochemical type was formed by recharging not only from the upper gradient area but also from the second Songhua River [50].

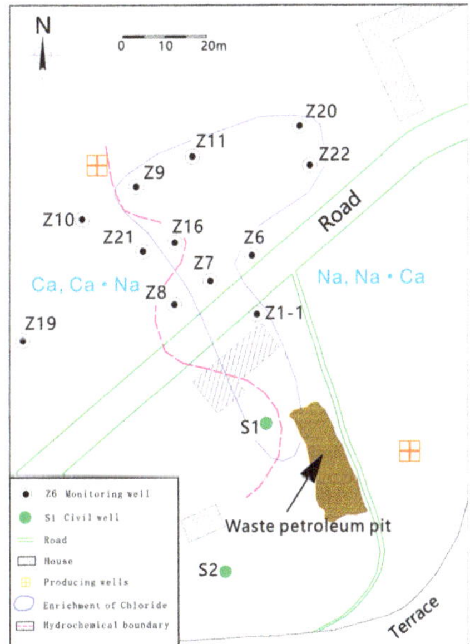

Figure 6. Chemical types and distributions of chlorine.

During the monitoring period, the concentration of dissolved oxygen (DO) was very low. DO concentrations in the high-TPH wells were lower than in the low-TPH wells, and DO concentrations typically were less than 1 mg/L. In the heavily contaminated area, especially near the WPP, the DO concentrations were less than 0.1 mg/L (Z1-1, Z6, Z7, and Z8). These data fully prove that aerobic biodegradation was occurring at this site.

The decreasing trends of nitrate concentrations also can be observed. Nitrate concentrations in the contaminated areas ranged from 1.2 to 14.0 mg/L (Table 1). Higher TPH and lower nitrate were detected near the source, and DO was depleted. The nitrate concentrations decreased from September 2009 to January 2010, and they were stable from January to March 2010 (1 mg/L). The ammonium concentrations increased from September to November 2009. The average ammonium concentrations ranged from 0.142 mg/L to 1.03 mg/L. This showed that denitrification was occurring on the site.

Table 1. Summary of water quality data intervals for monitoring wells in September 2009 to March 2010.

	September 2009	October 2009	November 2009	January 2010	March 2010
TPH (mg/L)	3.7–15.1	2.8–12.0	4.3–6.6	3.4–6.8	3.2–5.6
pH	6.8–7.6	6.8–8.3	7.9–8.6	7.3–9.7	7.2–7.9
ORP	−163.0–36.4	−137.6–45.7	−130.2−−45.4	−160.9−−77.4	−126.5−−21.5
NO_3^- (mg/L)	1.2–14.0	1.2–3.0	-	0.6–2.0	0.6–4.0
NH_4^+ (mg/L)	0.1–0.3	0.2–1.3	0.56–1.86	-	-
Mn^{2+} (mg/L)	0.68–2.98	0.05–2.27	0.90–4.89	0.85–5.23	0.36–3.52
Total Fe (mg/L)	1.7–27.8	0.3–20.9	0.1–10.9	2.0–13.3	0.8–47.1
Fe^{3+} (mg/L)	0.1–1.1	0.1–0.3	0.1–0.6	0.1–0.5	0.2–0.8
SO_4^{2-} (mg/L)	0.2–142.7	0.2–104.7	0.3–70.8	4.8–48.0	7.9–42.5

Parallel results were measurable for the distribution of manganese in groundwater in September 2009. Manganese is the production of manganese dioxide, which is a component of soil. Mn(II) occurs because of the metabolism by microbes [51]. Downstream of WPP, the average manganese concentrations increased (1.51 to 1.93 mg/L). The average iron(III) concentrations were as high as 0.337 to 7.79 mg/L in wells, while background levels were approximately 1.112 mg/L. Iron(III) concentrations can be used as an indicator of anaerobic degradation of the petroleum hydrocarbons at the site.

Sulfate can be reduced to sulfide by microbes. The background level of sulfate was 143 mg/L. In wells downstream of WPP (Z1-1, Z6, Z7, Z8, Z11, and Z16), the sulfate concentrations were less than 50 mg/L, lower than the farther wells.

Oxidation reduction potential (ORP) indicates that the groundwater environment was weak oxidation to weak restoration (36.4–163.0) in September 2009. ORP was lower downstream of WPP, and the lowest was Z1-1. This illuminates that petroleum contaminated Z1-1 first, and Z1-1 experienced natural attenuation for a long time. ORP decreased from September 2009 to January 2010 (36.81 dropped average) and increased from January to March 2010 (35.01 raised averagely) in some wells (Z1-1, Z6, Z7, Z8, etc.) downstream of WPP. This showed that electron acceptors were depleted, and biodegradation was unable to continue. Groundwater was in a neutral environment in September 2009 (the average pH was 7.23). Compared with the ORP, higher ORP and lower pH were detected near the source zones from September 2009 to January 2010. The pH decreased from January to March 2010 (1.24 dropped average).

3.2. Quantitative Assessment of Biodegradation Processes

3.2.1. Statistical Methods

The evaluation of MNA based on statistical methods was calculated by the TPH concentrations to confirm the flowline, the attenuation rate, and the half-life of petroleum pollutants. It is defined as

$$C(t) = Ae^{\frac{1}{\lambda}t}, \tag{1}$$

$C(t)$ is the concentration function, λ is half-life, t is time, and A is the initial concentration.

3.2.2. Mann–Kendall Test

The zero hypothesis H_0 to time series arrays $(x_1, \ldots, x_n)$ is n independent and the random variable distribution of the same sample, and alternative hypothesis H_1 is a bilateral examination. For the related $k, j \leq n$ and $k \neq j$, and x_k and x_j distribution is not the same; inspection of the statistical variable S is based on the following formula:

$$S = \sum_{i=1}^{n-1} \sum_{k=i+1}^{n} sgn(x_k - x_i), \tag{2}$$

x_k, x_i are particular values in a sequence of data; n is the length of data; and

$$sgn(\theta) = \begin{cases} 1, \theta > 0 \\ 0, \theta = 0 \\ -1, \theta < 0 \end{cases}, \tag{3}$$

$sgn(\theta)$ is the cumulative amount of time series. The Mann–Kendall test requires at least four independent samples of the time [52]. The normal statistical distribution of S is

$$Z_C = \begin{cases} \frac{S-1}{\sqrt{var(s)}}, S > 0 \\ 0, S = 0 \\ \frac{S+1}{\sqrt{var(s)}}, S < 0 \end{cases}, \tag{4}$$

Variance:

$$var[s] = \frac{[n(n-1)(2n+5) - \sum\limits_{t} t(t-1)(2t+5)]}{18}, \tag{5}$$

For the trend in the bilateral inspection, at a given confidence level α, if $|Z| \geq Z_{1-\alpha/2}$, then the original assumption is unacceptable. That means time-series data versus an obvious is upward or downward trend at a confidence level. Statistical variables can determine the trend of the plume, and based on the corresponding numerical credibility trends, have complete the plume assessment. For statistical variables Z, if greater than 0, it is and upward trend; if less than 0, it is a downward trend. The absolute values of Z were greater than or equal to 1.28, 1.64, and 2.32 at the time, showing that Z passed the reliability of the 90%, 95%, and 99% significance tests, respectively.

In this test, the plume stability is represented as decreasing, probably decreasing, stable, no trend, probably increasing, and increasing at given confidence levels.

3.2.3. Mass Flux

This method is based on four assumptions: (1) the monitoring wells of each CP could contain the plume, the diffusion effect in the longitudinal and vertical profile is negligible, and the TPH distribution in the direction is linear; (2) verified by multiple pumping tests, the TPH concentrations in groundwater and depth meet the power function; (3) adsorption and desorption are instantaneous, and the aquifer is homogeneous and isotropic, with the proportion of adsorption to desorption being 1:3; and (4) volatilization of TPH in groundwater is negligible [39].

In this article, three transects were drawn at distances of 38 m, 58 m, and 66 m from the main source location (Figure 1). The mass fluxes were calculated for the six different sampling periods. The development of the well capture area with time can be described as

$$r(t) = \sqrt{\frac{Qt}{\pi H n_e}}, \tag{6}$$

$r(t)$ is the radius (L) of the well capture zone at time t. Q is the pumping rate (L3 T1). T is the time (T). H is the aquifer thickness (L), and n_e is the effective porosity (–).

The Darcy law can be described as

$$Q = KAI, \tag{7}$$

K is the hydraulic conductivity (L T^{-1}), A is the area of the cross-section (L^2), and I is the hydraulic gradient (–).

The TPH concentration in groundwater and depth meet the power function:

$$C(h) = C_0 \times h^{-0.385}, \tag{8}$$

$C(h)$ is the concentration of TPH (mg/L) at depth (m), C_0 is the concentration at the surface of the aquifer(mg/L), and h is depth (m).

The mass flux of TPH Md_i (M T^{-1}) related to an individual sampling port located within each CP is then defined as

$$M_{di} = C_i q_i A_i, \tag{9}$$

C_i is the concentration measured at the sampling location (M L^{-3}), q_i is the specific discharge perpendicular to CP (L T^{-1}), and A_i is the capture area (m^2). The total mass discharge M_d (M T^{-1}) crossing a groundwater fence with n sampling ports can subsequently be calculated as

$$M_d = \sum_i^n M_{di} = \sum_i^n C_i(h)dh \times K_i I_i \times [2r_i(t)]^2 H, \tag{10}$$

K_i is the coefficient of permeability (m/d), I_i is hydraulic slope $(-)$, and H is the thickness of the confined aquifer (m).

The mass flux between two CPs can be described by the following equality:

$$M_{d_{CP1-CP2}} = \frac{M_{d_{CP1}} - M_{d_{CP2}}}{2} \times \frac{L}{V}, \tag{11}$$

$M_{dCP1-CP2}$, M_{dCP1}, and M_{dCP2} represent the measured compound-specific mass fluxes between CP1 and CP2 (g/d); at CP1 and CP2 (g), L is the distance between CP1 and CP2 (m), and V is the average groundwater velocity (m/d). The adsorbance can be described by the following equality:

$$M_a = k_f \times M_d, \tag{12}$$

M_a is adsorbance, and k_f is the absorption coefficient. The amount of microbial degradation is the decrement of M_a and M_d. Compound-specific total mass fluxes (M T^{-1}) are quantified at different distances from a contaminant source zone using either point scale or integral investigation approaches. If the average travel time Δt (T) between the two existing control planes is known, it is possible to quantify the compound-specific effective first-order natural attenuation rate constant (T^{-1}) [40]:

$$\lambda = -\ln\left(\frac{M_{d_{CP2}} + M_{a_{CP2}}}{M_{d_{CP1}} + M_{a_{CP1}}}\right)\frac{1}{\Delta t}, \tag{13}$$

λ (T^{-1}) is the effective natural attenuation rate constants.

4. Discussion

4.1. Statistical Methods

The results of statistical methods are shown in Table 2. Z1-1 is the nearest monitoring well to the WPP. Its natural attenuation rate is 0.101 mg/d, and 68% of TPH attenuated in 105 days, with its half-life being 87 days. The natural attenuation rate of Z7 is 0.043 mg/d, and 47% of TPH attenuated in 83 days; the natural attenuation rate of Z16 is 0.026 mg/d, and 61% TPH attenuated in 162 days; the natural attenuation rate of Z16 is 0.012 mg/d, and 31% of TPH attenuated in 162 days. The components of petroleum are complex. The longer the distance to the 344 source, the harder the attenuation (Figure 7). Attenuation rate curves followed the exponential distribution, $y = 0.33911e^{-0.0522x}$, $R_2 = 0.9868$. Without human intervention, the remediation time of S1 will cost 10.6 years, which is the longest time in the site. After 3 years, 53.8% of monitoring wells will achieve the remediate goals; after 5 years, 76.9% will achieve the remediate goals; and after 7 years, 92% of monitoring wells will achieve the remediate goals.

Table 2. Estimation from point concentrations.

Location	Initial Concentration (mg/L)	Distance to the Source (m)	Attenuation Time (d)	Attenuation Rate (mg/d)	Half-Life (d)	Remediation Time (a)
Z1-1	15.1	25.1	105	0.101	87	1.3
Z6	5.8	38.0	115	0.017	270	3.5
Z7	7.4	36.6	83	0.043	130	1.8
Z8	6.1	39.2	115	0.023	212	2.8
Z9	6.2	65.1	162	0.012	472	6.2
Z10	10.3	68.4	73	0.082	84	1.3
Z11	5.3	64.4	46	0.035	145	2.0
Z16	8.3	49.0	194	0.026	182	2.6
Z19	5.6	67.1	194	0.011	500	6.5
Z20	6.6	64.4	162	0.018	288	3.9
Z21	9.7	52.0	162	0.039	168	2.4
Z22	7.1	55.3	194	0.019	327	4.3
S1	7.3	8.8	194	0.009	775	10.6

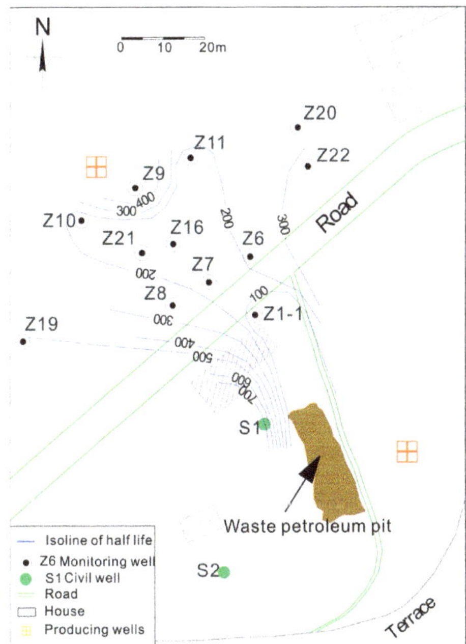

Figure 7. Distributions of half-life results of TPH concentrations in the monitoring wells.

4.2. Trend Analysis Based on the Mann–Kendall Test

Table 3 shows that none of the wells were found to be increasing or probably increasing, as well as showing eight decreasing or probably decreasing wells (69%). Only Z8 was stable; Z7, Z10, and Z11 had no trend, and attenuation in these monitoring wells was not apparent. Z10 and Z11 might have been contaminated, being near the producing well. The initial concentrations of Z6, Z7, and Z8 were relatively lower than in other wells. There was no decreasing trend downstream of WPP (Figure 8). These results support the conclusions that the TPH plume was stable or declining. We can also detect that the attenuation trend in the plume was more obvious than at the edge of the plume; for adsorption and dispersion, it was more efficient in the initial condition.

Table 3. Results of the Mann–Kendall trend analyses.

	Concentration Trend	Confidence Level (%)
Z1-1	Probably decreasing	93
Z6	Probably decreasing	97
Z7	No trend	69
Z8	Stable	83
Z9	Probably decreasing	97
Z10	No trend	79
Z11	No trend	56
Z16	Probably decreasing	97
Z19	Probably decreasing	93
Z20	Probably decreasing	96
Z21	Decreasing	99
Z22	Probably decreasing	97
S1	Probably decreasing	93

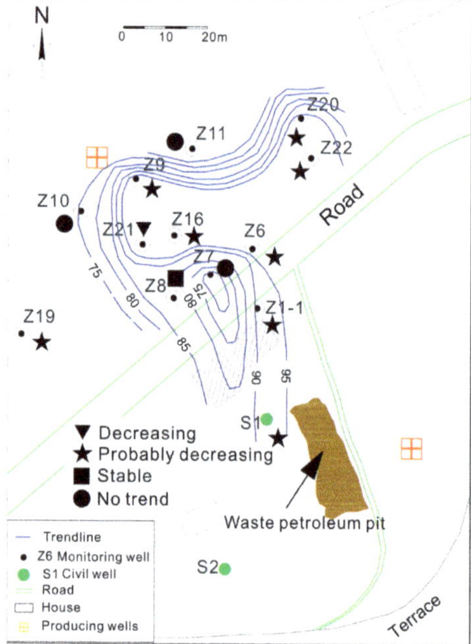

Figure 8. Distributions of the trend analysis results of TPH concentrations in the monitoring wells.

4.3. The Analysis Based on Mass Flux

The TPH concentrations were given by the data of six monitoring periods. Aquifer thickness was equal to the well depth minus the aeration zone depth. The hydraulic gradient was calculated by the water level. Mass flux results are shown in Figure 9. The mass flux of TPH generally reduced over time, only increasing slightly in January 2010. The fluctuation occurred mainly when the TPH concentration was relatively small. Test accuracy was impacted by the fluctuation. According to the changes of mass flux, it could be seen that CP1 > CP2 > CP3. The mass flux of CP1 was 1415 mg/d in September 2009, which reduced to 524 mg/d in January 2010, and then it rebounded slightly to 580 mg/d. CP2 and CP3 had the same trend (Figure 9). This indicated that the TPH concentrations through each section decreased. Natural attenuation was carried on in the contaminated plume, and TPH in the site reached the half-steady state from time to time.

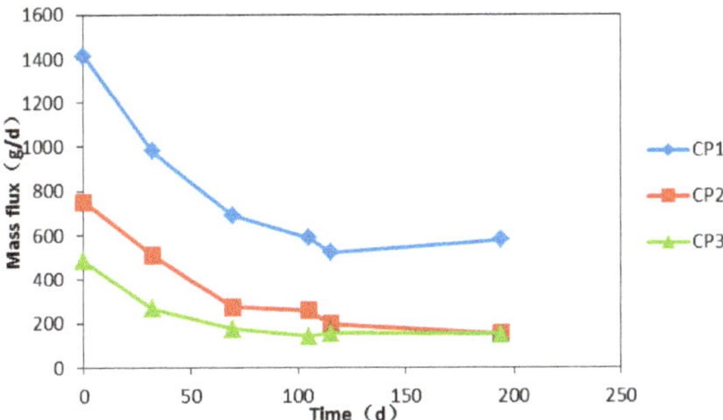

Figure 9. Changes in mass fluxes of each control plane.

The main attenuation processes were adsorption, volatilization, biodegradation, and dispersion. Based on the research of Suarez (2002) [39], the volatilization of TPH in groundwater is negligible. Once the medium and environment are fixed, adsorption and dispersion are stable, so the decrement of mass flux can be considered as the approximate contribution of biodegradation. We calculated the biodegradation using Formulas (9)–(11), and the results are listed in Table 4. Based on the mass flux data in time and space, biodegradation of TPH and natural attenuation were able to be analyzed. However, if the mass flux is reduced, the soil particle will resolve part of the petroleum hydrocarbons, and then the biodegradation of capacity will be the sum of reduction and desorption. We were able to calculate desorption by using Formula (12). From September 2009 to March 2010, biodegradation due to mass flux reduction between CP1 and CP2 was 215 g; total flux reduction was 242 g through calculus, and biodegradation occupied 89% of the total attenuation. This explained why microbial degradation was a major attenuation mechanism in the site. Similarly, from September 2009 to March 2010, biodegradation calculated by mass flux between CP2 and CP3 was 140 g; total mass reduction was 152 g, and biodegradation occupied 92% of the total attenuation. During this process, the adsorption capacity between CP1 to CP2 was from 81 g down to 27 g, and adsorption shrank. In the monitoring process, adsorption and biodegradation declined continuously, but the biodegradation ratio increased. The biodegradation ratio at the upstream was 64–89%, and the biodegradation rate downstream was 70–92%. The average biodegradation rate in the upstream was 0.4911 mg/d/m^3, and it was 0.338 mg/d/m^3 downstream. The attenuation rate upstream was higher than downstream.

In short, the following were found: (1) Calculations showed that attenuation stably occurred in the plume, the mass flux of each CP was reducing, and there were more contaminants upstream than downstream. (2) Adsorption and biodegradation were the main attenuation processes; biodegradation was more efficient. (3) Biodegradation increased, and adsorption decreased with time. The biodegradation capacity was 383 g.

The three findings above were able to be used to estimate the natural attenuation of petroleum-contaminated sites. Comprehensive results could reflect the intensity and trend of natural attenuation. The distribution and concentrations were able to verify the trend analysis results compared to the monitoring results. Natural attenuation occurred in the groundwater of the petroleum-contaminated site, and the TPH concentrations decreased continuously. Attenuation strength in the plume was higher than the edge, and attenuation patterns were different.

4.4. The Division of Functional Areas and the Utilization of Electron Acceptor

OPR is the parameter that indicates redox conditions of groundwater in the site. According to Norris and Diápela's research [32,53], when microorganisms consume electron acceptors, ORP indicates the utilization of an electron acceptor. If pH is known, ORP can be determined, as shown in Figure 10. When pH = 7 and temperature is 25 °C, if Eh is more than 740, it is an aerobic condition, and oxidation can occur in the area. If Eh is less than

740, it is an anaerobic condition, and there is no oxygen available; thus, denitrification can happen in the area [54]. However, biodegradation in the site is a complex process. pH and temperature are changeable, so Fetter, C.W [55] studied the relationship between Eh and various ions under different pH values, as shown in Figure 11. Thus, microbial functional areas of the site can be speculated.

Table 4. The result of mass flux and the contribution of each attenuation.

	Time (d)	0	32	69	105	115	194
	Mass flux (mg)	243,868	168,081	108,778	95,533	80,997	82,453
	Adsorbed mass (mg)	81,289	56,027	36,259	31,844	26,999	27,484
	Total mass (mg)	325,157	224,108	145,037	127,377	107,996	109,937
	Mass flux decreased (mg)	0	75,787	135,090	148,335	162,871	161,415
CP1–CP2	Adsorbed mass decreased (mg)	0	25,262	45,030	49,445	54,290	53,805
	Biodegradation (mg)	0	101,049	180,120	197,780	217,161	215,220
	Total attenuation (mg)	81,289	157,076	216,379	229,624	244,160	242,704
	Biodegradation ratio (%)	0	64	83	86	89	89
	Attenuation rate (%)	25	48	67	71	75	75
	Mass flux (mg)	139,415	87,636	50,803	45,171	39,551	34,223
	Adsorbed mass (mg)	46,472	29,212	16,934	15,057	13,184	11,408
	Total mass (mg)	185,887	116,848	67,737	60,228	52,734	45,631
	Mass flux decreased (mg)	0	51,780	88,613	94,244	99,865	105,192
CP2–CP3	Adsorbed mass decreased (mg)	0	17,260	29,538	31,415	33,288	35,064
	Biodegradation (mg)	0	69,040	118,150	125,659	133,153	140,257
	Total attenuation (mg)	46,472	98,251	135,084	140,716	146,336	151,664
	Biodegradation ratio (%)	0	70	87	89	91	92
	Attenuation rate (%)	25	53	73	76	79	82

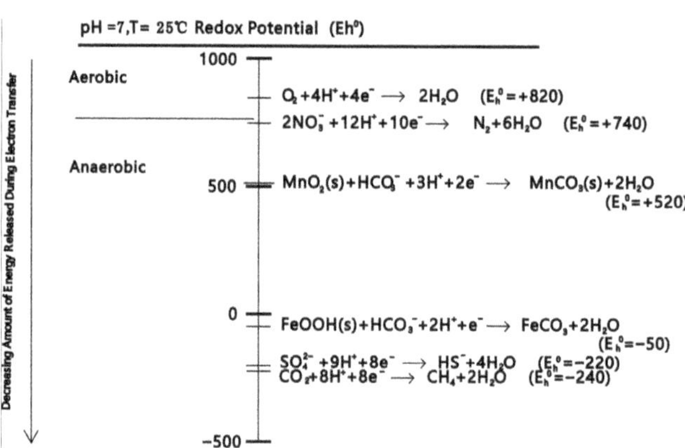

Figure 10. Redox potentials for various electron acceptors (based on Norris and Drápela) [32,53].

Furthermore, electron acceptor concentration is also an effective way to determine the microbial functional areas. Electron acceptors of the site have background values, and anomalies are usually caused by microorganisms. Therefore, concentrations of electron acceptors can reflect the microbial functional areas, at least those existing or taking place.

Based on Norris and Drápela's research, we can determine that the site was utterly anaerobic. Denitrification and manganese reduction occurred in all monitoring wells of the site. Until September 2009, iron reduction occurred in all of the wells except Z19 and S1, and sulfate reduction and methanogenesis did not occur during the monitored period. However, there was a zone whose sulfate concentration was lower than the background value; ORP was unable to prove it, so we considered that sulfate reduction had occurred before the monitored period.

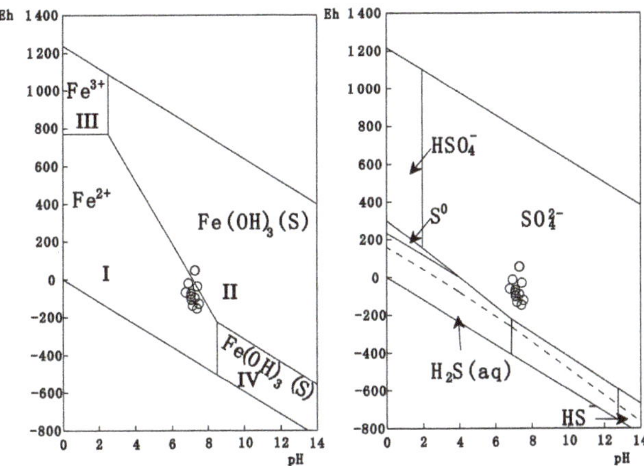

Figure 11. The relationship between Eh, pH, Fe, and S (based on Fetter) [55].

Suppose all electron acceptors in each monitoring well had an unchanged background value. Within the monitoring period, all the monitoring wells' concentrations were at a lower level, except S1. Manganese had a similar trend with iron, so the main boundary was unable to be distinguished, and manganese was assigned to the same boundary. An apparent low value area was able to be obtained through sulfate concentrations change analysis. This area is downstream of WPP and could maintain stability during the monitoring period. It could also prove that sulfate reduction had taken place before the monitored period. Through the analysis of electron acceptor concentrations, the boundaries of iron, manganese, and sulfate in the site were able to be identified, as shown in Figure 12.

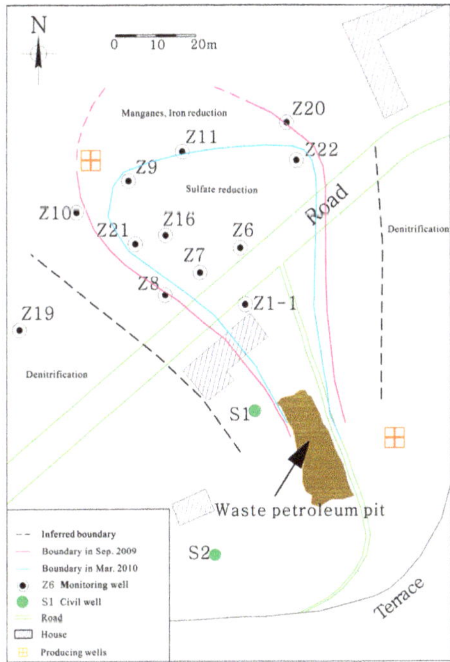

Figure 12. Electron acceptor zones in the groundwater.

By comparing the results, the boundary of the functional area remained unchanged. Only the reduction area of manganese and iron expanded slightly.

5. Conclusions

Based on the results monitored from petroleum-contaminated sites, we estimated MNA by statistical methods, the Mann–Kendall test, and mass flux, and thus we came to the following conclusions:

The contamination in the site was severe and needed remediation.

The TPH concentrations were decreasing in the monitoring period but still seriously exceeded the standard (Quality Standards for Groundwater (GB/T 14848-2017) [49] of China).

The attenuation rate in the site was from 0.00876 mg/L to 0.10095 mg/L, and the relationship between attenuation rate and distance met the exponential distribution. Remediation time was from 1.3 years to 10.6 years. Analysis of the TPH trend indicated that 62% of wells had decreasing or probably decreasing trends: these support the conclusion that the TPH plume is shrinking. Mass fluxes of the three CPs and changes in the plume also indicated that the TPH plume had reached a quasi-steady state and that mass flux was declining. The attenuation trend in the plume was more evident than the edge. The most essential process of MNA is biodegradation, and the second important is sorption. During the monitoring period, 393 g TPH attenuated, including 355 g TPH degraded by microbe. Biodegradation in the upstream is stronger. Iron(III) and manganese were the main electron acceptors utilized by microbes during the monitored period.

Based on the result of monitoring and estimation, we were able to come to the same result: MNA is in progress, and MNA can be an effective method to remediate the petroleum-contaminated site.

Author Contributions: Conceptualization, Y.X. and Y.Y.; methodology, Y.X.; validation, B.W.; formal analysis, M.Y.; investigation, Y.X. and M.Y.; resources, Y.Y. and X.D.; data curation, Y.X.; writing—original draft preparation, Y.X.; writing—review and editing, B.W.; supervision, X.D.; project administration, Y.Y.; funding acquisition, Y.Y. All authors have read and agreed to the published version of the manuscript.

Funding: This research was funded by Inner Mongolia Autonomous Region Department of Natural Resources, grant number 2022-TZH08; the China Geological Survey, grant number DD20230431; the National High-tech Research Development Program of China "863" project, grant number 2007AA06Z343; and the Guizhou Provincial Science and Technology projects, grant number QKHJC-ZK(2022)-General 186.

Institutional Review Board Statement: Not applicable.

Informed Consent Statement: Not applicable.

Data Availability Statement: Data are contained within the article.

Conflicts of Interest: The authors declare no conflict of interest.

References

1. Wang, L.H.; Li, M.M.; Zhang, Y.; Ma, W.M. Health risk assessment of groundwater organic pollution in an abandoned chemical plant. *Geol. Surv. Res.* **2012**, *35*, 4. (In Chinese) [CrossRef]
2. Zhao, Y.S.; Wang, B.; Qu, Z.H.; Zheng, W.; Jia, X.; Sun, M. Natural Attenuation of diesel pollution in sand layer of vadose zone. *J. Jilin Univ. (Earth Sci. Ed.)* **2010**, *40*, 389–393. (In Chinese) [CrossRef]
3. Li, Y.J.; Wang, S.J.; Zhang, M.; He, Z.; Zhang, W. Research progress of monitored natural attenuation remediation technology for soil and groundwater pollution. *China Environ. Sci.* **2018**, *38*, 1185–1193. (In Chinese) [CrossRef]
4. Sarkar, D.; Ferguson, M.; Datta, R.; Birnbaum, S. Bioremediation of petroleum hydrocarbons in contaminated soils: Comparison of biosolids addition, carbon supplementation, and monitored natural attenuation. *Environ. Pollut.* **2005**, *136*, 187–195. [CrossRef] [PubMed]
5. Brett, R.B.; Cindy, H.N.; Loring, N. Enumeration of aromatic oxygenase genes to evaluate monitored natural attenuation at gasoline-contaminated sites. *Water Res.* **2008**, *42*, 723–731. [CrossRef]
6. Choi, H.M.; Lee, J.Y. Groundwater contamination and natural attenuation capacity at a petroleum spilled facility in Korea. *J. Environ. Sci.* **2011**, *23*, 1650–1659. [CrossRef] [PubMed]

7. Declercq, I.; Cappuyns, V.; Duclos, Y. Monitored natural attenuation (MNA) of contaminated soils: State of the art in Europe—A critical evaluation. *Sci. Total Environ.* **2012**, *426*, 393–405. [CrossRef] [PubMed]

8. Li, T. Study on optimization of groundwater monitoring network for petroleum hydrocarbon pollution site based on natural attenuation process simulation. Master's Thesis, Jilin University, Changchun, China, 2021. (In Chinese) [CrossRef]

9. U.S. EPA. *Guidance for Developing Ecological Soil Screening Levels*; Office of Solid Waste and Emergency Response: Washington, DC, USA, 2005.

10. Sun, Y.; Liu, Y.; Yue, G.; Cao, J.; Li, C.; Ma, J. Vapor-phase biodegradation and natural attenuation of petroleum VOCs in the unsaturated zone: A microcosm study. *Chemosphere* **2023**, *336*, 139275. [CrossRef]

11. Song, Q.W.; Xue, Z.K.; Wu, H.J.; Zhai, Y.; Lu, T.T.; Du, X.Y.; Zheng, J.; Chen, H.K.; Zuo, R. The collaborative monitored natural attenuation (CMNA) of soil and groundwater pollution in large petrochemical enterprises: A case study. *Environ. Res.* **2023**, *216*, 114816. [CrossRef]

12. Lv, H.; Su, X.S.; Wang, Y.; Dai, Z.X.; Liu, M.Y. Effectiveness and mechanism of natural attenuation at a petroleum-hydrocarbon contaminated site. *Chemosphere* **2018**, *206*, 293–301. [CrossRef]

13. U.S. NRC (National Research Council). *Natural Attenuation for Groundwater Remediation*; National Academy Press: Washington, DC, USA, 2000.

14. Yan, X.X.; An, J.; Zhang, Y.Z.; Wei, S.H.; He, W.X.; Zhou, Q.X. Photochemical degradation in natural attenuation of characteristics of petroleum hydrocarbons (C_{10}-C_{40}) in crude oil polluted soil by simulated long term solar irradiation. *J. Hazard. Mater.* **2023**, *460*, 132259. [CrossRef]

15. Zito, P.; Bekins, B.A.; Martinović-Weigelt, D.; Harsha, M.L.; Humpal, K.E.; Trost, J.; Cozzarelli, I.; Mazzoleni, L.R.; Schum, S.K.; Podgorski, D.C. Photochemical mobilization of dissolved hydrocarbon oxidation products from petroleum contaminated soil into a shallow aquifer activate human nuclear receptors. *J. Hazard. Mater.* **2023**, *459*, 132312. [CrossRef] [PubMed]

16. Blum, P.; Hunkeler, D.; Weede, M.; Beyer, C.; Grathwohl, P.; Morasch, B. Quantification of biodegradation for o-xylene and naphthalene using first order decay models, Michaelis-Menten kinetics and stable carbon isotopes. *J. Contam. Hydrol.* **2009**, *105*, 118–130. [CrossRef] [PubMed]

17. Agnello, A.C.; Bagard, M.; Hullebusch, E.D.; Esposito, G.; Huguenot, D. Comparative bioremediation of heavy metals and petroleum hydrocarbons co-contaminated soil by natural attenuation, phytoremediation, bioaugmentation and bioaugmentation-assisted phytoremediation. *Sci. Total Environ.* **2016**, *563–564*, 693–703. [CrossRef] [PubMed]

18. Balseiro-Romero, M.; Monterroso, C.; Casares, J.J. Environmental Fate of Petroleum Hydrocarbons in Soil: Review of Multiphase Transport, Mass Transfer, and Natural Attenuation Processes. *Pedosphere* **2018**, *28*, 833–847. [CrossRef]

19. Kristensen, A.H.; Henriksenm, K.; Mortensen, L.; Scow, K.M.; Moldrup, P. Soil physical constraints on intrinsic biodegradation of petroleum vapors in a layered subsurface. *Vadose Zone J.* **2010**, *9*, 1–11. [CrossRef]

20. Wang, B.; Xia, Y.B.; Qu, Z.H.; Zhao, Y.S. Research on natural attenuation of diesel in the aquifer. *Hydrogeol. Eng. Geol.* **2014**, *41*, 138–143. (In Chinese) [CrossRef]

21. Bauer, S.; Beyer, C.; Kolditz, O. Assessing measurement uncertainty of first-order degradation rates in heterogeneous aquifers. *Water Resour. Res.* **2006**, *42*. [CrossRef]

22. Yang, Y.S.; Li, P.; Zhang, X.; Li, M.J.; Lu, Y.; Xu, B.; Yu, T. Lab-based investigation of enhanced BTEX attenuation driven by groundwater table fluctuation. *Chemosphere* **2017**, *169*, 678–684. [CrossRef]

23. Chen, K.F.; Kao, C.M.; Chen, C.W.; Surampalli, R.Y.; Lee, M.S. Control of petroleum-hydrocarbon contaminated groundwater by intrinsic and enhanced bioremediation. *J. Environ. Sci.* **2010**, *22*, 864–871. [CrossRef]

24. Li, G.H.; Zhang, X.; Huang, W. Characteristics of degrading microorganism distribution in polluted soil with petroleum hydrocabons. *Environ. Sci.* **2000**, *21*, 61–64. (In Chinese) [CrossRef]

25. Eulãlia, M.P.; Tim, G.; Anna, M.S.; Marc, V. Coupling chemical oxidation and biostimulation: Effects on the natural attenuation capacity and resilience of the native microbial community in alkylbenzene-polluted soil. *J. Hazard. Mater.* **2015**, *300*, 135–143. [CrossRef]

26. Alakendra, N.R.; Greg, L.M. Redox pathways in a petroleum contaminated shallow sandy aquifer: Iron and sulfate reductions. *Sci. Total Environ.* **2006**, *366*, 262–274. [CrossRef]

27. Guarino, C.; Spada, R.; Sciarrillo, R. Assessment of three approaches of bioremediation (natural attenuation, landfarming and bioagumentation-assistited landfarming) for a petroleum hydrocarbons contaminated soil. *Chemosphere* **2017**, *170*, 10–16. [CrossRef]

28. Chen, K.F.; Kao, C.M.; Wang, J.Y.; Chen, T.Y.; Chien, C.C. Natural attenuation of MTBE at two petroleum-hydrocarbon spill sites. *J. Hazard. Mater.* **2005**, *125*, 10–16. [CrossRef]

29. Wang, B.; Zhao, Y.S.; Qu, Z.H.; Zheng, W.; Long, B.S.; Jiao, L.N.; Xu, C. Impact of depth and moisture to diesel degradation in sand layer of vadose zone. *Environ. Sci.* **2011**, *32*, 227–232. (In Chinese) [CrossRef]

30. Xia, Y.B.; Wang, B.; Yang, Y.S.; Du, X.Q.; Yang, M.X. Evaluation effect of monitored natural attenuation in groundwater of petroleum-contaminated site. *Hydrol. Eng. Geol.* **2013**, *40*, 85–90. (In Chinese) [CrossRef]

31. Chiu, H.Y.; Hong, A.; Lin, S.L.; Surampalli, R.Y.; Kao, C.M. Application of natural attenuation for the control of petroleum hydrocarbon plume: Mechanisms and effectiveness evaluation. *J. Hydrol.* **2013**, *505*, 126–137. [CrossRef]

32. Drápela, K.; Drápelová, I. Application of Mann-Kendall test and the Sen's slope estimates for trend detection in deposition data from Bílý Kříž (Beskydy Mts. the Czech Pepublic) 1997–2010. *Časopis Beskydy* **2011**, *4*, 133–146. [CrossRef]

33. Kumar, K.S.; Rathnam, E.V. Analysis and Prediction of Groundwater Level Trends Using Four Variations of Mann Kendall Tests and ARIMA Modelling. *J. Geol. Soc. India* **2019**, *94*, 281–289. [CrossRef]

34. Yue, S.; Wang, C. The Mann-Kendall Test Modified by Effective Sample Size to Detect Trend in Serially Correlated Hydrological Series. *Water Resour. Manag.* **2004**, *18*, 201–218. [CrossRef]

35. Schäfer, D.; Hornbruch, G.; Schlenz, B.; Dahmke, A. Contaminant spreading assuming different kinetic approaches to simulate microbial degradation. *Grundwasser* **2007**, *12*, 15–25. [CrossRef]

36. Wilson, R.D.; Thornton, S.F.; Mackay, D.M. Challenges in monitoring the natural attenuation of spatially variable plumes. *Biodegradation* **2004**, *15*, 459–469. [CrossRef] [PubMed]

37. Newell, C.J.; Rifai, H.S.; Wilson, J.T.; Connor, J.A.; Aziz, J.A.; Suarez, M.P. *Calculation and Use of First-Order Rate Constants for Monitored Natural Attenuation Studies*; U.S. EPA Ground Water Issue; EPA/540/S-02/500; U.S. EPA, Office of Research and Development: Washington, DC, USA, 2002.

38. Metvalf, M.M.; Stevens, G.J.; Robbins, G.A. Application of first order kinetics to characterize MTBE natural attenuation in groundwater. *J. Contam. Hydrol.* **2016**, *187*, 47–54. [CrossRef] [PubMed]

39. Suarez, M.P.; Rifai, H.S. Evaluation of BTEX remediation by natural attenuation at a coastal facility. *Ground Water Monit. Remediat.* **2002**, *22*, 62–77. [CrossRef]

40. Bockelmann, A.; Zamfirescu, D.; Ptak, T. Quantification of mass fluxes and natural attenuation rates at an industrial site with a limited monitoring network: A case study. *J. Contam. Hydrol.* **2003**, *60*, 97–121. [CrossRef] [PubMed]

41. Verginelli, I.; Pecoraro, R.; Baciocchi, R. Using dynamic flux chambers to estimate the natural attenuation rates in the subsurface at petroleum contaminated sites. *Sci. Total Environ.* **2018**, *619–620*, 470–479. [CrossRef]

42. Vasudevan, M.; Nambi, M.I.; Kumar, S.G. Scenario-based modelling of mass transfer mechanisms at a petroleum contaminated field site-numerical implications. *J. Environ. Manag.* **2016**, *175*, 9–19. [CrossRef]

43. Gutierrez-Neri, M.; Ham, P.A.S.; Schotting, R.J.; Lerner, D.N. Analytical modelling of fringe and core biodegradation in groundwater plumes. *J. Contam. Hydrol.* **2009**, *107*, 1–9. [CrossRef]

44. Medici, G.; Lorenzi, V.; Sbarbati, C.; Manetta, M.; Petitta, M. Structural classification, discharge statistics, and recession analysis from the springs of the Gran Sasso (Italy) carbonate aquifer; comparison with selected analogues worldwide. *Sustainability* **2023**, *15*, 10125. [CrossRef]

45. Steelman, C.M.; Meyer, J.R.; Parker, B.L. Multidimensional investigation of bedrock heterogeneity/unconformities at a DNAPL-impacted site. *Groundwater* **2017**, *55*, 532–549. [CrossRef] [PubMed]

46. King, M.W.G.; Barker, J.F.; Devlin, J.T.; Butler, B.J. Migration and natural fate of a coal tar creosote plume: 2. Mass balance and biodegradation indicators. *J. Contam. Hydrol.* **1999**, *39*, 281–307. [CrossRef]

47. Bayer-Raich, M.; Jarsjö, J.; Liedl, R.; Ptak, T.; Teutsch, G. Average contaminant concentration and mass flow in aquifers from time-dependent pumping well data: Analytical framework. *Water Resour. Res.* **2004**, *40*. [CrossRef]

48. Chiu, H.Y.; Verpoort, F.; Liu, J.K.; Chang, Y.M.; Kao, C.M. Using intrinsic bioremediation for petroleum-hydrocarbon contaminated groundwater cleanup and migration containment: Effectiveness and mechanism evaluation. *J. Taiwan Inst. Chem. Eng.* **2017**, *72*, 53–61. [CrossRef]

49. *GB/T 14848-2017*; Quality Standard for Ground Water. General Administration of Quality Supervision, Inspection and Quarantine of the People's Republic of China, Standardization Administration of China. Standards Press of China: Beijing, China, 2017.

50. Xia, Y.B.; Yang, Y.S.; Du, X.Q.; Yang, M.X. Evaluation of in-situ MNA remediation potential and biodegradation efficiency of petroleum-contaminated shallow groundwater. *J. Jilin Univ. Earth Sci. Ed.* **2011**, *41*, 831–839. [CrossRef]

51. Wang, J.L.; Zhang, Y.L.; Ding, Y.; Song, H.W.; Liu, T.; Zhang, Y.; Xu, W.Q.; Shi, Y.J. Comparing the indigenous microorganism system in typical petroleum-contaminated groundwater. *Chemosphere* **2023**, *311*, 137173. [CrossRef]

52. Lee, J.Y.; Lee, K.K. Viability of natural attenuation in a petroleum-contaminated shallow sandy aquifer. *Environ. Pollut.* **2003**, *126*, 201–212. [CrossRef]

53. Norris, R.D.; Hinchee, R.E.; Brown, R. *Handbook of Bioremediation*; Lewis Publishers, Inc.: Boca Raton, FL, USA, 1994; 257p.

54. Jiang, W.; Sheng, Y.; Wang, G.; Shi, Z.; Liu, F.; Zhang, J.; Chen, D. Cl, Br, B, Li, and noble gases isotopes to study the origin and evolution of deep groundwater in sedimentary basins: A review. *Environ. Chem. Lett.* **2022**, *20*, 1497–1528. [CrossRef]

55. Fetter, C.W. *Contaminant Hydrogeology*; Macmillan: New York, NY, USA, 1993; pp. 285–291.

applied sciences

MDPI

The Role of Geological Methods in the Prevention and Control of Urban Flood Disaster Risk: A Case Study of Zhengzhou

Shuaiwei Wang [1,2], Weichao Sun [1,2,3], Xiuyan Wang [1,2], Lin Sun [1,2,*] and Songbo Liu [1,2]

[1] Institute of Hydrogeology and Environmental Geology, Chinese Academy of Geological Sciences, Shijiazhuang 050061, China; tairan_w@163.com (S.W.); 17730568088@163.com (W.S.); wxiuyan9948@163.com (X.W.); l20240110@163.com (S.L.)

[2] Fujian Provincial Key Laboratory of Water Cycling and Eco-Geological Processes, Xiamen 361021, China

[3] School of Chinese Academy of Geological Sciences, China University of Geosciences (Beijing), Beijing 100086, China

* Correspondence: lin_sun166@163.com

Abstract: The frequent occurrence of urban flood disasters is a major and persistent problem threatening the safety of cities in China and elsewhere in the world. As this issue is so pervasive, exploring new methods for more effective risk prevention and urban flood disaster control is now being prioritized. Taking the case of the city of Zhengzhou as an example, this paper proposes using geological, hydrogeological, ecological, and environmental conditions together with appropriate engineering designs to address the problem of urban flooding. The strategy includes integrating urban sponge–hydrogeological conditions, ecological engineering, and the construction of deep underground water storage facilities. Field investigations, data collection and analysis, in situ observations, testing, and laboratory experiments, are analyzed to explain the formation mechanism and means to mitigate flood disasters in Zhengzhou. Our results suggest that the appropriate use of geological, ecological, and hydrogeological aspects, combined with effective engineering practices, can significantly improve the city's flood control capacity. These measures can solve the problem of the "once-in-a-millennium" occurrence of torrential rain disasters such as the "720" torrential rainstorm that has affected the city of Zhengzhou.

Keywords: flood disaster; Zhengzhou; sponge city; underground drainage; ecological engineering

Citation: Wang, S.; Sun, W.; Wang, X.; Sun, L.; Liu, S. The Role of Geological Methods in the Prevention and Control of Urban Flood Disaster Risk: A Case Study of Zhengzhou. *Appl Sci.* **2024**, *14*, 1839. https://doi.org/10.3390/app14051839

Academic Editor: Nicola Magnavita

Received: 10 January 2024
Revised: 6 February 2024
Accepted: 10 February 2024
Published: 23 February 2024

1. Introduction

With the high pace of urban environment development, many cities worldwide face risks from urban floods, which is evolving as a major threat to life and property. According to the statistics of the World Meteorological Organization (WMO), around 7870 hydrometeorological disasters occurred in the world from 1970 to 2016. These floods have resulted in 1.86 million deaths and an economic loss of USD 1.9 trillion [1]. Chinese cities are no exception in this regard. Based on the available statistics, more than 400 Chinese cities with a wide geographical distribution were affected by floods from 2007 to 2018. These flood-related disasters have claimed 21,720 lives and resulted in a staggering economic loss of CNY 3163.9 billion [2,3] (p. 57; pp. 661–662). The problem of floods has greatly affected safe operations in cities, disrupted normal life, and has become a matter of serious concern. Flood-related disasters have become a severe impediment to the high-quality development of cities in China [4].

Urban waterlogging disaster is an extremely complex engineering problem with many dimensions [5–9] (pp. 1–2; pp. 661–662). The processes that lead to such urban disasters are interactive and complicated, involving many natural and human-induced factors. They can be understood only by understanding the state of natural systems, social conditions, and developmental activities [10–15]. Experts and scholars worldwide have been exploring ways to prevent and control flood disasters. They have put forward many suggestions based on ground conditions and theoretical methods that can be used effectively for flood control. Researchers and engineers based in China are constantly engaged in developing flood control strategies; they have also learned techniques practiced in foreign countries, which they successfully adapt for effective flood control [16–22].

In the face of the rapid development of cities built on regions of diverse geological environments, climatic zoning, and other conditions, preventing and controlling urban floods are challenging. However, after decades of continuous exploration, China has successfully developed a suitable strategy to manage and mitigate urban flood disasters. It has been proposed that by 2025, China will build a "sixteen-character" urban flood control and drainage system. This system works on a combination of parameters such as source emission reduction, well-designed storage and drainage facilities, risk elimination, and emergency response if stipulated standards are exceeded. Together, various elements of this system significantly improve the urban drainage system and effectively prevent waterlogging and flooding [23].

Fundamentally, the "sixteen-character" strategy aims at an efficient disposal capacity by building artificial reservoirs and lakes to improve flood detention and storage. Laying water pipelines, dredging rivers to improve flood discharge, and building sponge cities to enhance urban rainwater flood infiltration and storage are the other steps suggested to prevent flooding. On the mitigation front, steps are taken for emergency flood discharge, evacuation of residents, providing emergency aid, transportation of materials and implementation of traffic control. Undoubtedly, these measures have played a vital role in urban flood control in China. However, in the face of pervasive floods that are becoming more and more severe and seriously restricting the high-quality development of its cities, there is a need to develop innovative ideas and theoretical and practical methods for more effective solutions.

So, in this study, we use the rainfall during the "720" torrential rainstorm (in three days, from 17 to 20 July 2021, Zhengzhou received 617.1 mm of rainfall. On 19 July alone, the rainfall was 552.5 mm. This torrential rain disaster flooded all the urban areas of Zhengzhou and paralyzed the city's traffic. More than 2000 roads were damaged, and many infrastructure facilities were destroyed. The flood disaster killed 380 people and caused direct economic losses as high as CNY 40.9 billion. The intense rainfall received in this city during a short period broke the 70-year historical records of hourly and single-day rainfall since 1951. The probability and return period of hourly and daily precipitation rates leading to such events is estimated to be more than 1000 years.) as the upper limit value [24,25], and we take the example of Zhengzhou city (also spelt as Chengchow), the capital and largest city of Henan Province in the central part of the People's Republic of China (Figure 1), to discuss the role of geological methods in solving urban flood disasters, from the aspects of integrating urban sponge engineering with hydrogeological conditions, guiding engineering construction under the guidance of ecological geological theory, and the use of underground space for water storage projects.

This study will analyze and explain the importance of geological methods for the prevention and control of urban flood disasters according to the following logic, as shown in Figure 2.

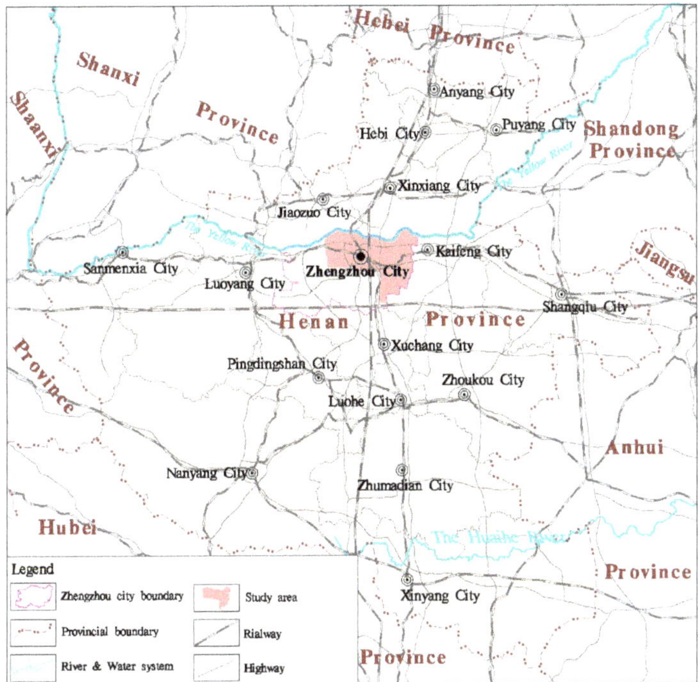

Figure 1. Location map of the study area.

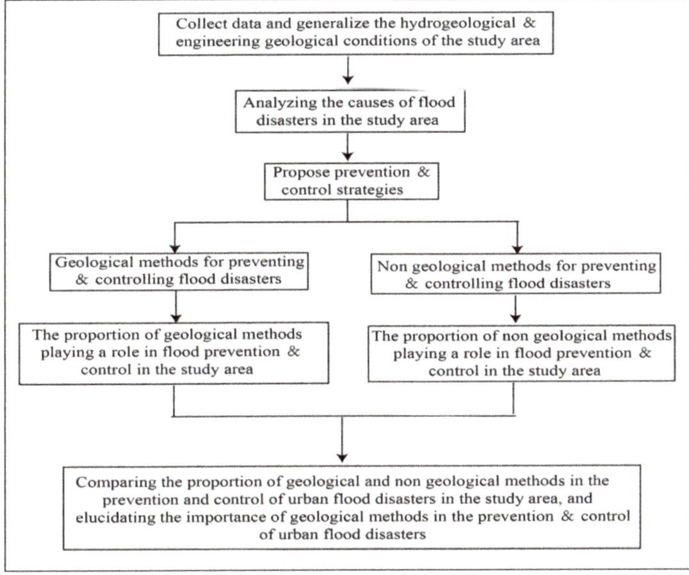

Figure 2. Thesis research logic map.

2. Materials and Methods

2.1. Methods

We collected and analyzed a variety of data that form the basis for the development of the city of Zhengzhou. The data included basic information on geography, geology, hydrology, social-economic development, and the area of urban land space. On the planning and administrative side, the data covered aspects of urban engineering, including sponge city construction, ecological environment aspects, river system governance, urban flood control etc. Many relevant kinds of literature based on in-depth studies at home and abroad that have discussed the developmental status and layout of cities prone to such disasters were consulted to develop the background of our study. These research papers have helped gather the concepts of advanced flood control, technologies and methods practiced for flood control at home and abroad.

Regional geology data were obtained from regional geological maps with accuracies of 200,000 and 50,000. In addition, the map with an accuracy of 50,000 was used to identify regions prone to geological disasters. Hydrogeological data for farmland water conservation and geotechnical data required for underground constructions were collected using maps with an accuracy of 50,000. Evaluation of urban environmental and geological problems and investigations of the section of the lower Yellow River in the city of Zhengzhou was conducted to gather the background data. Geological and environmental investigations were conducted in various mines in the area, and the data were analyzed in detail to evaluate the Quaternary geological conditions in the area. The large amount of data collected is useful for an integration of the regional geology, hydrogeology, environmental and engineering geology and assessing how their characteristics influence urban flooding.

A comprehensive geological survey was carried out with an accuracy of 50,000 scale to evaluate hydrogeology, engineering geology and environmental geology. The study area also carried out geophysical exploration, engineering, geological drilling and in situ testing, hydrogeological drilling, and pumping tests. The regional hydrologic environment, geological conditions and their spatial distribution were documented by the above investigations. Maps presenting the relevant parameters of hydrogeology, engineering and environmental geology were compiled.

Further, maps of groundwater depth were prepared based on water level measurements, water quality analysis, and tests conducted during high- and low-water periods. Through in situ tests such as water seepage test and pumping test carried out on-site, parameters such as soil seepage rate, rainfall infiltration replenishment coefficient, and permeability coefficient were obtained. Water, rock, and soil samples were collected, water quality and geotechnical characteristics were tested indoors, and the related parameters of hydrogeology, engineering geology and environmental geology were also obtained. The detailed work deployment is shown in Figure 3.

2.2. Materials

2.2.1. Geographical Setting

With a built-up area of 1181.61 km^2, Zhengzhou is a megacity in the Henan Province in the central region of China. By 2021, Zhengzhou registered a permanent population of 12.742 million, an urbanization rate of 79.1%, and a GDP of CNY 1269.1 billion, ranking 15th in China [26]. The city is located on the top of the fluvial alluvial plain of the Yellow River, with well-developed alluvial terraces. It falls in the north temperate continental monsoon climate zone and is also adjacent to the south subtropical monsoon climate zone, sharing the climatic characteristics of both zones. Precipitation occurs mainly from June to September, and the annual average rainfall is 542.15 mm [18].

2.2.2. Distribution of Watersheds and Water Systems

Zhengzhou city, with an area of 7446 km^2, spans two major river basins: the Yellow River and the Huaihe River. The city is built on the lower reaches of the Yellow River, the upper reaches of the Huaihe River and the upper and middle reaches of the Jialu River. It is adjacent to the Yellow River in the north, the Songshan Mountain in the west, and the Huang Huai plain in the east and south. The western part of the city, located in the Yellow River Basin, accounts for 24.6% of the city's total area. The remaining 75.40% of the city is in the Huaihe River Basin. The part of the city located in the Jialu River Basin of the Huaihe River covers 2750 km^2, of which the urban planning and construction area is 1945 km^2, which is the key study area. Located in the upper and middle reaches of the tributary Jialu River Basin, it also covers a small part of the Yellow River Basin, as shown in Figure 4. The Jialu River in the urban area mainly has 11 tributaries, such as Suoxu, Jiayu, Jinshui and Qili rivers. All tributaries flow into the Jialu River and flow southeast to Zhoukou outside the city, through Zhongmou County. It can be seen from Figure 1 that the Jialu River plays a critical role in flood control and drainage in the urban area of Zhengzhou [27].

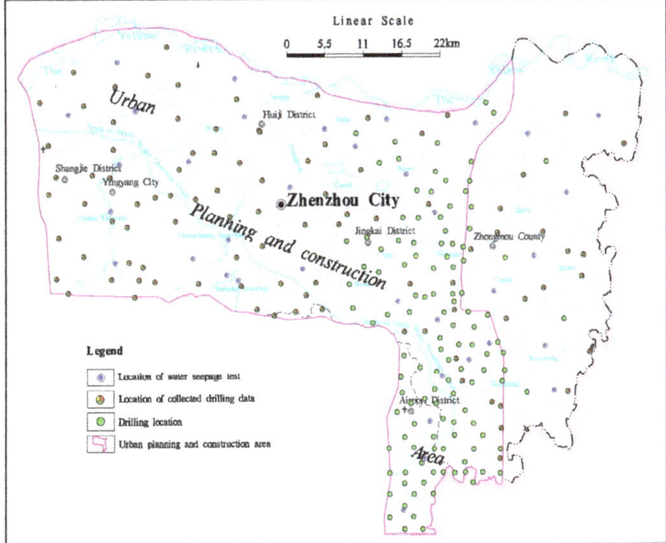

Figure 3. Work deployment in study area.

2.2.3. Geological and Engineering Geological Conditions

As discussed above, the study area is in the hinterland of China, bordering the Yellow River in the north, Songshan Mountain in the west, and Huang Huai plain in the east and south. The overall terrain is high in the southwest and low in the northeast, descending in a ladder shape. From the low and medium mountains eroded by the structure in the west and southwest, it gradually declines and transits to the structural denudation hills, loess hills, inclined (hill) plains and alluvial plains.

The inclined (hill) plain is located in front of the hills and is distributed in the central area in a strip form near the south and north. The terrain elevation is 100 to 150 m. From the west to the east, it is inclined from the front of the hills to the downstream longitudinally, with a gradient of 3 to 10 degrees. From south to north, it occurs in a wavy undulating form of hillocks. The well-developed and distributed strata are composed mainly of quaternary loess, loess-like soil, silt, and silty clay. The alluvial plains are widely distributed in the eastern region, formed by the Yellow River alluvium. The terrain is flat, the ground elevation is 80 to 100 m, and the slope is from the northwest to the southeast. The quaternary loose silt, silty clay and sandy soil, which define the strata, have good engineering properties and are suitable for construction, including underground water storage facilities (Figure 5). There are few active faults in the study area, and the peak ground-shaking acceleration is 0.10 to 0.15 g. There are no earthquakes of medium intensity or above, and the region is considered to fall in the seismic zone of intensity VII, which is conducive to urban safety. In conclusion, the geological conditions of the study area are generally suitable for urban engineering construction, including underground water storage projects.

Figure 4. Distribution map of water system in the study area.

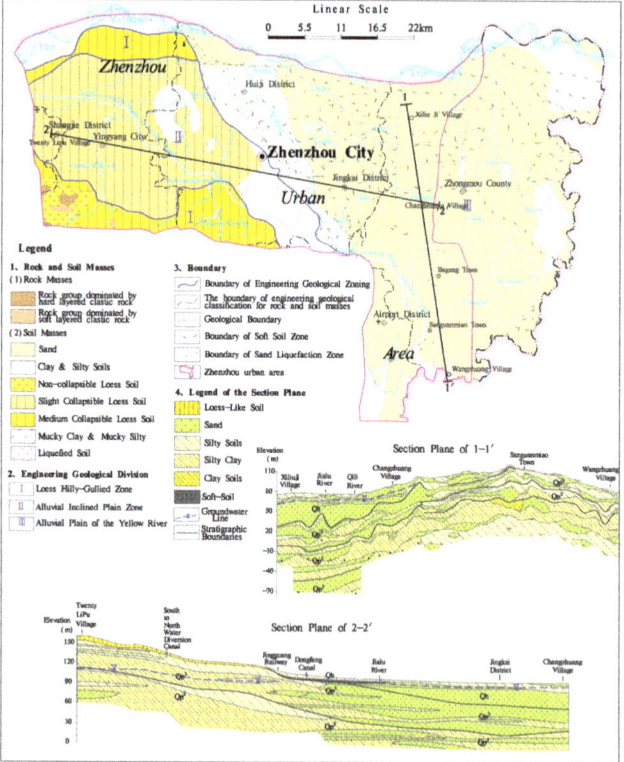

Figure 5. Engineering geological map of the study area.

2.2.4. Hydrogeological Conditions

There are four main types of groundwater in the study area: pore water, pore fissure water, carbonate fissure karst water, and bedrock fissure water. However, the key study area is mainly composed of loose rock pore water, which is widely stored in quaternary loose sediments. Among them, the shallow water (buried water level depth less than 80 m) with a single well inflow of 1000 to 5000 m^3/d is a good water supply source. The medium and deep water (buried water level depth is 80 to 400 m) is mainly distributed in the sand, and silty fine sand aquifer in the plain area, with a single well inflow of 500 to 1000 m^3/d. This is currently the primary water supply source for the city. Low impermeable silty clay layers with a thickness of 5 to 12 m widely distributed between the shallow, middle and deep aquifers are very suitable for constructing the deep tunnel water storage projects.

As shown in Figure 6, the buried depth of the shallow groundwater level in the urban area generally increases first and then decreases from the front of the mountain to the middle. The buried depth increases from 15 to 20 m to more than 30 m, then gradually decreases to 5–10 m eastward. The buried depth of 15–20 m has the largest regional distribution area. The part with a buried depth of more than 6 m accounts for 97.68% of the study area. The part with a buried depth of less than 6 m, with the smallest distribution area, accounting for only 2.32% of the area of the study area. The vadose zone in most of the urban area is composed of silt or fine sand, with good permeability, which is conducive for infiltration. The shallow groundwater is buried deep (as shown in Figure 6), and the soil layer in the vadose zone has enough space to store the rainwater. Building a sponge city can make full use of this hydrogeological feature.

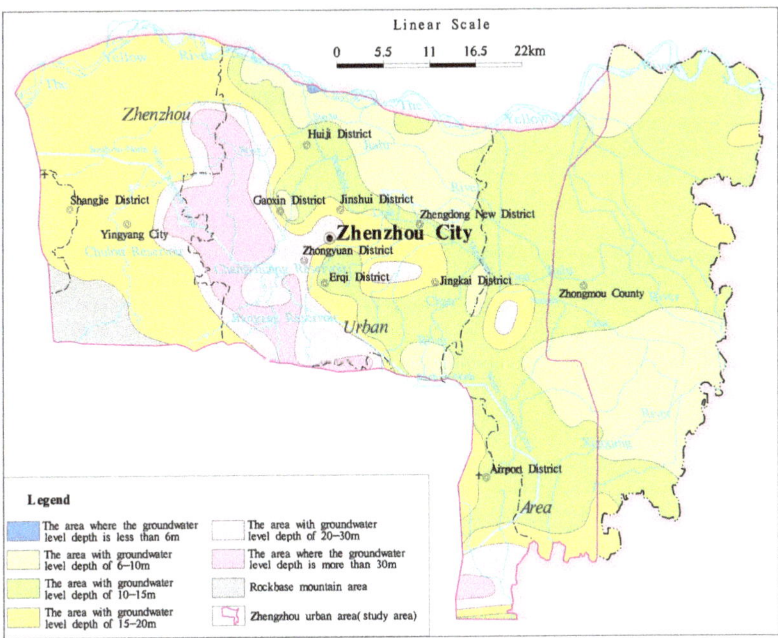

Figure 6. Buried depth map of groundwater level in the study area.

3. Results and Discussions

3.1. Analysis on the Causes of Flood Disasters in the Study Area

Generally speaking, the urban rain-flood is caused mainly by the rainfall and the river water flowing through the urban area from the upstream rivers. As shown in Figure 7, the drainage paths of precipitation mainly include the surface runoff generated by the rainwater. After falling to the ground, the rainwater follows multiple paths. It flows into the underground drainage pipe network, surface rivers, ditches, lakes, and wetlands. The precipitation also infiltrates the vadose zone soil layer or the urban sponge. Some rainwater will enter the sewage treatment plant designed for treatment and reuse. Thus, the stormwater will eventually enter the rivers, except for the loss through evaporation and ecological and human consumption.

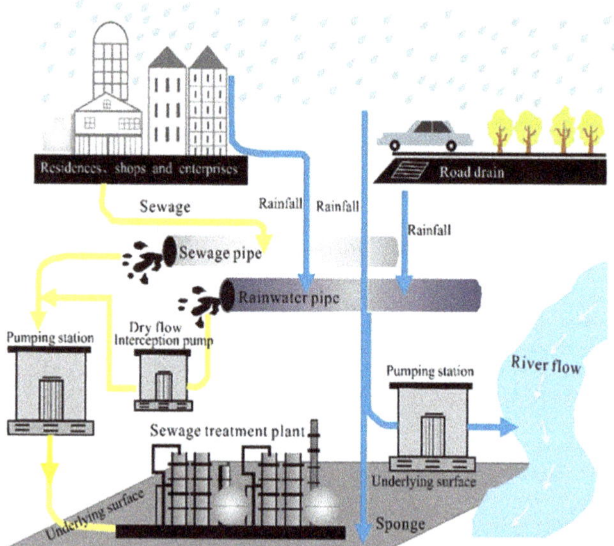

Figure 7. Pathways of rainwater and flood discharge after building sponge city.

If the urban rainwater drainage is not smooth, the situation may lead to urban flood disasters. For the study area, the mechanism of flood disasters is controlled by the following aspects of the rainfall as well as the ground conditions:

(1) Because of the abnormal superposition of the temperate monsoon climate and global climate change, the rainfall gets concentrated from June to September. During this period, it is easy to form heavy or extremely heavy rainstorms, which is the direct cause of flood disasters in the study area.

(2) The water and soil erosion in the upper reaches and source of the Jialu River Basin is quite intense. By the end of 2015, there was still a total area of 881 km^2 of soil erosion, and the total annual soil loss was 2.16 million tons [28]. Soil erosion upstream leads to siltation, and the flood interception capacity of the reservoirs is reduced substantially. The sediment in the middle and lower reaches can easily cause silting, which greatly reduces the flood-carrying capacity of the rivers.

(3) With rapid urbanization, the urban area has increased from 65 km^2 in 1981 to 1181.51 km^2 in 2019, an increase of 18 times. As a result, the impervious underlying layers are being rapidly exposed, altering the original geological conditions. Subsequently, the ability of the rainwater to penetrate the ground is reduced, and the surface runoff coefficient is increased from 0.25–0.45 to 0.65–0.85 [29]. An increase in surface runoff by about 60% results in poor drainage conditions, leading to flood disasters.

(4) During the construction of water ecological projects or the sponge city, the hydrogeological conditions are not fully utilized. For example, during the construction of projects such as wetlands, river treatment and other aquatic ecological projects, anti-seepage treatment is implemented at the bottom, which reduces the infiltration capacity. During the construction of sponge city, engineering measures such as pervious pavements are emphasized, but the natural rainwater absorption capacity of the

aeration zone soil layer is not fully considered. This situation, which lacks integration between the two important hydrologic parameters, reduces the sponge city's rainwater and flood absorption capacity.

(5) In the process of urban construction, the natural law of the urban hydrological cycle is not fully understood or paid attention to. As the flood plains are blindly developed and the river course is occupied, the drainage ditches are either buried or filled, and the open spaces are channelized. Thus, during urban construction on both riverbanks, the space for discharge space is not fully conserved. This misappropriation of the flood plain discharge space results in a significant decline in the flood discharge capacity of the river.

(6) The main rainwater drainage facilities, such as the underground drainage pipe network, roads, and river systems, have not been considered as a whole, and some urban areas have no drainage facilities, resulting in a substantial decline in urban drainage capacity.

Among the causes of the above-mentioned flood disasters, Articles 2, 3, and 4 are directly related to the underutilization of geological conditions.

3.2. Prevention and Control Strategies for Flood Disasters in Study Area

Urban waterlogging disasters mainly occur in highly developed urban areas. Although Zhengzhou spans the two river basins of the Yellow River and the Huaihe River, the waterlogging disasters mainly occur in the tributaries of the Jialu River in the Huaihe River Basin. These disasters include the "720" torrential rainstorm. Therefore, the study area of this paper is mainly 2750 km^2 in Zhengzhou, of which the key area is the urban planning and construction area of 1945 km^2.

With the purpose of flood risk prevention and control, and based on a comprehensive understanding of the actual geological, hydrological, and climatic conditions and the current urban development situation, a general idea for flood risk prevention and control in the study area is proposed as follows: taking the Jialu River basin where the city is located as a complete system, following the idea of "combining ecological geological methods with engineering construction to control flood risk; coordinating upstream, middle and downstream, upstream conservation and storage, and developing both midstream (urban) storage and drainage, with downstream emphasis on drainage and coordination above ground and underground", so the risk of flood disasters occurring in the study area can be minimized.

It is evident that to fully implement a composite plan for flood control; it is necessary to combine the geological and non-geological methods. Non-geological methods include planning of the flood discharge, design and use of underground drainage pipe networks, and storage of reservoirs, lakes, and wetlands. The geological method is based on the geological conditions of the study area, fully following and applying hydrogeological laws in sponge city engineering and water ecological engineering construction, fully utilizing the ecological geological theory in upstream or source ecological restoration engineering, and utilizing engineering geological conditions to build underground water storage engineering, in order to improve the capacity of rainwater and flood absorption, and achieve the goal of reducing the risk of urban flood disasters.

3.3. The Role of Geological Methods in Urban Flood Control

3.3.1. The Role of the Geological Method-Integrating Sponge City Construction with Hydrogeology in the Prevention and Control of Urban Flood Disasters

The concept of "sponge city" refers to the method of "strengthening the management of urban planning and construction, paying full attention to the absorption, storage, slow-release and drainage of rainwater by ecosystems such as buildings, roads, green spaces, and water systems, to effectively control rainwater runoff and realize natural accumulation, penetration, and purification of the urban environment". The concept regards cities as living organisms and uses the local conditions to improve drainage and waterlogging, with the goal of controlling urban floods [30]. Obviously, the construction of "sponge city" has become one of the important means of checking urban waterlogging in China. The main purpose of a sponge city in solving urban flood disasters is to improve the function of "water absorption, storage and seepage". Its construction guide proposes more than 10 engineering measures for the development of the sponge city (see Figure 8 for details).

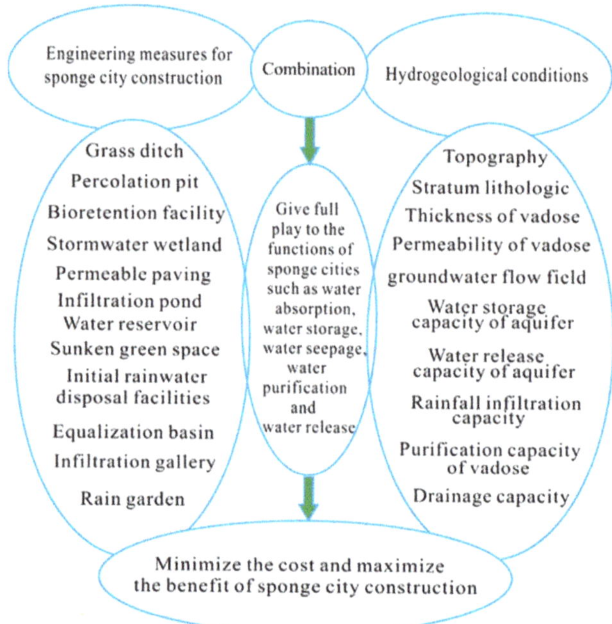

Figure 8. Engineering–hydrogeology integration concept for sponge city construction.

However, it must be regarded that each city is built under specific geological conditions and has its own unique physical and hydrogeological conditions, such as topography, lithology, rock and soil thickness of vadose zone, rainfall infiltration capacity, underground water storage performance, etc. (Figure 8). These hydrogeological conditions play an extremely important role in the "water absorption, storage and seepage" of urban rainwater. In the planning and construction of a sponge city, the "engineering measures" are organically integrated with the natural "hydrogeological conditions". The goal is to form a whole "sponge" of the city, which can reduce the construction and make the city functionally more efficient. This is the concept of the integration of engineering measures with hydrogeology, as illustrated in Figure 8.

For urban flood control, the construction of sponge cities not only needs to consider whether rainwater can infiltrate into the ground, but also whether there is space underground to store rainwater, which means that both the infiltration capacity and water storage capacity of the vadose zone should be considered.

(1) Rainfall infiltration capacity of the soil layer in the vadose zone

The infiltration capacity of the soil layer in the vadose zone in the absence of surface runoff is an important parameter to be evaluated. This paper suggests several water seepage, infiltration and geotechnical experiments to evaluate the infiltration capacity of the vadose zone soil layers in response to different intensities of rainfall. These experiments are based on previously collected hydrogeological data by water seepage, infiltration and geotechnical experiments to obtain parameters such as the infiltration rate, the permeability coefficient, and the void ratio of the soil layer in the vadose zone. Infiltration in the various soil layers of the vadose zone corresponding to the intensity of rainfall, calculated when there is no surface runoff, is shown in Table 1.

Table 1. Permeability parameters and corresponding rainfall of different soil layers in the vadose zone (not considering surface runoff).

Soil Mass Types	Seepage Velocity (mm/d)	Permeability Coefficient (cm/s)	Rainfall Intensity That Can Be Absorbed	Rainfall (mm/d)
Silty clay	6.2	7.1×10^{-6}–6.4×10^{-5}	light rain	<10
Silt	56.75	6.6×10^{-5}–2.1×10^{-4}	heavy rain	25–50
Loess-like soils	62.374	5.4×10^{-5}–9.7×10^{-5}	heavy rain	25–50
Silty sand or silty-fine sand	122.60	1.8×10^{-4}–7.2×10^{-4}	rainstorm	50–100
Medium sand	2583.2	5.0×10^{-3}–1.4×10^{-2}	extremely heavy rainstorm	>250
Coarse sand	7251.2	1.8×10^{-2}–6.8×10^{-2}	extremely heavy rainstorm	>250

As shown in Table 1, the infiltration capacity of silty clay can meet the requirements for most rainfalls of light rain intensity. The infiltration capacity of silt or loess can meet the requirements for heavy rain intensity. The infiltration capacity of silty sand or silty fine sand can meet the infiltration requirements for a rainstorm. The infiltration capacity of medium sand and coarse sand can meet the infiltration requirements for an extremely heavy rainstorm. Therefore, when the vadose zone is composed of fine-grained soils such as silty clay, silt, and loess, its ability to infiltrate rainfall is low, and it is not an ideal stratum for the construction of the sponge city. Coarse-grained soils such as medium sand and coarse sand have a strong rainwater infiltration capacity. If there is enough storage space—that is, if the thickness of the vadose zone soil layer is sufficient—it is an ideal stratum for building a sponge city.

(2)   Water storage capacity of the vadose zone

The water storage capacity of the vadose zone refers to the amount of rainwater that can be stored in the effective pores of the soil layer. The porosity, natural saturation and maximum saturation obtained by engineering geological drilling, geotechnical testing and other methods, are used to calculate the effective influent porosity Ne of each soil layer by Equation (1) (as shown in Table 2). And then, the thickness H_t of the vadose zone soil layer required to store rainfall of different intensities was calculated by Equation (2) (as shown in Table 3). The thickness of this soil layer is an indicator of underground water storage capacity.

$$N_e = (S_{max} - S_0) \times n, \tag{1}$$

where S_{max} is the maximum saturation, S_0 is the natural saturation, and n is the porosity.

$$H_t = (q \times S_t \times N_e)/S_t = q \times N_e \tag{2}$$

where q is the upper limit value for different rainfall intensities, and S_t is both the rainfall area and the infiltration area of the vadose zone.

Table 2. The effective influent porosity of different vadose zone soil.

Soil Mass Types	Porosity (%)	Natural Saturation (%)	Maximum Saturation (%)	The Effective Influent Porosity (%)
silty clay	44.32	76.36	92.34	7.08
silt or loess-like soils	40.76	59.51	91.73	13.13
silty-fine sand	42.23	62.11	90.03	11.79
medium sand	36.14	61.19	94.72	12.12
coarse sand	35.66	56.29	93.24	13.18

Table 3. Thickness of vadose zone soil required by different strata for different rainfall intensity.

Soil Mass Types	Light Rain (10 mm/d)	Moderate Rain (25 mm/d)	Heavy Rain (50 mm/d)	Rainstorm (100 mm/d)	Heavy Rainstorm (250 mm/d)	Extremely Heavy Rainstorm (>250 mm/d)
silty clay	141	353	706	1412	3531	>3531
silt or loess-like soils	76	190	381	762	1904	>1904
silty-fine sand	85	212	424	848	2120	>2120
medium sand	83	206	412	825	2062	>2062
coarse sand	76	190	379	759	1879	>1879

The area of the key study area is 1945 km^2, and its vadose zone is composed of mixed sands and clay. Fine silt makes up 47.65%, and 18.72% is composed of medium-coarse sand. Silt makes up 31.22%, and the rest, 2.41%, is silty clay. For the water storage capacity of a heavy rainstorm, the required thickness of soil layers such as silt, silty sand and medium-coarse sand is 1.88 to 2.12 m, while the required thickness of silty clay is 3.53 m. The rainfall of the "720" torrential rain that lasted for three days in Zhengzhou was 617.1 mm. Based on this calculation, the thickness of soil layers such as silt, silty sand, and medium-coarse sand to accommodate the rainfall was no more than 5.23 m. The required thickness of the clay is 8.72 m. Judging from the depth distribution of the groundwater level in study area, if the rainwater can seep into the ground, most of the vadose soil layers within the study area have the ability to store the "720" torrential rain.

(3) Assessment of rainwater infiltration and storage capacity in the study area based on the geological method-integrating sponge city construction with hydrogeology

The rainwater storage capacity of the vadose zone soil layer is closely related to the rainfall intensity, the permeability coefficient of the vadose zone soil layer, the thickness of the vadose zone, the depth of rainfall infiltration, the effective water inflow rate, and the distribution area of each soil layer. Among them, the permeability coefficient determines whether rainwater under different intensities can infiltrate into the vadose zone, and the thickness of the vadose zone determines whether there is enough space to accommodate rainfall; the depth of rainfall infiltration, effective water inflow rate, and distribution area of each soil layer determine whether the soil layer in the vadose zone can accommodate rainfall of different intensities.

Based on the previous experimental data, we have obtained the permeability coefficient, vadose zone depth, effective water inflow rate and other parameters of soil layers with different rock types in the study area. Through analysis, we have found that as long as there is sufficient rainfall infiltration depth, the study area has extremely strong rainfall storage capacity. Next, this article will evaluate the rainwater storage capacity of the study area based on the concept of sponge city engineering and hydrogeological integration by restoring the rainfall intensity and infiltration depth at different return periods.

① Rain intensity and return periods

Since many aspects of urban waterlogging risk prevention and control involve parameters such as rainfall intensity and its return period, the methods used for their calculation are very important. This paper uses the following formula selected by the "720" disaster investigation group of the State Council:

$$R = ((2479.78 + 2388.2762 \times \lg p)/(t + 15.3)^{0.775}) \times 8.64 \tag{3}$$

where R is the is the rainfall (mm/d) at different return periods (rainfall intensity), p is the return period (year), and t is the rainfall duration (1440 min). The rain intensities of different return periods calculated by the above formula are given in Table 4.

Table 4. Geological methods for assessing the drainage capacity of rainwater and floods in the study area.

Rainfall Intensity/Return Period	Rainfall (mm/d)	Total Rainfall (106 m^3/d)	Suburban Storage Capacity (10^6 m^3/d)	Storage Capacity of Urban "Sponge" Areas (10^6 m^3/d)	Bottom Storage Capacity of Aquatic Ecological Engineering (10^6 m^3/d)	The Increased Storage Capacity of the Ecological Restoration Reservoir at the Source (10^6 m^3/d)	Total (10^6 m^3)
Light rain (upper limit)	10	27.50	5.62	2.69	14.53	3.28	26.12
Moderate rain (upper limit)	25	68.75	12.47	6.38	14.53	3.28	36.66
Heavy rain (upper limit)	50	137.5	23.89	12.51	14.53	3.28	54.21
Once a year	76	209.00	35.77	18.90	14.53	3.28	72.48
Rainstorm (upper limit)	100	275.00	46.74	24.79	14.53	3.28	89.34
Once in 3 years	111	305.25	51.77	27.49	14.53	3.28	97.07
Once in 5 years	127	349.25	59.07	31.42	14.53	3.28	108.30
Once in a decade	149	409.75	69.12	36.82	14.53	3.28	123.75
Once in 30 years	184	506.00	85.11	45.41	14.53	3.28	148.33
Once in 50 years	200	550.00	92.42	49.34	14.53	3.28	159.57
Once in a century	222	610.50	102.47	54.74	14.53	3.28	175.02
Once in 200 years	244	671.00	112.52	60.14	14.53	3.28	190.47
Heavy rainstorm (upper limit)	250	687.50	115.26	61.61	14.53	3.28	194.68
Once in 500 years	273	750.75	125.77	67.26	14.53	3.28	210.84
Once in a millennium	295	811.25	135.82	72.66	14.53	3.28	226.29
The 3-day rainfall of the "720"	617.1	1697.03	285.08	152.21	43.59	3.28	484.16

② Infiltration depth of rainfall and rainfall storage capacity of soil layer in vadose zone

According to the existing research results [31], the formula for calculating the rainfall infiltration depth (H) in the study area is as follows:

$$H = 0.8905R + 0.79 \text{ (medium − coarse sand)} \tag{4}$$

$$H = 0.4951R + 4.47 \text{ (silt, silty clay)} \tag{5}$$

$$H = 0.7647R + 0.74 \text{ (silty fine sand)} \tag{6}$$

The formula for calculating the rainwater infiltration and storage capacity Q of the vadose zone soil layer is as follows:

$$Q = H \times S \times N_e \tag{7}$$

where Q is the infiltration and storage capacity of the vadose zone soil layer (m^3); H is the rainfall infiltration depth (m), and S is the distribution area of soil layers with different lithologies (m^2). N_e is the effective porosity of the soil, which we define here as the difference between porosity and soil saturation.

③ The rainwater storage capacity of the soil layer in the vadose zone of key study area

The area of the key study area is 1945 km^2, of which 14% is green spaces and squares [32], and the area that can be used to build a sponge city is 272.30 km^2. According to the concept of integrating sponge city engineering and hydrogeological conditions, the soil layer in the aeration zone of green space and square land can be generalized as fine sand and silt layer, with an effective inflow porosity of 11.79%. Equation (7) is used to estimate the rainwater seepage storage capacity of the vadose zone soil layer, as shown in Table 4.

④ The rainwater storage capacity of the soil layer in the vadose zone of the suburban area

After investigation, it was found that the suburban area of the study area is 805 km², and rainfall cannot penetrate areas such as roads, urban buildings and factories, accounting for 22.375% of the suburban area. The underlying surface of other areas in the suburb is farmland, forest, and grasslands with good permeability. The area of vadose zone lithology with silty fine sand accounts for 48.89%, medium-coarse sand accounts for 15.88%, and silt and silty clay accounts for 35.23% (see Figure 9 for details). According to the method discussed earlier, the rainwater infiltration and storage capacity of the soil layer in the vadose zone in the suburbs were obtained (Table 4).

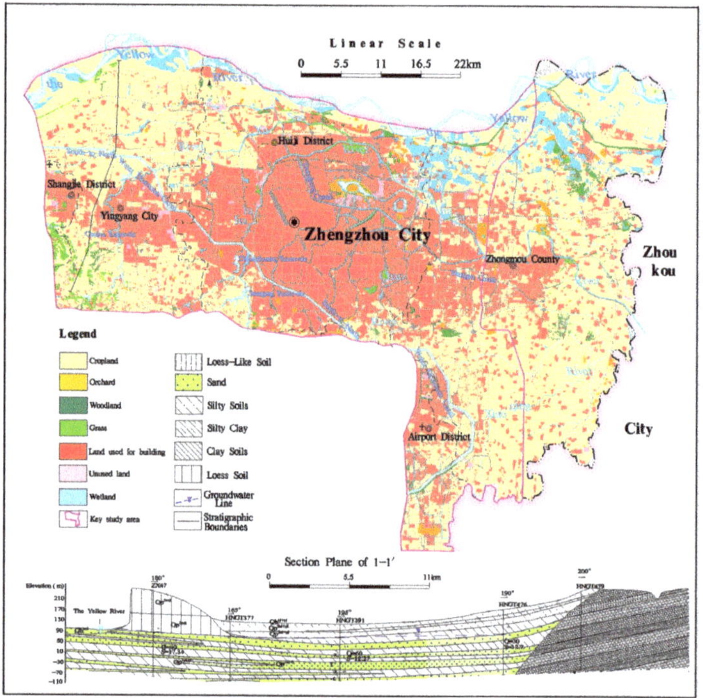

Figure 9. Land use types in study area.

3.3.2. The Role of the Geological Method-Integrating Water Ecological Engineering Construction with Eco-Geological in the Prevention and Control of Urban Flood Disasters

It has been recognized that urban water ecological engineering construction can greatly improve the city's ability to prevent flood disasters. By controlling soil erosion, the amount of sediment entering the reservoir or river can be substantially reduced, thereby maintaining the capacity of these water bodies to intercept floods. The water and soil loss in the upper reaches or source of the Jialu River Basin is quite significant. By the end of 2015, the total water and soil loss area was 881 km², and the average annual total soil loss was 2.16 million tons [28]. On the one hand, the loss of water and soil at the upstream or the source will lead to sedimentation and weaken the flood interception capacity of the reservoir, and on the other, sedimentation at the middle and lower reaches will lead to the blockage of the river channel and will greatly reduce its flood discharge capacity. It is estimated that the upstream siltation of the reservoir due to soil erosion has reduced the flood interception capacity of the reservoir from 260.91 million m^3 to 195.68 million m^3, which is about 25.00%.

Developing green patches by giving due consideration to ecological geology can address this issue quite successfully. Thus, areas suitable for planting trees or grass need to be selected based on the local geological conditions. By following scientific practices in ecological restoration and management, the sediment deposition in reservoirs and river channels can be greatly reduced to maintain the capacity of the reservoir to intercept floods and the river, and to discharge flood water. At present, the amount of annual soil erosion is 3.281 million m^3. After ecological management, the reservoir can reduce the sediment deposition volume by 3.281 million m^3 every year [28], and this number is temporarily used as the standard for the improvement of the flood-holding capacity in this paper.

The goal of increasing infiltration storage and river discharge capacity are achieved through the proper implementation of urban ecological engineering. Zhengzhou plans to build 54 water ecological projects within the city. These projects are planned for various areas, including the Jialu River, Lianhu Lake, Dongfeng Canal, Ruyi River, Xiushui River, Xiliu Lake, Yuema Lake, and Yanming Lake. After the completion of these projects, the water-covered area of Zhengzhou city will increase from 156.9 km^2 to 211.66 km^2. The urban water surface area will increase from 0.93% to 3.9% by 2020 [33]. Concept-wise, the construction of water ecological engineering is to follow "Tao follows nature" and not artificially destroy the law of water cycle. Geologically speaking, it is necessary to follow the laws of ecological hydrogeology. Thus, there should be no seepage prevention at the bottom of lakes and rivers, so that water can naturally seep into the ground. The area of the study area is 2750 km^2, and the water surface area is 3.9% × 2750 = 107.25 km^2. If the bottom of the basin (area: 107.25 km^2) is not treated with anti-seepage, the situation would be drastically different. Considering the hydrogeological parameters such as the permeability coefficient of the soil layer in the local vadose zone, the buried depth of the groundwater level, and the water level of lakes and rivers, the amount of seepage water is calculated to be 14.53 million m^3/d. The results show that the water ecological engineering based on this concept is very beneficial to the prevention and control of flood disasters.

3.4. The Importance of Non-Geological Methods in Urban Flood Control

In addition to the geological methods mentioned above, there are non-geological methods such as river drainage, reservoir and lake water storage and underground drainage pipe network system storage to prevent and control flood disasters. These methods applicable to the study area are discussed below.

(1) Flood discharge capacity of the river channel

The flood is mainly discharged in the Jialu River and its tributaries, which eventually flow into its main stream and discharge outward in the downstream Zhongmu County. The flood discharge capacity can be characterized by the overcurrent capacity of the Zhongmu hydrological station. According to the calculation of the warning water level of the Zhongmu Hydrological Station of the Jialu River at 77.5 m, the total drainage capacity of the river channel for floods is 323 m^3/s [34], and the flood drainage capacity is 27.91 million m^3/d, as shown in Table 5.

(2) Water storage capacity of reservoirs, lakes, and wetlands

The water storage capacity includes the storage capacity of reservoirs and lakes after the completion of the Jialu River source or upstream reservoir and the water ecological projects. After the completion of the urban water ecological project, the area of lakes and wetlands will increase to 107.25 km^2. In addition to the storage by this increased area of rivers, ditches, etc., and the ecological water demand for other lakes and wetlands under normal circumstances, about 62.57 million m^3 of flood control water storage can be reserved (see Table 5 for details).

The total catchment area of the Jialu River source or upstream reservoir is 786 km^2, and the total storage capacity is about 260.9 million m^3. But, after years of sediment deposition, the current storage capacity has been reduced to 195.68 million m^3. In fact, the storage capacity of the reservoir for rainwater is positively correlated with the rainfall and the catchment area. Only when the rainfall intensity reaches a certain level will the reservoir reach the maximum storage capacity, which is 195.68 million m^3. The calculations for the storage capacity of the reservoir for rainwater under different rainfall intensities are shown in Table 5.

Table 5. Non geological methods for assessing the drainage capacity of rainwater and floods in the study area.

Rainfall Intensity/Return Period	Rainfall (mm/d)	Total Rainfall (10^6 m^3/d)	Storage Capacity of Upstream Reservoirs (10^6 m^3)	Flood Discharge Capacity of Jialu River (10^6 m^3/d)	Reserved Storage Capacity for Lakes, Wetlands, etc. (10^6 m^3)	Storage Capacity of Underground Pipeline Network (10^6 m^3)	Total (10^6 m^3)
Light rain (upper limit)	10	27.50	7.86	27.91	62.57	7.79	106.13
Moderate rain (upper limit)	25	68.75	19.65	27.91	62.57	7.79	117.92
Heavy rain (upper limit)	50	137.5	39.30	27.91	62.57	7.79	137.57
Once a year	76	209.00	59.74	27.91	62.57	7.79	158.01
Rainstorm (upper limit)	100	275.00	78.60	27.91	62.57	7.79	176.87
Once in 3 years	111	305.25	87.25	27.91	62.57	7.79	185.52
Once in 5 years	127	349.25	99.82	27.91	62.57	7.79	198.09
Once in a decade	149	409.75	117.11	27.91	62.57	7.79	215.38
Once in 30 years	184	506.00	144.62	27.91	62.57	7.79	242.89
Once in 50 years	200	550.00	157.20	27.91	62.57	7.79	255.47
Once in a century	222	610.50	174.49	27.91	62.57	7.79	272.76
Once in 200 years	244	671.00	191.78	27.91	62.57	7.79	290.05
Heavy rainstorm (upper limit)	250	687.50	195.68	27.91	62.57	7.79	293.95
Once in 500 years	273	750.75	195.68	27.91	62.57	7.79	293.95
Once in a millennium	295	811.25	195.68	27.91	62.57	7.79	293.95
The 3-day rainfall of the "720"	617.1	1697.03	195.68	83.73	62.57	7.79	349.77

(3)   Storage capacity of underground drainage network system

The study area plans to build a new underground pipe network of 2576 km. With the completion of this project, the total length of the pipe network will increase from 1831 km to 4407 km. With that, the water storage capacity of the underground pipe network will increase from 3.24 million m^3 to 7.79 million m^3, as shown in Table 5.

The flood control capacity of Zhengzhou using geological and non-geological methods shown in Tables 4 and 5 are estimations based on the results obtained after the completion of the various projects for the Jialu River. These include the comprehensive treatment project, construction of the sponge city, and implementation of various projects that employ the concepts of ecological and geological aspects. It is to be noted that the geological method does not include the flood control capacity obtained by using underground space to build water storage projects. Based on the relevant data in Tables 4 and 5, the proportion of the geological method's ability in flood control was calculated. The calculation was based on the formula expressed as the proportion of the ability of geological methods to remove floods = (the amount of flood water removed by geological methods)/(the amount of flood water removed by geological methods + the amount of flood water removed by non-geological methods) $\times 100\%$. The results of the calculations are shown in Table 6.

Table 6. The important role of geological prevention and control methods of flood and waterlogging disasters in the study area.

Rainfall Intensity/Return Period	Rainfall (mm/d)	Combining Two Methods Can Solve the Proportion of Rainfall (%)	The Proportion of Geological Methods Solved (%)	Geological Conditions Not Fully Utilized	
				Reduced Rainwater Storage Capacity (10^6 m^3)	Combining Two Methods Can Solve the Proportion of Rainfall (%)
Light rain (upper limit)	10	480.91	19.75	20.50	406.36
Moderate rain (upper limit)	25	224.84	23.72	24.19	189.66
Heavy rain (upper limit)	50	139.48	28.27	30.32	117.43
Once a year	76	110.28	31.45	36.71	92.72
Rainstorm (upper limit)	100	96.80	33.56	42.60	81.31
Once in 3 years	111	92.57	34.35	45.30	77.74
Once in 5 years	127	87.73	35.35	49.23	73.63
Once in a decade	149	82.77	36.49	54.63	69.43
Once in 30 years	184	77.32	37.91	63.22	64.82
Once in 50 years	200	73.35	38.47	67.15	63.25
Once in a century	222	73.35	39.09	72.55	61.46
Once in 200 years	244	71.61	39.64	77.95	59.99
Heavy rainstorm (upper limit)	250	71.07	39.84	79.42	59.52
Once in 500 years	273	67.24	41.77	85.07	55.90
Once in a millennium	295	64.13	43.50	90.47	52.98
The 3-day rainfall of the "720"	617.1	49.14	58.06	199.08	37.40

Table 6 shows that the combined action of geological and non-geological methods can almost solve the rain and flood-related problems caused by rainstorms in the study area, and geological methods can take a 33.56% share of the prevention and control of floods. However, these methods are insufficient to meet the requirements if the rainstorms are larger. In the case of preventing floods caused by heavy rainstorms, geological methods account for more than 30%, and the greater the rainfall intensity, the greater the proportion of drainage capacity of geological methods. For the "720" torrential rainstorm in the study area, the two methods taken together account for only a combined share of 49.14% in controlling the flood disasters.

If the geological, ecological, and hydrogeological conditions are not fully utilized, and if the engineering aspects of sponge city construction are not followed, the situation could be different for the study area. If so, the impounding capacity at the source of the Jialu River will be reduced by 3.28 million m^3, and the seepage volume will be reduced by 14.53 million m^3 per day. This will lead to a sharp decline in the amount of rainwater and flood infiltration in the sponge construction area of the city, reducing its flood control capacity. As shown in Table 6, the flood control capacity achievable through a combination of geological and non-geological methods will be reduced from rainstorm to heavy rain. For example, for the urban flood disaster caused by the "720" rainstorm in study area, the combined efforts of the two methods can only remove 49.14% of the rain and flood. This is the reason behind the extremely serious consequences caused by the "720" torrential rainstorm in Zhengzhou.

4. Conclusions and Suggestions

4.1. Prevention and Control Strategies for Flood Disasters in Study Area

Our analysis shows that, by making full use of Zhengzhou's geological conditions and by considering the ecological and hydrogeological conditions, the city's flood control capacity can be greatly improved. Proper engineering measures such as drainage and the development of sponge cities would further facilitate flood control.

By taking advantage of the favorable engineering geological conditions of the city, the construction of deep underground water storage projects can almost solve the problem of floods, including those caused by the extremely heavy rainstorm that occurred once in a millennium. The ability to store the excess water and use it as a resource to alleviate the problem of water shortage in the city is an added advantage.

Construction of pipelines connecting deep underground storage tunnels, using the space available below the basements of subways, shopping malls, and parking lots, as well as those beneath the low-lying areas of high flood risk, is effective in controlling the flood situation. The pipeline network will divert the flood waters that enter the subway or other high-risk areas into the deep-water storage tunnel. This can eliminate the risk of flood disasters and greatly improve the emergency response capability of urban flood disasters.

4.2. Suggestion

Conduct targeted hydrogeological, ecological, engineering, and environmental geological surveys in the flood-prone areas, including the city of Zhengzhou, from the point of urban flood control. Based on the survey results, make full use of geological conditions to address the problem of urban flood disasters. The specific suggestions are as follows:

In the urban area, the sponge city engineering construction needs to be closely combined with the urban hydrogeological conditions, giving full attention to the water seepage and storage performance of the vadose zone strata. This would improve the urban sponge's storage capacity for rainwater.

Consider the laws of hydrogeology while constructing water ecological projects such as lakes, wetlands, or other ecological management projects. Through the proper use of a hydrological system, the infiltration capacity of water bodies to penetrate deeper into the strata can be improved.

In accordance with the laws of engineering geology and urban construction, the deep underground space or underground goaf can be used to build deep water storage projects. This would help to improve flood absorption and emergency response capabilities.

It is necessary to execute steps for the control of soil erosion and improve the soil and water conservation capacity while considering the ecological and geological conditions. This would reduce the accumulation of sediments in the reservoirs and rivers and maintain their flood retention capacity.

In summary, through a studied application of geological, ecological, and hydrological conditions and by the proper design and implementation of engineering practices, the flood situation in Zhengzhou can be controlled. The potential for recycling the flood water for the city's consumption is a positive fallout of the city's flood disaster management.

Author Contributions: S.W. contributed to conception and design of the study. S.W., L.S., X.W. and W.S. carried out the experiments. W.S., S.L. and L.S. compiled the drawings. S.W. and W.S. performed the statistical analysis and wrote the first draft of the manuscript. All authors have read and agreed to the published version of the manuscript.

Funding: This research was supported by the Hebei Natural Science Foundation: D2021504034, E2021210072; Basal Research Fund of IHEG, CAGS: SK202106.

Institutional Review Board Statement: Not applicable.

Informed Consent Statement: Not applicable.

Data Availability Statement: The original contributions presented in this study are included in the article, and further inquiries can be directed to the corresponding author.

Acknowledgments: This research was supported by the Basal Research Fund of Iheg, CAGS: SK202314, SK202319. Geological Survey Project: DD20211309.

Conflicts of Interest: The authors declare that the research was conducted in the absence of any commercial or financial relationships that could be construed as a potential conflict of interest.

References

1. World Meteorological Organization. *Reducing and Managing Risks of Disasters in a Chancing Climate*; WMO: Geneva, Switzerland, 2018; pp. 23–31.
2. Ministry of Water Resources of the PRC. *China Flood and Drought Disaster Bulletin 2017*; Ministry of Water Resources of the PRC: Beijing, China, 2018.
3. Ministry of Water Resources of the PRC. *China Flood and Drought Disaster Bulletin 2018*; Ministry of Water Resources of the PRC: Beijing, China, 2019.
4. Guha-Sapir, D.; Hoyois, P. *Annual Disaster Statistical Review 2016*; CRED: Brussels, Belgium, 2016.
5. Liu, S. *Analysis and Evaluation of Risk of Flood Disasters in Zhengzhou*; Henan University: Zhengzhou, China, 2023.
6. Zhang, H. *Research on the Urban Flood Disaster Carrying Capacity Evaluation Model Based on Remote Sensing Technology—Taking Wuhan City as an Example*; China University of Geosciences: Wuhan, China, 2020.
7. Xu, Z.; Ye, C.; Liao, R. Integrated management technology for urban flooding/waterlogging disaster: Research progress and case study. *Adv. Earth Sci.* **2023**, *38*, 1107–1120. [CrossRef]
8. Zhou, M.; Shang, Z.; Cai, Z. Assessment of Urban Flood Disaster Resiliencein Guangdong Province Based on VIKOR Method. *J. Catastrophology* **2023**, *38*, 206–212. [CrossRef]
9. Hou, J.; Dong, M.; Li, D. Drainage effect of urban drainage-pipe network under extreme rainstorms—Taking Fengxi New City in Xi'an City, China as an example. *J. Earth Sci. Environ.* **2023**, *45*, 427–436.
10. Liu, Y.; Tang, W.; Zhang, W. A review of flood risk analysis based on disaster chain. *Water Resour. Prot.* **2021**, *37*, 8. [CrossRef]
11. Yan, X.; Wang, J.; Fan, L. Research on subway flood disasternfrom the perspective of Resilient city-based on Bow-Tie_Bayesian network model. *J. Catastrophology* **2022**, *37*, 36–43. [CrossRef]
12. Wang, X.; Leng, C. Risk Analysis of Flood Disaster in the Middle Reaches of the Yangtze River. *Sci. Technol. Rev.* **2008**, *26*, 61–66.
13. Li, J.; Gao, J.; Li, N.; Yao, Y.; Jiang, Y. Risk Assessment and Management Method of Urban Flood Disaster. *Water Resour. Manag.* **2023**, *37*, 2001–2018. [CrossRef]
14. Yitnaningtyas, T.B.K.; Hasibuan, H.S.; Tambunan, R.P. Strengthening resilience to flood disaster in Depok urban areas. *IOP Conf. Ser. Earth Environ. Sci.* **2021**, *802*, 012043. [CrossRef]
15. Prashar, N. Urban Flood Resilience: A comprehensive review of assessment methods, tools, and techniques to manage disaster. *Prog. Disaster Sci.* **2023**, *20*, 100299. [CrossRef]
16. Ding, L.; Wang, H. Evolution of deep stormwater storage tunnel projects in the United States and its reference value to China. *China Water Wastewater* **2016**, *32*, 35–41.
17. Ishikawa, T.; Akoh, R. Assessment of flood risk management in lowland Tokyo areas in the seventeenth century by numerical flow simulations. *Environ. Fluid Mech.* **2019**, *19*, 1295–1307. [CrossRef]
18. National Research Council of United States. *Urban Stormwater Management in the United States*; National Academies Press: Washington, DC, USA, 2009.
19. Marco, C.; Antonio, M.-G. Flood Risk Evaluation in Urban Spaces: The Study Case of Tormes River (Salamanca, Spain). *Int. J. Environ. Res. Public Health* **2018**, *16*, 5.
20. Karthikeyan, N.; Gugan, I.; Kavitha, M.S.; Karthik, S. An effective ontology-based query response model for risk assessment in urban flood disaster management. *J. Intell. Fuzzy Syst.* **2023**, *44*, 5163–5178. [CrossRef]
21. Park, K.; Oh, H.; Won, J.-H. Analysis of disaster resilience of urban planning facilities on urban flooding vulnerability. *Environ. Eng. Res.* **2021**, *26*, 76–85. [CrossRef]
22. Wu, M.; Wu, Z.; Ge, W.; Wang, H. Identification of sensitivity indicators of urban rainstorm flood disasters: A case study in China. *J. Hydrol.* **2021**, *599*, 126393. [CrossRef]
23. General Office of the State Council. *Implementation Opinions of the General Office of the State Council on Strengthening Urban Waterlogging Management (State Council Office [2021] No. 11)*; General Office of the State Council: Beijing, China, 2021.
24. Wang, L.; Huang, H. Spatio-temporal Dynamics analysis of 720 flood disaster in Henan Province. *J. Catasrophology* **2022**, *37*, 205–211.
25. The Disaster Investigation Group of the State Council. *Investigation Report on "July 20" Extremely Heavy Rainstorm Disaster in Zhengzhou*; The Disaster Investigation Group of the State Council: Zhengzhou, China, 2022.
26. Zhengzhou Municipal Bureau of Statistics. Statistical Bulletin on National Economic and Social Development of Zhengzhou City in 2021. Available online: https://www.henan.gov.cn (accessed on 15 March 2022).
27. Lv, D.; Lin, C. *Report on the Results of Zhengzhou Urban Geological Survey*; China Geological Survey: Beijing, China, 2021; pp. 37–100.
28. Zhengzhou Water Bureau. *Zhengzhou City Soil and Water Conservation Plan (2016–2030)*; Zhengzhou Water Bureau: Zhengzhou, China, 2017.

29. Liang, W. Research on rainwater utilization and control planning in the central urban area of Zhengzhou. *Urban Roads Bridges Flood Control.* **2011**, 95–97, 242.
30. Ministry of Housing and Urban-Rural Development. *Technical Guidelines for Sponge City Construction—Construction of Low Impact Development Rainwater Systems (Trial)*; Ministry of Housing and Urban-Rural Development: Beijing, China, 2014.
31. Ye, M.; Xue, C. Research on the Method of Calculating Penetration Depth during a Rainfall Event. *Chin. J. Agrometeorol.* **2010**, *A1*, 66–69.
32. Zhengzhou Natural Planning Bureau. *Zhengzhou Sponge City Construction Special Plan (2017–2030)*; Zhengzhou Natural Planning Bureau: Zhengzhou, China, 2016.
33. Zhengzhou Municipal People's Government of the Communist Party of China. *Implementation Plan for Water Ecological Construction in Zhengzhou City in 2020*; Zhengzhou Municipal People's Government of the Communist Party of China: Zhengzhou, China, 2020.
34. Gao, Y.; Zhang, H. Study on the Enhancing the Capacity of Jialu River System to Withstand Rainfall. *Yellow River* **2014**, *36*, 7–12.

Article

Groundwater Vulnerability Assessment and Protection Strategy in the Coastal Area of China: A GIS-Based DRASTIC Model Approach

Qian Zhang [1,2,*], **Qiang Shan** [3,4,*], **Feiwu Chen** [1,2], **Junqiu Liu** [1,2] **and Yingwei Yuan** [1,2]

[1] College of Hydraulic Engineering, Tianjin Agricultural University, Tianjin 300384, China; 2208028105@stu.tjau.edu.cn (F.C.); jqliu86@tjau.edu.cn (J.L.); 2108028115@stu.tjau.edu.cn (Y.Y.)

[2] Joint Tianjin Agricultural University-China Agricultural University Smart Water Conservancy Research Center, Tianjin 300384, China

[3] Hebei Key Laboratory of Geological Resources and Environment Monitoring and Protection, Shijiazhuang 050021, China

[4] Hebei Geo-Environment Monitoring Centre, Shijiazhuang 050021, China

* Correspondence: cathy_zhang@tjau.edu.cn (Q.Z.); 15102533329@163.com (Q.S.)

Abstract: Groundwater vulnerability reflects the risk level of groundwater contamination and its self-repairing ability, as well as its sustainability for use. Therefore, it provides significant scientific support for implementing measures to prevent groundwater contamination, especially in coastal areas. In this study, considering the lithology of vadose in valley plains and the extent of karst subsidence areas, a GIS-based DRASTIC model was employed to assess groundwater vulnerability in Tangshan City, a coastal area in China. The assessment results were presented and mapped using GIS, based on a comprehensive evaluation of seven parameters, including "Depth of groundwater, Vertical net recharge, Aquifer thickness, Soil media, Topography, Impact of vadose zone, and Hydraulic conductivity". The identified groundwater vulnerability zones included the highest, higher, moderate, low vulnerability those four zones, which accounted for 4%, 53%, 25%, and 18%, respectively. In addition, according to the results of field investigation, the karst subsidence area and the mined-out coastal area were directly classified as the highest vulnerable areas and covered 1.463 km^2; more attention is required here in subsequent groundwater protection processes and strategies. Finally, the groundwater pollution index was used to validate the groundwater vulnerability distribution results, and these two were in high agreement, with an R^2 coefficient of 0.961. The study is crucial for the rational utilization and protection of water resources in Tangshan City.

Keywords: groundwater; vulnerability assessment; GIS-based DRASTIC model; protection strategy; coastal areas

Citation: Zhang, Q.; Shan, Q.; Chen, F.; Liu, J.; Yuan, Y. Groundwater Vulnerability Assessment and Protection Strategy in the Coastal Area of China: A GIS-Based DRASTIC Model Approach. *Appl. Sci.* **2023**, *13*, 10781. https://doi.org/10.3390/app131910781

Academic Editor: Dino Musmarra

Received: 29 August 2023
Revised: 24 September 2023
Accepted: 25 September 2023
Published: 28 September 2023

1. Introduction

Water resources are a vital and irreplaceable natural asset for human survival and development. With the rapid progress of national economic construction and the improvement of living standards, the demand for groundwater resources continues to increase, leading to an increasingly prominent imbalance between supply and demand. Simultaneously, the issue of water pollution has become more pronounced due to the continuous development of industry and agriculture, posing a serious threat to human development. Shallow groundwater systems are highly susceptible to changes in precipitation levels since their recharge primarily comes from surface sources [1]. Groundwater, as a crucial water source, has garnered significant attention.

Under recent climate change, groundwater resources have been increasingly exploited and utilized worldwide, resulting in pollutant issues, particularly in highly vulnerable areas [2]. Groundwater extraction, which serves as a widespread and accessible method for obtaining high-quality freshwater, accounts for up to 30% of total global freshwater utilization [3,4]. However, excessive groundwater exploitation often leads to a continuous decline in the water table [5–7], posing risks of land subsidence and collapse, especially in densely populated urban areas such as Tianjin, Jakarta, Konya, Mashhad, and Hanoi [8–12]. Moreover, groundwater exploitation significantly contributes to vulnerability associated with high levels of human activity [13,14]. The exploitation of groundwater and its impact on ecological systems pose challenges to global economic and social development [15,16].

Groundwater vulnerability reflects both the extent of contamination and the ability of groundwater to self-repair and maintain sustainability. As a complex black box system, groundwater vulnerability factors are diverse, and various research methods are available. For example, Chaves et al. [17], Van Dijck et al. [18], and Cui et al. [19] employed Mann–Kendall's Test, Poisson Distribution, End-member mixing analysis, and other relevant models based on land use change and soil infiltration. These studies revealed that groundwater responses to precipitation weaken or even disappear due to the thickened vadose zone and reduced permeability caused by declining groundwater levels and the conversion of natural grasslands to artificial farmland [20]. In a recent study in Nigeria, GLSI and LC models were used to conduct a case study of the Ijero mining site, examining the comparative effect of lateritic shield on groundwater vulnerability [21,22]. Groundwater vulnerability assessment provides significant scientific support for implementing measures to prevent groundwater pollution [23,24]. Vaezihir and Tabarmayeh [25] and Wen et al. [26] selected environmental parameters such as vadose zone impact, hydraulic conductivity, population density, and river recharge, utilizing the DRASTIC model to evaluate aquifer vulnerability.

The DRASTIC model aims to use seven factors such as "Depth to water, Net recharge, Aquifer media, Soil media, Topography, Impact of vadose zone, and Hydraulic Conductivity", and allocates the ratings to all factors according to the range of factors [27], which is simple and widely used in evaluating groundwater contamination vulnerability. However, previous studies tend to assess groundwater vulnerability subjectively and refer only to a small number of factors, or have reported only a weak correlation between water pollutant concentrations (commonly nitrate) and the calculated DRASTIC index [28,29], which provided difficulties in obtaining conclusion accurately. This research integrated DRASTIC with additional methods like the GIS techniques, PCA (Principal Component Analysis), and other statistical approaches, which improve the evaluation systems to be modified to obtain more accurate results based on local hydrogeological conditions [30–32]. Furthermore, regression analysis is also applied in the DRASTIC model which has developed a new system for evaluating groundwater vulnerability, thereby enhancing scientific rigor and accuracy in utilization.

Groundwater vulnerability studies are crucial foundational work for the rational development, utilization, and protection of groundwater resources. To establish timely zoning systems for groundwater pollution prevention and control, it is necessary to evaluate the pollution resistance of groundwater systems. In this study, a GIS-based DRASTIC model was employed to assess groundwater vulnerability, with emphasis on the lithology of the vadose in the valley plain and the extent of karst subsidence areas. Comprehensive evaluation results based on seven parameters were presented and mapped using GIS to reflect the typical characteristics of coastal areas, enabling the efficient formulation of protection strategies. The aim of this study is to construct a set of groundwater vulnerability evaluation index systems and methods in the coastal area, to characterize the spatial distribution of vulnerability, and to propose corresponding protection measures and environmental protection strategies for highly vulnerable areas.

2. Study Area and Data

2.1. Location and Meteorology

Tangshan City is situated in the northeast of Hebei Province, with geographical coordinates ranging from 117°31′ to 119°19′ east longitude and 38°55′ to 40°28′ north latitude. The total land area of Tangshan City is 13,472 km^2, comprising 5131 km^2 (38.09%)

of mountainous terrain and 8341 km^2 (61.91%) of plains. Additionally, it has a sea area spanning 4472 km^2, with a coastline stretching 229.72 km in length. Tangshan experiences a typical warm temperate sub-humid continental monsoon climate characterized by cold and dry winters with northerly winds, and hot and rainy summers with southerly winds. The annual average temperature is 10.6 °C, while the average precipitation from 1956 to 2020 amounts to 608.7 mm.

2.2. Hydrogeology

Based on the occurrence conditions of groundwater and the characteristics of aqueous media in Tangshan, three types of groundwater can be identified: pore water in loose rocks, carbonate karst water, and bedrock rock fissure water. Pore water is primarily found in the piedmont plain, coastal plain, inter-mountain basin, and valley zones of the Yan Mountain region (Figure 1). The main source of recharge for this groundwater type is atmospheric precipitation, and its dynamics closely follow the annual distribution of rainfall. Karst water is predominantly present in the southern foothills of the Yan Mountain, concealed within sloping plains, piedmont areas, and intermountain basins. The degree of karst development varies and is closely associated with the local geological structure and geomorphology. Individual well yields typically range from 40 to 25 m^3/h·m, while spring water flow rates vary between 0.3 and 3.0 L/s. Fissure water is located in the northern hilly area, characterized by a recorded fracture rate of 1.3%. The weathered zone has an approximate depth of 50 m, with some localized fractures exceeding 100 m due to significant geological influences [33].

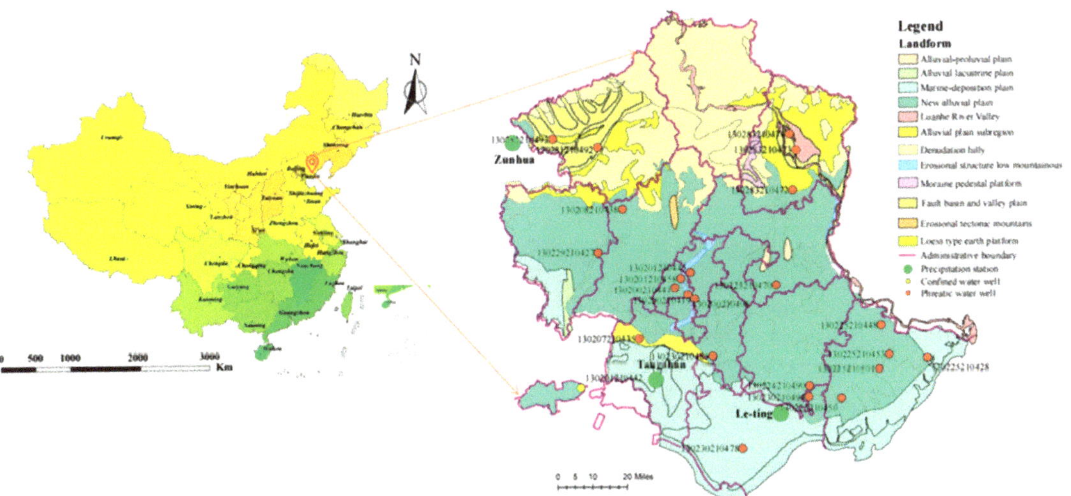

Figure 1. The location of study area and its hydrogeological condition.

2.3. Data Availability

This study is a significant component of the zoning and technical research for groundwater pollution prevention and control in Tangshan City, encompassing two main aspects: the analysis of groundwater dynamics and the evaluation of groundwater vulnerability.

In the investigation of groundwater dynamics, a total of 75 government-monitored wells with long-term observation data were carefully selected to ensure completeness, control, and representativeness. For each district or county level, 1–2 monitoring wells were chosen for analysis, ensuring a uniform distribution across the study area. Specifically, this study focused on analyzing the response of unconfined groundwater, directly recharged by precipitation. Based on these principles, data from 24 monitoring wells were ultimately determined. To complement this analysis, three precipitation stations in Zunhua, Tangshan, and Leting were selected to match similar monitoring wells based on their proximity.

In the evaluation of groundwater vulnerability, data acquisition was conducted through indoor data collection and gathering relevant information about the study area's geology, hydrogeology, and drilling. This included parameters such as groundwater depth, vertical net recharge, aquifer thickness, soil medium, terrain slope, vadose medium type, and permeability coefficient. A total of 1599 boreholes and associated information were gathered from various sources, including survey reports on land pollution from key industries in Hebei Province, China (974 cases), hydrogeology and engineering geology borehole construction and in-situ tests (28 cases), comprehensive geological surveys of major cities and towns (19 cases), the national important geological borehole database service platform (55 cases, https://zk.cgsi.cn/ (accessed on 28 August 2023)), groundwater level surveys (164 cases), and geotechnical investigation reports (359 cases). Aquifer permeability data were obtained by combining the aquifer media type with empirical values from the Hydrogeological Manual [34]. All data collected are accurate, providing strong support for the reliability of this study.

3. Methodology

3.1. Description of the GIS-Based DRASTIC Model Framework

In this study, a GIS-based DRASTIC model was employed to assess the vulnerability of groundwater in Tangshan City, a coastal area in China. The DRASTIC model, initially developed by the US Environmental Protection Agency (EPA) to evaluate groundwater pollution potential across the United States [35], considers intrinsic vulnerability based on hydrogeological characteristics while disregarding the specific contaminants. The use of a GIS-model interface facilitates data input, and the GIS can also serve as a common platform for transferring information between different examples of software. By combining various data sets within the GIS framework, thematic maps can be generated for utilization by the DRASTIC model. Furthermore, the GIS enables the conversion of these maps into raster mode, dividing them into pixels with dimensions of 300 m × 300 m. Each pixel is assigned a numerical value corresponding to the resulting grid, which serves as a basis for the seven parameters of the DRASTIC model, as detailed in the following equation:

$$DRASTIC_{\text{index}} = W_D R_D + W_R R_R + W_A R_A + W_S R_S + W_T R_T + W_I R_I + W_C R_C \quad (1)$$

where, the parameters in Equation (1) include the following: depth of groundwater (D), vertical net recharge of groundwater (R), aquifer thickness of media (A), soil media (S), topography (T), impact of vadose zone (I), and hydraulic conductivity (C).

Consequently, vulnerability is calculated at a high resolution for each individual pixel. The GIS is further utilized to overlay the seven thematic maps, thereby generating a groundwater vulnerability map. The DRASTIC index computation involves assigning weights (W) to each parameter and using the assigned values (R).

3.2. Description of the GIS-Based DRASTIC Model Parameters

The seven parameters in Equation (1) that affect and control groundwater flow and pollutant transport are adopted in DRASTIC to constitute the factor system for vulnerability assessment. Based on the "Delineation of Priority Areas for Groundwater Pollution Prevention and Control Technical Guideline" published by the China Geological Survey in 2022 [36], different parameter weights have been assigned by investigation experience and field experiment. According to the influence of each parameter on groundwater vulnerability, different weight values are assigned to them in Table 1.

Table 1. Description of DRASTIC model parameters and its weights.

Parameter	Description	Relative Weight
Depth of groundwater (D)	Depth to groundwater table is the distance from the surface to the submerged surface; unit is m.	5
Vertical net recharge (R)	Approximate using precipitation infiltration recharge instead of vertical net recharge; unit is mm/a.	4
Aquifer thickness (A)	Aquifer thickness can be analyzed from borehole data; refers to the saturated zone material properties, which controls the pollutant attenuation processes; unit is m.	3
Soil media (S)	The soil media is a weathered layer with a thickness of 2 m or less at the surface, which controls the amount of recharge that can infiltrate downward.	2
Topography (T)	Slope values can be automatically generated in GIS after DEM extraction from 1:50,000 or 1:10,000 topographic maps.	1
Impact of Vadose Zone (I)	The unsaturated zone material: it controls the passage and attenuation of the contaminated material to the saturated zone.	5
Hydraulic conductivity (C)	Indicates the ability of the aquifer to transmit water, and hence determines the rate of flow of contaminant material within the groundwater system.	3

Based on the aforementioned seven parameters, an evaluation level is assigned to each parameter. As the local groundwater depth increases, vertical net recharge decreases, aquifer thickness increases, soil and vadose zone cutoff particles become finer, topographic slope increases, and aquifer permeability coefficient decreases, the scores for each parameter will decrease. A lower score indicates a smaller contribution of that parameter to the groundwater vulnerability assessment index, suggesting a reduced susceptibility to pollution. Conversely, higher scores indicate a greater contribution of the parameter to the groundwater vulnerability assessment index, indicating a higher vulnerability to pollution. The scores for each indicator range from 1 to 10, with specific classifications provided in Table 2.

Table 2. The parameter Ranking and Assignment in DRASTIC.

Parameter	Grade									
	1	2	3	4	5	6	7	8	9	10
D (m)	>30	(25, 30]	(20, 25]	(15, 20]	(10, 15]	(8, 10]	(6, 8]	(4, 6]	(2, 4]	≤2
R (mm/a)	0	(0, 51]	(51, 71]	(71, 92]	(92, 117]	(117, 147]	(147, 178]	(178, 216]	(216, 235]	>235
A (m)	>50	(45, 50]	(40, 45]	(35, 40]	(30, 35]	(25, 30]	(20, 25]	(15, 20]	(10, 15]	≤10
S	rock	clay loam	silt loam	loam	sandy loam	swelling or condensing clay	silt-sand/ fine sand	medium sand/coarse sand	gravel-cobble	thin layer or missing
T(%)	>10	(9, 10]	(8, 9]	(7, 8]	(6, 7]	(5, 6]	(4, 5]	(3, 4]	(2, 3]	≤2
I	clay	loam	sandy loam soil	silt-sand	silty, fine sand	fine sand	medium sand	coarse sand	sand gravel	gravel-cobble
C (m/d)	[0, ≤4]	(4, 12]	(12, 20]	(20, 30]	(30, 35]	(35, 40]	(40, 60]	(60, 80]	(80, 100]	>100

A step wise flow diagram representing the data processing, preparations of layers, and assigning of weights has been demonstrated below in Figure 2.

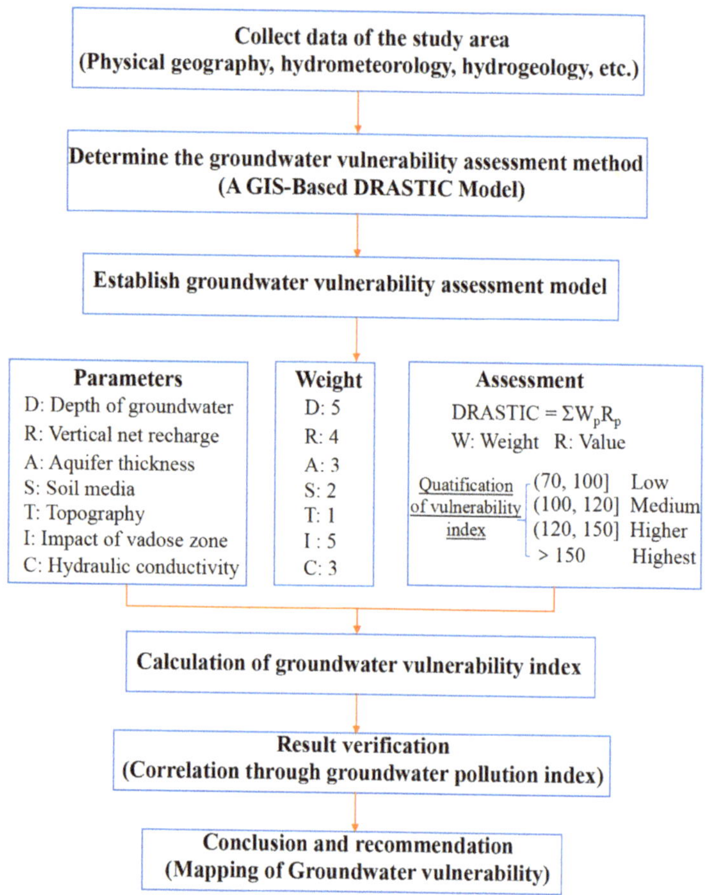

Figure 2. Flowchart of groundwater vulnerability assessment.

4. Results and Discussion

4.1. Results of Single Parameter Evaluation

4.1.1. Depth of Groundwater

A total of 1867 data points were collected for groundwater depth. The depth at which groundwater is located primarily determines the depth of the transmission medium through which pollutants migrate to the aquifer [37]. Generally, a greater groundwater depth requires more time for pollution to reach the aquifer. This results in a smaller amount of pollution entering the aquifer and weaker levels of contamination within the aquifer, with the opposite being true as well.

The method for analyzing the zoning map of buried depths differs between mountainous and plain areas. In the plain and valley regions, water level data collected and supplementary surveys are interpolated and adjusted based on hydrogeological and river conditions. In mountainous areas, characterized by significant topographic variations and complex geological conditions, underground water depths are mainly classified according to aquifer types. The same type of water-bearing medium within an area is assigned an average value based on measured groundwater levels. Additionally, for valleys within mountainous regions where groundwater exists within loose rock formations, score values are determined by interpolating the buried depths observed in plain areas.

Once the above analysis is completed, the mountainous and plain areas are superimposed using GIS, ensuring uniform layer attributes. Gradation mapping (Figure 3a) is then assigned based on Table 2.

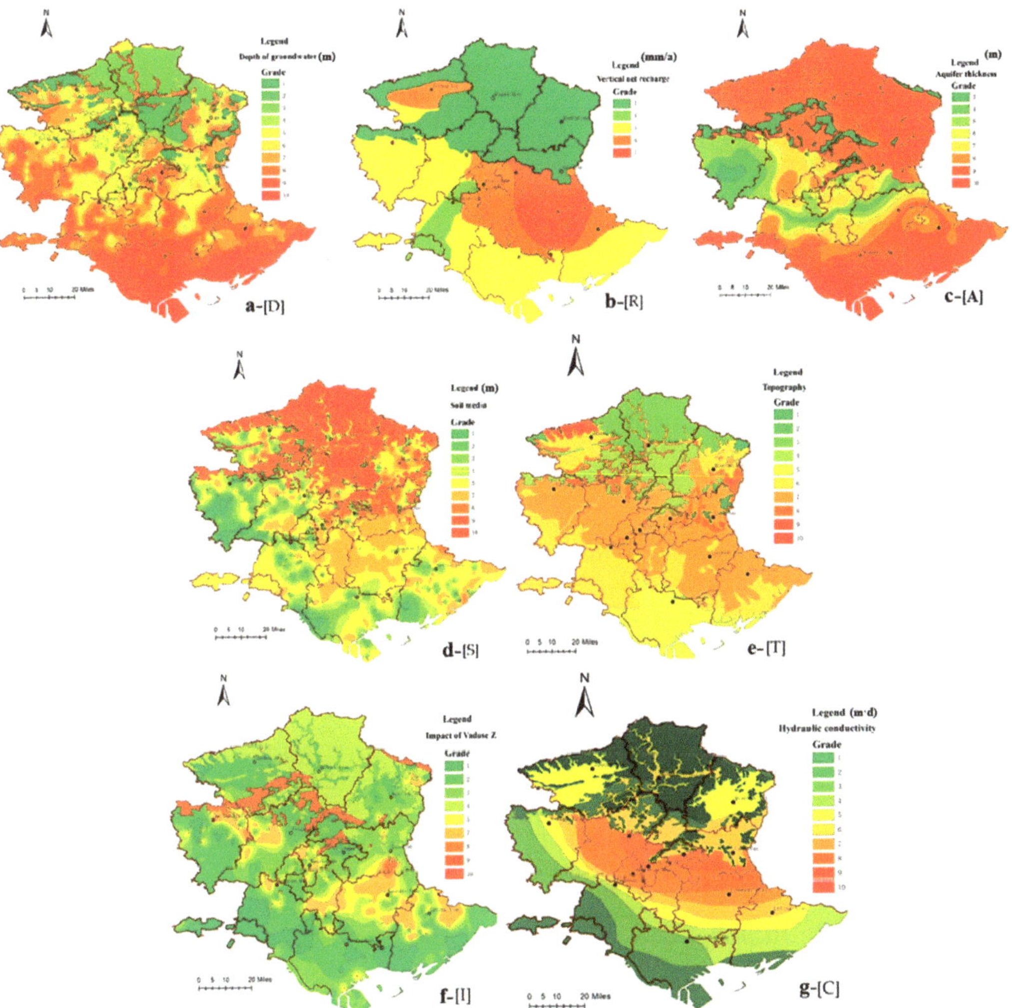

Figure 3. Mapping of DRASTIC parameters ((**a**) Depth of groundwater; (**b**) Vertical net recharge; (**c**) Aquifer thickness; (**d**) Soil media; (**e**) Topography; (**f**) Impact of vadose zone; (**g**) Hydraulic conductivity).

4.1.2. Vertical Net Recharge

Net recharge refers to the amount of water per unit area that infiltrates the surface and reaches the aquifer. Contaminants can enter the aquifer vertically through this recharge water [38]. In the DRASTIC model, net recharge represents the total amount of water applied to the surface and infiltrated into the aquifer. In Tangshan City, atmospheric precipitation serves as the primary source of regional recharge, and therefore, the vertical net recharge can be estimated using precipitation infiltration [39]. Generally, the equation

$R = P \times \alpha$ is used, where R denotes the vertical net recharge, P represents the precipitation infiltration coefficient, and α represents the annual precipitation.

The annual precipitation data for the region were obtained from the 2020 precipitation records of four gauging stations. In 2020, Tangshan City experienced rainfall ranging from 480–720 mm, with higher precipitation observed in the northern mountainous area compared to the southern plain area. Notably, Luannan County in the southern plain area received relatively higher precipitation. In the study area, the values of α range from 0.15–0.2 in the plain area, 0.09 in the igneous and metamorphic rock area, and 0.08 in the carbonate rock area. The empirical values of the precipitation infiltration coefficient were adjusted based on field seepage test results (30 groups) to ensure accuracy. The resulting vertical net recharge is illustrated in Figure 3b.

4.1.3. Aquifer Thickness

Aquifer thickness (Figure 3c) is determined by combining groundwater depth measurements with borehole data [40]. The data sources include the national important geological borehole database service platform (55 points), Geo-technical Investigation Reports (359 points), and borehole data collected during field investigations (30 holes), totaling 444 points.

In the bedrock mountain areas, the thickness of the weathered fissure layer is primarily considered, and it is obtained by subtracting the groundwater depth. The average value is then calculated for different regions. In the metamorphic rock and igneous rock mountain areas of Tangshan City, the thickness of the weathering layer ranges from 40–50 m, with some areas reaching up to 60 m. For the bare leaky karst mountains, the thickness of the strong water-bearing zone identified through borehole data is used to determine the aquifer thickness. Again, the average value is calculated for different regions. For example, in the northern part of Yutian County, Tangshan City, the thickness of the Jixian dolomite strong water-bearing zone is 40–45 m. In the plain area, the shallow aquifer thickness in the coastal region is relatively thin, mostly less than 10 m. This is determined based on calculations using borehole data and groundwater depth measurements.

4.1.4. Soil Media

Soil refers to the biologically active upper layer of the seepage zone. The influence of soil on groundwater vulnerability primarily relies on the physicochemical properties of different soil types [41]. In the study area, there are various soil media, with sandy loam and swelling or condensing clay being widely distributed. In the northern mountainous area of Tangshan City, the surface soil layer is very thin in some areas, while the lower part consists of weathered bedrock crust. This results in predominantly thin or deficient soil media. In the middle region, sandy loam, silty sand, and fine sand are widely distributed. In the coastal area, clay loam (clay), and silty loam are prevalent. The soil media were classified into different grades as shown in Figure 3d.

4.1.5. Topography

Topographic slope plays a significant role in groundwater migration and subsequently affects groundwater vulnerability [42]. In this study, a digital elevation model (DEM) of the research area was acquired through a measurement system. The DEM had an accuracy of 12.5 m, which met the guidelines' requirements. Using the GIS's slope tool, the slope was calculated based on the projected DEM map. Then, according to the scoring standard, each zone's attributes were determined through reclassification (Figure 3e). Due to the high resolution of the data, fragmentation zones were formed considering the impact of rivers and dams in the plain area on slope classification. This evaluation utilized the geomorphic zones of the study area as a reference, and the average terrain slope within each zone was considered the representative slope value.

The study area exhibits significant variations in topographic slope, generally decreasing from north to south. High topographic slope values are concentrated in the northern mountainous area, while the intermountain basin, piedmont plain, and coastal plain exhibit relatively low slopes. The highest topographic slope values are found along the northern edge of Yutian County, the southern part of Zunhua City, the northern part of Fengrun District, the southern part of Qianxi County, the western part of Qian'an City, and the northwestern part of Luanzhou City. Conversely, the lowest slope values are primarily distributed in Caofeidian District along the southern coast.

4.1.6. Impact of Vadose Zone

The unsaturated zone, also known as the vadose zone, refers to the area between the Earth's surface and the water table. The type of vadose zone medium plays a crucial role in controlling water exchange between the soil layer and aquifer, the migration and transformation of pollutants, as well as various physicochemical and biological processes [43]. Additionally, the vadose zone medium determines the length and path of percolation. Therefore, it serves as an important indicator for evaluating groundwater vulnerability.

In the plain area of the study region, the vadose zone mainly consists of Quaternary sediments, which can be categorized into nine types. When assessing the self-purification capacity of the vadose zone, factors such as particle size and permeability of the Quaternary sediments are primarily considered. The assigned values range from 1 to 10. In the mountain aquifers, the vadose zone comprises regolith with varying degrees of metamorphic rocks and igneous rocks, as well as dissolution layers of dolomites and limestone in karst areas. Based on vulnerability assessments conducted in different mountainous regions in China, the medium score for the vadose zone in bedrock regolith is determined to be 4 points.

This determination takes into account the extent of fracturing and dissolution observed in the surface regolith of the Tangshan mountain area, finally showed in Figure 3f. For dolomite or limestone vadose zones, the medium score is set at 8 points.

4.1.7. Hydraulic Conductivity

The permeability of an aquifer is influenced by the type of aquifer medium, and the permeability coefficient partially reflects the lithological composition of the aquifer [44]. Generally, larger particle sizes or more voids in the aquifer medium indicate greater permeability, lower dilution capacity, and higher potential for pollution [45]. The aquifer's permeability coefficient is determined based on the type of aquifer medium, combined with empirical values, and adjusted using data obtained from field exploration and pumping tests (30 groups). The classification of aquifer medium types refers to the data source of aquifer thickness.

In the plain area, the permeability coefficient was obtained by interpolating the collected drilling data using Kriging interpolation with software such as Golden Surfer. For mountain aquifers, the permeability coefficient is classified based on weathered fissure aquifers in metamorphic rocks, weathered fissure aquifers in igneous rocks, and karst fissure aquifers. The assignment of permeability coefficients is performed according to the aquifer types derived from pumping test results from hydrologic boreholes. GIS technology is utilized to overlay the mountain and plain areas, harmonize layer attributes, and assign values to each partition following evaluation guidelines.

In the northern mountainous area of Tangshan City, the permeability coefficient mostly ranges from 0.5–5 m/d, with some areas reaching up to 10 m/d. In the southern coastal area, where silt is the predominant aquifer type, the permeability coefficient is relatively low, mostly below 12 m/d. The Shanqian area in central Tangshan exhibits higher permeability coefficients, with the lowest values observed in the north of Zunhua City, Qianxi County, and the northwest of Qian'an City. Conversely, the highest values are found in the east of Fengrun District, the south of Kaiping District, the south of Guye District, and the south of Luanzhou City. The assignment result for hydraulic conductivity in the DRASTIC parameter is presented in Figure 3g.

4.2. Results of Comprehensive Vulnerability Assessment and Its Validation

4.2.1. Results of Groundwater Vulnerability

Vulnerability mapping was conducted using the Geographic Information System (GIS) based on hydrogeological data of the study area and with reference to the DRASTIC model, as shown in Figure 4. The groundwater vulnerability index values for Tangshan City range from 85 to 173, classified into four levels according to the DRASTIC classification principle (refer to Table 3). Higher vulnerability assessment index values indicate a higher susceptibility to pollution, while lower values suggest a lower vulnerability and reduced susceptibility to pollution. Additionally, karst collapse areas and regions affected by mining subsidence are directly classified as the highest vulnerability areas.

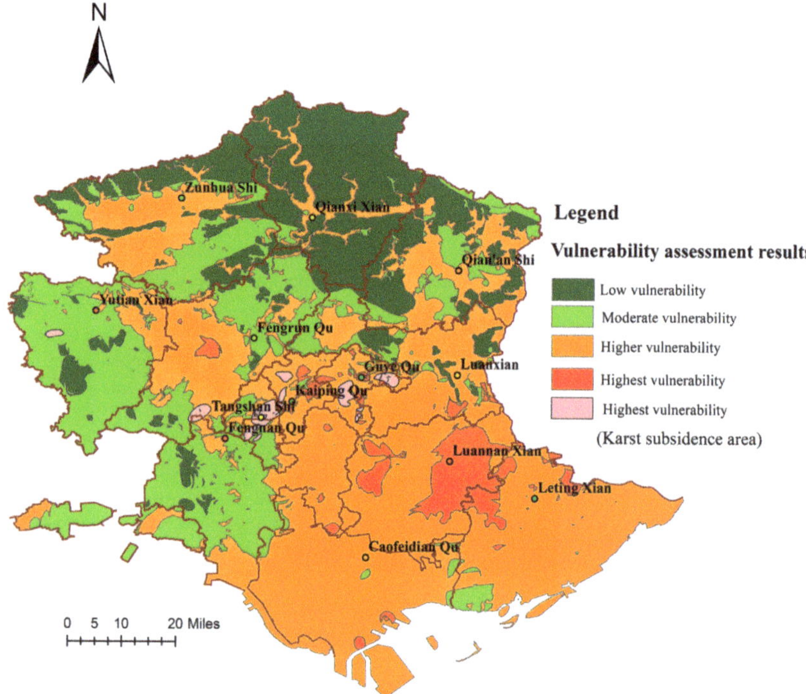

Figure 4. Comprehensive evaluation of vulnerability of groundwater in Tangshan.

Table 3. Classification of vulnerability of shallow groundwater in Tangshan City.

Groundwater Vulnerability Composite Index Value	Vulnerability	Level
(70, 100]	Potentially contaminated	Low
(100, 120]	Easily contaminated	Moderate
(120, 150]	highly prone to be contaminated	Higher
>150	Particularly vulnerable to be contaminated	Highest

The vulnerability assessment map of shallow groundwater in Tangshan City reveals the following distribution results. Firstly, the shallow groundwater in Tangshan City exhibits high vulnerability, with the highest vulnerability area accounting for 4% of the total study area, the higher vulnerability area accounting for 53%, and the moderate vulnerability area accounting for 25%. Together, these three categories account for 82% of the area, while the low vulnerability area only accounts for 18%. This highlights the weak protection performance of shallow groundwater in the region, making it susceptible to pollution.

Secondly, the spatial variability and causes of groundwater vulnerability in the study area are complex. The area around Luannan County has relatively shallow groundwater depth, significant vertical recharge, and a thick aquifer, leading to its classification as an area with high underground vulnerability. The valley area in the study region is also classified as a high vulnerability area due to its homogeneous lithology, simple rock structure, and strong permeability. These areas experience frequent agricultural activities and have well-developed aquaculture, resulting in a high pollution load and easy pollution of groundwater [46]. Similarly, highly vulnerable areas are also found in the central and southeastern regions of Tangshan, characterized by high vertical net recharge, relatively high permeability coefficients, good water permeability, and limited self-purification capacity. In addition, the karst subsidence area and the mined-out coastal area were directly classified as the highest vulnerable areas, covering 1.463 km^2. These factors contribute to weak anti-pollution capabilities of groundwater, making it prone to pollution. Additionally, these areas are marked by intense human activities, including urban industrial and domestic sewage discharge from Tangshan City, as well as excessive fertilization and sewage irrigation in agricultural regions. Therefore, they should be prioritized as key areas for groundwater protection.

Thirdly, low-vulnerability areas are scarce and mainly concentrated in the northern part of Tangshan, such as the northern edge of Zunhua City, Qianxi County, and the northwest of Qian'an City. In these areas, groundwater depth is significant, the vadose zone exhibits fine lithology, and recharge amounts are limited, reducing vulnerability to external pollutants.

4.2.2. Validation for the Mapping of Groundwater Vulnerability

Groundwater with high vulnerability is more easily contaminated and thus more likely to have poorer water quality conditions. Currently, there are fewer methods to validate the vulnerability of groundwater, and the pollution distribution of "NO_3^-, NO_2^-, NH_4^+" is generally used to compare with the vulnerability distribution of groundwater [47,48]. Xu et al. utilized single and combined factors to evaluate the water quality of groundwater, and then examined the correlation between the evaluation level and the vulnerability index [49]. Chao et al. chose typical water quality indicators characterizing groundwater taste, color, and scaling, which were selected to calculate the value of the groundwater pollution index according to the "Groundwater Quality Evaluation Standards" [50], and then compared them with the spatial distribution of groundwater vulnerability. Since this study also addresses the vulnerability of shallow groundwater, and there are similarities with the above studies, similar validation methods were selected.

We selected 11 water quality indicators (in Table 4) from nearly 30 well sites to calculate the Groundwater Pollution Index (GPI), which summed every index up after assigned number based on different categorization criteria in Table 4. The GPIs, ranging from 26 to 72, were used to validate the groundwater vulnerability and mapped in Figure 5, which demonstrates that high groundwater vulnerability corresponded to a high groundwater pollution index. These two were in high agreement, with a R^2 coefficient of 0.961, as shown in Figure 6. This certificated that the GIS-based DRASTIC model for groundwater vulnerability assessment was scientific and reliable, and could provide reliable experience for groundwater pollution prevention and management.

Table 4. Assignment criteria of groundwater pollution index.

Assigned Index	pH	TH mg/L	TDS mg/L	F⁻ mg/L	Cl⁻ mg/L	SO₄²⁻ mg/L	NO₃⁻ mg/L	NO₂⁻ mg/L	Fe mg/L	Cu mg/L	Mn mg/L
1	6.5–8	150	300	0.2	50	50	2	0.005	0.1	0.01	0.01
3	6–6.5; 8–8.5	300	500	0.5	150	150	5	0.01	0.2	0.05	0.05
5	5.5–6.0; 8.5–9	450	1000	1	250	250	20	1	0.3	1	0.1
7	1–5.5; 9–13	650	2000	2	350	350	30	4.8	2	1.5	1.5
10	0–1; 13–14	>650	>2000	>2	>350	>350	>30	>4.8	>2	>1.5	>1.5

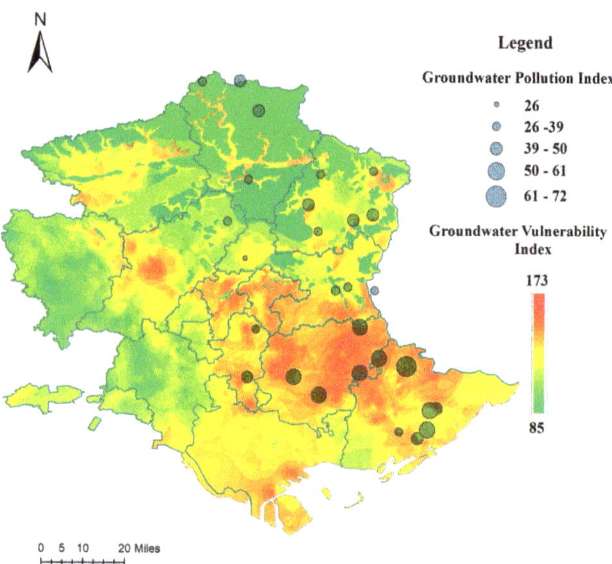

Figure 5. The mapping of groundwater vulnerability index and groundwater pollution index.

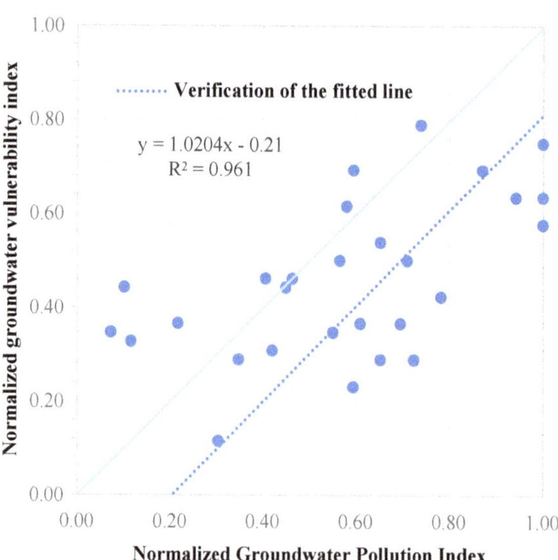

Figure 6. Groundwater vulnerability assessment and verification index correlation.

4.3. Groundwater Protection Strategy in Coastal Area

4.3.1. Groundwater Protection Objectives Based on Groundwater Vulnerability

According to the results of the vulnerability assessment of shallow groundwater in Tangshan City, the following medium- and long-term objectives for groundwater protection have been formulated in view of the current water quality and water environment.

1. To guarantee water safety and prevent groundwater pollution. Groundwater is one of the important sources of drinking water, and protecting groundwater means guaranteeing people's drinking water safety and basic usage of water.
2. To maintain ecological balance on this basis. Protecting groundwater can maintain the water supply of wetlands, keep the stability of the ecological environment, and promote the maintenance of species diversity and the balance of the natural ecosystem.
3. To pay attention to the protection of water sources for agricultural irrigation to ensure the sustainability of agricultural irrigation and maintain the normal growth of crops, paying particular attention to reducing the adverse effects on soil salinization in the coastal area.
4. To prevent the decline in groundwater level and ground subsidence. Over-exploitation of groundwater will lead to the decline in groundwater level and ground subsidence, which will in turn lead to geological disasters and ground subsidence. One of the objectives of groundwater protection is to avoid these problems and maintain the stability of the groundwater system.

To realize the sustainable management and utilization of groundwater and to support the sound economic development of Tangshan City, groundwater resources are non-renewable resources, and the protection and rational utilization of groundwater is an important part of realizing the sustainable development of the region, which is also an important goal of this study.

4.3.2. Groundwater Protect Measures in Study Area

Based on the results of vulnerability assessment of groundwater in Tangshan City and the above protection objectives, we have clarified that the protection performance of shallow groundwater is weak, and the current status of shallow groundwater pollution is serious, so we propose the following protection measures for the current situation characteristics.

1. Reasonable land use and control of new highly polluting enterprises to reduce groundwater pollution: Urban planning, especially industrial zoning, should consider vulnerable areas and restrict the establishment of industrial zones in areas vulnerable to groundwater pollution, such as the center of Tangshan City, Kaiping District, Guye District, and Luannan County. In addition, wastewater discharge from industrial zones must be strictly managed to ensure that standards are met and groundwater pollution is controlled [51]. By controlling the sources of pollution, restricting the discharge of hazardous substances, and rationally managing groundwater extraction, it is possible to ensure that the quality of groundwater meets safety standards.
2. Balanced control of groundwater exploitation: Over-exploitation of groundwater will lead to increased water hardness, groundwater pollution, and even ground subsidence. At this stage, deep wells with serious over-exploitation should be closed, the amount of groundwater extraction should be limited, and the number of new wells should be controlled.
3. Regulate sewage irrigation and fertilizer use: It is important to control the amount of fertilizers used in agricultural areas and promote their effective use in order to minimize the amount that enters groundwater. Additionally, effluent used for irrigation should be analyzed and tested to ensure that it does not adversely affect soil and groundwater quality [52].
4. The establishment of a comprehensive groundwater monitoring network and early warning system throughout the region is essential. Regular water quality monitoring should be carried out to detect early warning signs in time for timely interventions

to mitigate serious consequences and to report data to ecological and environmental authorities.

4.3.3. Rigorous Implementation Plans for Groundwater Exploitation in High Vulnerability Area

Based on the above groundwater protection objectives and implementation methods, this study suggests that the Tangshan area should follow the following most stringent groundwater exploitation plan to ensure the sustainable development of groundwater resources.

(1). Key groundwater pollution sources such as chemical and metal products industrial clusters, landfills, hazardous waste disposal sites, and other key groundwater pollution sources should be investigated and evaluated as soon as possible to map out the status of groundwater pollution, establish a system of seepage and leakage prevention for key groundwater pollution sources, carry out seepage and leakage prevention inspections every year, formulate seepage prevention and renovation plans for seepage prevention measures that do not satisfy the corresponding seepage prevention specification requirements, and take technical and management measures in a timely manner to eliminate hidden dangers.

(2). For areas where groundwater pollution exists, detailed investigation and assessment of groundwater shall be further carried out, and where groundwater treatment and remediation or risk control is required after detailed investigation and assessment, control shall be strengthened and appropriate management shall be carried out in a timely manner. In areas where the health risk of groundwater pollution is unacceptable, the use of groundwater should be prohibited, and drainage methods such as pits and ponds should be restricted to reduce the disturbance of the polluted area; if drainage is really needed, it should be discharged in compliance with the standard after treatment.

(3). Penalties should be established for enterprises and individuals who do not comply with the rules on groundwater protection, and they should be ordered to rectify the situation thoroughly before continuing production.

(4). In areas where karst is strongly developed and where there are many fallout holes and karst funnels, construction projects that may cause groundwater pollution shall not be newly built, altered, or expanded.

5. Conclusions

Being a typical coastal city, Tangshan City exhibits a high level of vulnerability in its shallow groundwater system. The highly vulnerable area accounts for 4% of the total study area, while the areas classified as moderately and highly vulnerable account for 25% and 53%, respectively. Collectively, these three categories encompass 82% of the study area, leaving only 18% categorized as having low or minimal vulnerability. The groundwater pollution index was used to validate the groundwater vulnerability distribution results, and these two were in high agreement, with an R^2 coefficient of 0.961.

The study area, characterized by complex hydrogeological conditions and high vulnerability, is situated along the coast. The valley regions, with their homogeneous lithology, simple rock structure, and high permeability, are particularly vulnerable according to the assessment results. Additionally, the karst subsidence area and the mined-out coastal area were directly classified as the highest vulnerable areas, covering 1.463 km^2, where more attention is required in subsequent groundwater protection processes and strategies.

Based on the vulnerability assessment outcomes for shallow groundwater in Tangshan City, it is evident that the current state of shallow groundwater pollution is severe, indicating a weak performance in safeguarding the quality of shallow groundwater resources. Therefore, it is recommended that relevant departments address these concerns by taking appropriate actions informed by the evaluation results. These actions should focus on improving and developing groundwater resources in a rational manner, implementing effective measures to control groundwater pollution, and undertaking remediation efforts to restore polluted groundwater.

Author Contributions: Q.Z.: Writing—Original Draft; Q.S.: Writing—Review and Editing; F.C.: Drawing Pictures; J.L.: Writing—Review and Editing; Y.Y.: Data. All authors have read and agreed to the published version of the manuscript.

Funding: This study was supported by [the National Natural Science Foundation of China] (NO. 41907149), [Open Funding from Hebei Key Laboratory of Geological Resources and Environment Monitoring and Protection] (NO. JCYKT202210), [the China Postdoctoral Science Foundation] (NO. 2018M631732), and [Tianjin Graduate Research Innovation Project (No. 2022SKY195)].

Institutional Review Board Statement: Not applicable.

Informed Consent Statement: Not applicable.

Data Availability Statement: All the data in this manuscript are derived from the field surveys, reports and references.

Conflicts of Interest: The authors declare no conflict of interest.

References

1. Singh, S.K.; Taylor, R.W.; Rahman, M.M.; Pradhan, B. Developing robust arsenic awareness prediction models using machine learning algorithms. *J. Environ. Manag.* **2018**, *211*, 125–137. [CrossRef]
2. Houria, B.; Mahdi, K.; Zohra, T.F. Hydrochemical characterisation of groundwater quality: Merdja plain (Tebessa town, Algeria). *Civ. Eng. J.* **2020**, *6*, 318–325. [CrossRef]
3. Döll, P.; Hoffmann-Dobrev, H.; Portmann, F.; Siebert, S.; Eicker, A.; Rodell, M.; Strassberg, G.; Scanlon, B. Impact of water withdrawals from groundwater and surface water on continental water storage variations. *J. Geodyn.* **2012**, *59*, 143–156. [CrossRef]
4. Giordano, M. Global Groundwater? Issues and solutions. *Annu. Rev. Environ. Resour.* **2009**, *34*, 153–178. [CrossRef]
5. Castellazzi, P.; Martel, R.; Galloway, D.L.; Longuevergne, L.; Rivera, A. Assessing groundwater depletion and dynamics using GRACE and InSAR: Potential and limitations. *Groundwater* **2016**, *54*, 768–780. [CrossRef] [PubMed]
6. Çelik, R. Temporal changes in the groundwater level in the Upper Tigris Basin, Turkey, determined by a GIS technique. *J. Afr. Earth Sci.* **2015**, *107*, 134–143. [CrossRef]
7. Narany, T.S.; Aris, A.Z.; Sefie, A.; Keesstra, S. Detecting and predicting the impact of land use changes on groundwater quality, a case study in Northern Kelantan, Malaysia. *Sci. Total Environ.* **2017**, *599*, 844–853. [CrossRef]
8. Bui, L.K.; Le, P.V.V.; Dao, P.D.; Long, N.Q.; Pham, H.V.; Tran, H.H.; Xie, L. Recent land deformation detected by Sentinel-1A InSAR data (2016–2020) over Hanoi, Vietnam, and the relationship with groundwater level change. *GISci. Remote Sens.* **2021**, *58*, 161–179. [CrossRef]
9. Chaussard, E.; Amelung, F.; Abidin, H.; Hong, S.-H. Sinking cities in Indonesia: ALOS PALSAR detects rapid subsidence due to groundwater and gas extraction. *Remote Sens. Environ.* **2013**, *128*, 150–161. [CrossRef]
10. Khorrami, M.; Abrishami, S.; Maghsoudi, Y.; Alizadeh, B.; Perissin, D. Extreme subsidence in a populated city (Mashhad) detected by PSInSAR considering groundwater withdrawal and geotechnical properties. *Sci. Rep.* **2020**, *10*, 11357. [CrossRef]
11. Orhan, O. Monitoring of land subsidence due to excessive groundwater extraction using small baseline subset technique in Konya, Turkey. *Environ. Monit. Assess.* **2021**, *193*, 174. [CrossRef] [PubMed]
12. Tang, W.; Zhan, W.; Jin, B.; Motagh, M.; Xu, Y. Spatial Variability of Relative Sea-Level Rise in Tianjin, China: Insight from InSAR, GPS, and Tide-Gauge Observations. *IEEE J. Sel. Top. Appl. Earth Obs. Remote Sens.* **2021**, *14*, 2621–2633. [CrossRef]
13. Putranto, T.T. Determining the groundwater vulnerability using the aquifer vulnerability index (AVI) in the Salatiga groundwater basin in Indonesia. *AIP Conf. Proc.* **2018**, *2021*, 030016. [CrossRef]
14. Sarkar, M.; Pal, S.C. Application of DRASTIC and modified DRASTIC models for modeling groundwater vulnerability of Malda District in West Bengal. *J. Indian Soc. Remote Sens.* **2021**, *49*, 1201–1219. [CrossRef]
15. Konikow, L.F.; Kendy, E. Groundwater depletion: A global problem. *Hydrogeol. J.* **2005**, *13*, 317–320. [CrossRef]
16. Kumar, V.; Setia, R.; Pandita, S.; Singh, S.; Mitran, T. Assessment of U and As in groundwater of India: A meta-analysis. *Chemosphere* **2022**, *303*, 135199. [CrossRef]
17. Chaves, J.; Neill, C.; Germer, S.; Neto, S.G.; Krusche, A.; Elsenbeer, H. Land management impacts on runoff sources in small Amazon watersheds. *Hydrol. Process.* **2008**, *22*, 1766–1775. [CrossRef]
18. Van Dijck, S.J.; Laouina, A.; Carvalho, A.V.; Loos, S.; Schipper, A.M.; Van der Kwast, H.; Nafaa, R.; Antari, M.; Rocha, A.; Borrego, C. Desertification in northern Morocco due to effects of climate change on groundwater recharge. In *Desertification in the Mediterranean Region. A Security Issue*; Springer: Berlin/Heidelberg, Germany, 2006; pp. 549–577. [CrossRef]
19. Cui, Y.; Liao, Z.; Wei, Y.; Xu, X.; Song, Y.; Liu, H. The Response of Groundwater Level to Climate Change and Human Activities in Baotou City, China. *Water* **2020**, *12*, 1078. [CrossRef]

20. Mohammed, O.A.; Sayl, K.N. A GIS-based multicriteria decision for groundwater potential zone in the west desert of Iraq. In *IOP Conference Series: Earth and Environmental Science*; IOP Publishing: Bristol, UK, 2021; Volume 856, p. 012049. [CrossRef]

21. Mohammed, O.A.; Sayl, K.N. Determination of groundwater potential zone in arid and semi-arid regions: A review. In Proceedings of the 2020 13th International Conference on Developments in eSystems Engineering (DeSE), Virtual, 14–17 December 2020; IEEE: Piscataway, NJ, USA, 2020; pp. 76–81. [CrossRef]

22. Falade, A.O.; Oni, T.E.; Oyeneyin, A. Comparative effect of lateritic shield in groundwater vulnerability assessment using GLSI and LC models: A case study of Ijero mining site, Ijero-Ekiti. *Model. Earth Syst. Environ.* **2023**, *9*, 3253–3262. [CrossRef]

23. Pavlis, M.; Cummins, E. Assessing the vulnerability of groundwater to pollution in Ireland based on the COST-620 Pan-European approach. *J. Environ. Manag.* **2014**, *133*, 162–173. [CrossRef]

24. Prasad, R.K.; Singh, V.S.; Krishnamacharyulu, S.K.G.; Banerjee, P. Application of drastic model and GIS: For assessing vulnerability in hard rock granitic aquifer. *Environ. Monit. Assess.* **2011**, *176*, 143–155. [CrossRef] [PubMed]

25. Vaezihir, A.; Tabarmayeh, M. Total vulnerability estimation for the Tabriz aquifer (Iran) by combining a new model with DRASTIC. *Environ. Earth Sci.* **2015**, *74*, 2949–2965. [CrossRef]

26. Wen, X.; Wu, J.; Si, J. A GIS-based DRASTIC model for assessing shallow groundwater vulnerability in the Zhangye Basin, northwestern China. *Environ. Geol.* **2009**, *57*, 1435–1442. [CrossRef]

27. Aller, L.; Thornhill, J. *DRASTIC: A Standardized System for Evaluating Ground Water Pollution Potential Using Hydrogeologic Settings*; Robert, S., Ed.; Kerr Environmental Research Laboratory, Office of Research and Development, US Environmental Protection Agency: Washington, DC, USA, 1987.

28. Panagopoulos, G.P.; Antonakos, A.K.; Lambrakis, N.J. Optimization of the DRASTIC method for groundwater vulnerability assessment via the use of simple statistical methods and GIS. *Hydrogeol. J.* **2006**, *14*, 894–911. [CrossRef]

29. Kwon, E.; Park, J.; Park, W.-B.; Kang, B.-R.; Hyeon, B.-S.; Woo, N.C. Nitrate vulnerability of groundwater in Jeju Volcanic Island, Korea. *Sci. Total Environ.* **2022**, *807*, 151399. [CrossRef]

30. Rama, F.; Busico, G.; Arumi, J.L.; Kazakis, N.; Colombani, N.; Marfella, L.; Hirata, R.; Kruse, E.E.; Sweeney, P.; Mastrocicco, M. Assessment of intrinsic aquifer vulnerability at continental scale through a critical application of the drastic framework: The case of South America. *Sci. Total Environ.* **2022**, *823*, 153748. [CrossRef]

31. An, Y.; Lu, W. Assessment of groundwater quality and groundwater vulnerability in the northern Ordos Cretaceous Basin, China. *Arab. J. Geosci.* **2018**, *11*, 118. [CrossRef]

32. Bai, L.; Wang, Y.; Meng, F. Application of DRASTIC and extension theory in the groundwater vulnerability evaluation. *Water Environ. J.* **2012**, *26*, 381–391. [CrossRef]

33. Chen, S.M.; Liu, F.T.; Zhang, Z.; Zhang, Q.; Wang, W. Changes of groundwater flow field of Luanhe River Delta under the human activities and its impact on the ecological environment in the past 30 years. *China Geol.* **2021**, *4*, 455–462. [CrossRef]

34. Hydrogeological Manual, China Geological Survey. 2023. Available online: https://kns.cnki.net/KCMS/detail/detail.aspx?dbname=SNAD&filename=SNAD000001542056 (accessed on 28 August 2023).

35. US EPA (Environmental Protection Agency). *DRASTIC: A Standard System for Evaluating Groundwater Potential Using Hydrogeological Settings*; Oklahoma WA/EPA Series, Ada; US EPA (Environmental Protection Agency): Washington, DC, USA, 1985; p. 163.

36. China Geological Survey. *Delineation of Priority Areas for Groundwater Pollution Prevention and Control Technical Guideline*; China Geological Survey: Beijing, China, 2022.

37. Yu, H.; Wu, Q.; Zeng, Y.; Zheng, L.; Xu, L.; Liu, S.; Wang, D. Integrated variable weight model and improved DRASTIC model for groundwater vulnerability assessment in a shallow porous aquifer. *J. Hydrol.* **2022**, *608*, 127538. [CrossRef]

38. Abu-Bakr, H.A.E.-A. Groundwater vulnerability assessment in different types of aquifers. *Agric. Water Manag.* **2020**, *240*, 106275. [CrossRef]

39. Hayashi, M.; Farrow, C.R. Watershed-scale response of groundwater recharge to inter-annual and inter-decadal variability in precipitation (Alberta, Canada). *Hydrogeol. J.* **2014**, *22*, 1825–1839. [CrossRef]

40. Nair, A.M.; Prasad, K.R.; Srinivas, R. Groundwater vulnerability assessment of an urban coastal phreatic aquifer in India using GIS-based DRASTIC model. *Groundw. Sustain. Dev.* **2022**, *19*, 100810. [CrossRef]

41. Taghavi, N.; Niven, R.K.; Kramer, M.; Paull, D.J. Comparison of DRASTIC and DRASTICL groundwater vulnerability assessments of the Burdekin Basin, Queensland, Australia. *Sci. Total Environ.* **2023**, *858*, 159945. [CrossRef]

42. Rodriguez-Galiano, V.; Mendes, M.P.; Garcia-Soldado, M.J.; Chica-Olmo, M.; Ribeiro, L. Predictive modeling of groundwater nitrate pollution using Random Forest and multisource variables related to intrinsic and specific vulnerability: A case study in an agricultural setting (Southern Spain). *Sci. Total Environ.* **2014**, *476*, 189–206. [CrossRef]

43. Goyal, D.; Haritash, A.; Singh, S. A comprehensive review of groundwater vulnerability assessment using index-based, modelling, and coupling methods. *J. Environ. Manag.* **2021**, *296*, 113161. [CrossRef]

44. Khosravi, K.; Sartaj, M.; Tsai, F.T.-C.; Singh, V.P.; Kazakis, N.; Melesse, A.M.; Prakash, I.; Bui, D.T.; Pham, B.T. A comparison study of DRASTIC methods with various objective methods for groundwater vulnerability assessment. *Sci. Total Environ.* **2018**, *642*, 1032–1049. [CrossRef]

45. Jiang, W.; Sheng, Y.; Wang, G.; Shi, Z.; Liu, F.; Zhang, J.; Chen, D. Cl, Br, B, Li, and noble gases isotopes to study the origin and evolution of deep groundwater in sedimentary basins: A review. *Environ. Chem. Lett.* **2022**, *20*, 1497–1528. [CrossRef]

46. Jin, G.; Shimizu, Y.; Onodera, S.; Saito, M.; Matsumori, K. Evaluation of drought impact on groundwater recharge rate using SWAT and Hydrus models on an agricultural island in western Japan. *Proc. Int. Assoc. Hydrol. Sci.* **2015**, *371*, 143–148. [CrossRef]
47. Zhuang, Y.; Tao, W.; Jun, L. Evaluation of special vulnerability of groundwater in Guangzhou based on fuzzy comprehensive judgment. *Mod. Geol.* **2011**, *25*, 796–801. (In Chinese)
48. Hua, J.; Ke, W.; Ying, Q. Special vulnerability of groundwater in the Guanzhong Basin and its evaluation. *J. Jilin Univ.* **2009**, *39*, 1106–1116. (In Chinese)
49. Yuan, X.; Yang, Y.; Lu, L. Modeling and validation of groundwater pollution prevention performance zoning in reclaimed water irrigation areas. *J. Agric. Eng.* **2010**, *26*, 57–63. (In Chinese)
50. *GB14848-2017*; China Environmental Quality Standards for Groundwater. PRC State Administration of Quality Supervision and Quarantine: Beijing, China, 2017.
51. Balacco, G.; Alfio, M.R.; Fidelibus, M.D. Groundwater Drought Analysis under Data Scarcity: The Case of the Salento Aquifer (Italy). *Sustainability* **2022**, *14*, 707. [CrossRef]
52. Kruseman, G.P.; De Ridder, N.A.; Verweij, J.M. *Analysis and Evaluation of Pumping Test Data*; International Institute for land Reclamation and Improvement: Wageningen, The Netherlands, 1983; Volume 11, p. 200. Available online: https://www.researchgate.net/publication/284969758 (accessed on 28 August 2023).

Article

Research on Suitability Evaluation of Urban Engineering Construction Based on Entropy Weight Hierarchy-Cloud Model: A Case Study in Xiongan New Area, China

Yi-Hang Gao [1,2,3,4], Bo Han [1,2,3,4], Jin-Jie Miao [1,2,3,4], Shuang Jin [5,*] and Hong-Wei Liu [1,2,3,4,*]

[1] Tianjin Center, China Geological Survey, Tianjin 300170, China; gaoyihang1988@163.com (Y.-H.G.); hanbo1984@126.com (B.H.); tjmiaojj@163.com (J.-J.M.)
[2] North China Center of Geoscience Innovation, Tianjin 300170, China
[3] Xiongan Urban Geological Research Center, China Geological Survey, Tianjin 300170, China
[4] Tianjin Key Laboratory of Coast Geological Processes and Environmental Safety, Tianjin 300170, China
[5] Fifth Geological Brigade, Hebei Bureau of Geology and Mineral Resources, Tangshan 063000, China
* Correspondence: m18133550867@163.com (S.J.); liuhenry022@163.com (H.-W.L.)

Abstract: The development of Xiongan New Area in Hebei Province, China, as a significant national choice, has considerable strategic significance for the integrated growth of Beijing, Tianjin, and Hebei. This paper proposes a cloud model for the suitability evaluation of the construction of Xiongan New Area based on entropy weight analysis, taking into account the geological conditions, groundwater environment, environmental geological problems, and other factors of the suitability of image city development. According to the research, the suitability evaluation findings for the project building employing the cloud model are in strong accord with those of the traditional model and have some application potential. The evaluation's findings indicate that the project construction in Xiongan New Area is acceptable, with suitable and relatively suitable sites making up 81.4% of the total area and excellent circumstances for project development, construction, and usage. This study offers helpful direction for Xiongan New Area's urban land-space design and serves as a useful point of comparison for studies looking at the viability of other deep Quaternary Plain region engineering buildings.

Keywords: Xiongan New Area; entropy weight–analytic hierarchy process; cloud model; suitability evaluation

Citation: Gao, Y.-H.; Han, B.; Miao, J.-J.; Jin, S.; Liu, H.-W. Research on Suitability Evaluation of Urban Engineering Construction Based on Entropy Weight Hierarchy-Cloud Model: A Case Study in Xiongan New Area, China. *Appl. Sci.* **2023**, *13*, 10655. https://doi.org/10.3390/app131910655

Academic Editor: Dino Musmarra

Received: 3 August 2023
Revised: 15 September 2023
Accepted: 20 September 2023
Published: 25 September 2023

1. Introduction

Urban environmental geological issues have gradually garnered attention as China's urbanization process has accelerated in recent years. Urban development is reliant on the geological environment, which has a direct impact on city planning and development. At the same time, urban development responds to the geological environment. Numerous engineering, geological, and environmental issues with urban development include surface deformation, groundwater contamination, the instability of the rock surrounding subsurface caves, harm to neighboring structures, and the degradation of the ecological environment. Therefore, it is crucial to use the geological environment in urban design and construction in a scientifically sound manner in order to change it and make it compatible with the geological environment [1]. The development and construction of Xiongan New Area as a new ecological metropolis will inevitably face constraints from engineering geology and environmental geology. The suitability study of urban construction can mitigate or even prevent various geological problems caused by engineering construction. It can also facilitate and maintain the coordination between the urban environment and development to the greatest extent possible, based on a comprehensive understanding of the engineering geology and environmental geological conditions in the new area [2–7].

Although there is currently no scientifically unified understanding of the application of engineering construction suitability evaluation methods among academics at home and

abroad, the mathematical models and analytical theories of various widely used methods can be roughly categorized into five main categories. Through the use of the spatial data superposition analysis method in geographic information systems (GIS), the artificial neural network method, and the analytic hierarchy process, Sterling [8] developed an assessment system to evaluate the suitability of engineering construction in Minneapolis, Minnesota based on topography, engineering geological circumstances, hydrogeological conditions, and other indicators. He then utilized a complete index model for evaluation purposes. Professor Mario Mejia-Navarro [9,10] devised a geological disaster risk assessment system for the Glenwood Springs area on a GIS platform. The system evaluated the region's geological disaster risk and provided a framework for regional urban planning and construction. Based on the fuzzy comprehensive assessment approach, Gan Xin and coworkers [11] conducted a zoning evaluation study on the acceptability of site engineering construction of various building kinds in a plain area. A waste treatment plant in Israel's Kurdistan Province is used by Mozafar [12] as the research object. He employs the analytic hierarchy approach to systematically analyze and assess the site suitability of the plant. German academic Youssef Ahmed and others [13] included elements like topography, geological conditions, and environmental geological issues into their analytic hierarchy approach to thoroughly assess the viability of future urban development regions. Liu Hanqiang et al. [14] conducted a systematic and comprehensive evaluation of the suitability of the composite foundation of construction land in their study area through in-depth analysis and research on the engineering geological conditions of the construction land in a certain city. With the aid of the MapGIS spatial analysis function and the unit multi-factor grading weighted index method, Wang Wentao [15] and others investigated the suitability of urban construction land in the Yellow River alluvial plain. They used the geological environment and other elements to evaluate the suitability of construction land development in the plain area.

Due to differences in their mathematical models and analytical logic, each of these traditional methods has its own advantages and limitations (Table 1). However, the factors affecting the suitability of engineering construction are often complex, fuzzy, and random. Despite this complexity, most of these methods still rely on qualitative analysis and subjective evaluation to determine factor weights or membership functions. The evaluation of the adequacy of an engineering construction must consider its inherent imprecision and unpredictability. Academician Lee Deyi [16] made the cloud model his own towards the end of the 1990s. This model, which is based on fuzzy set theory and probability theory, fulfills the natural conversion between quantitative language values and quantitative values through consistently describing the randomness and fuzziness between uncertain linguistic values and exact values. It can explain uncertain issues and has greater universality than traditional fuzzy membership functions [17].

Little research has been done so far on the applicability of cloud-based urban engineering construction. Based on this, this paper uses Xiongan New Area as an example, attempts to apply the cloud model to the suitability evaluation of urban engineering construction in a plain area, resolves the uneven subjective and objective weights of conventional evaluation models, reflects some fuzziness and randomness, provides a new idea for the suitability evaluation of urban engineering construction, and serves as a foundation for Xiongan New Area's overall planning.

Table 1. Comparison of advantages and disadvantages of suitability evaluation methods.

Evaluation Methodology	Advantages	Disadvantages
Spatial data superposition analysis of GIS	The operation is simple and the portability is strong	The research of secondary development of software is insufficient, and the analysis of complex geological environment is still lacking
Artificial neural network method	Nonlinear correlation function, intelligent processing ability for wrong results; It has the ability to react quickly to different indicators	For local minimization problem, the convergence speed of the algorithm is slow; the choice of network structure is different, and there is no unified and perfect theoretical guidance
Analytic hierarchy process (AHP)	Simple application, hierarchical division system, few quantitative parameters; It is suitable for complex problems with multiple indicators	Less quantitative data, more qualitative components, and more subjective factors; when there are too many evaluation factors, the weight is difficult to determine
Fuzzy comprehensive evaluation method	The combination of qualitative and quantitative analysis of the selected index system; the evaluation accuracy is more accurate, and the evaluation result is close to the reality	The calculation is complicated and it is difficult to collect each factor quantitatively. There are subjective factors in determining the index weight
Composite index method	The method is simple, easy to understand and easy to calculate; reliable, practical, clear economic meaning	When selecting indicators, we should pay attention to select the same direction indicators; difficult to quantify accurately

2. Materials and Methods

The cloud model is utilized for the appropriateness evaluation of urban engineering construction, which establishes a grade standard for suitability evaluation using the cloud model. The digital features of the second-level index cloud are generated through utilizing the reverse cloud generator, and then obtaining the digital feat. Firstly, the suitability evaluation system for engineering construction is established through identifying single influencing factors. Then, the index weight is determined through a combination of the analytic hierarchy process and entropy weight method. Subsequently, the complete cloud map is generated using a forward cloud generator and compared with a standard cloud map to obtain the results of the suitability evaluation. The overall method flow is shown in Figure 1.

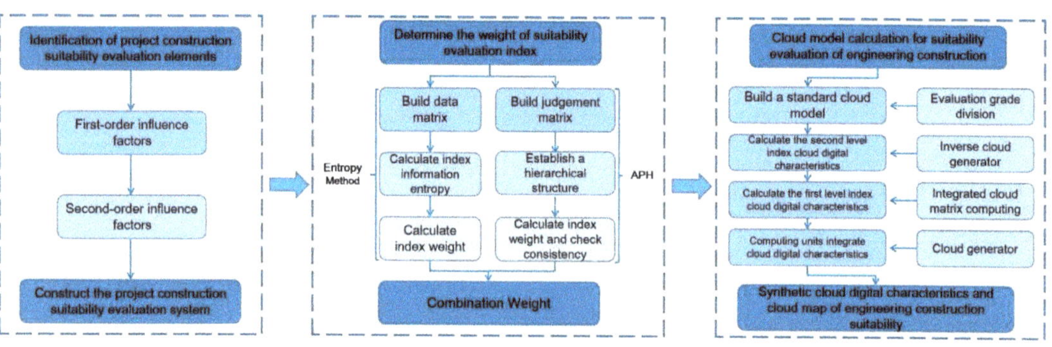

Figure 1. Flow chart of cloud model method for suitability evaluation of engineering construction.

2.1. Entropy Weight Hierarchy Combination Weight

2.1.1. Entropy Weight Method Weight Calculation

A material system's degree of disorder can be quantified according to its entropy, which objectively measures the amount of information contained in known data without subjective bias. This method is based on a solid mathematical foundation and involves several stages for calculating weight using the entropy weight method.

(1) Setting up the initial data matrix An original data matrix is created, presuming an evaluation index system has n evaluation indexes and m evaluated items:

$$A = \left(a_{ij}\right)_{m \times n} = \begin{bmatrix} a_{11} & \cdots & a_{in} \\ \vdots & \ddots & \vdots \\ a_{m1} & \cdots & a_{mn} \end{bmatrix} \tag{1}$$

(2) The original data is dimensionlessly processed:

$$r_{ij} = \frac{a_{ij} - min\{a_{ij}\}}{max\{a_{ij}\} - min\{a_{ij}\}} \tag{2}$$

After dimensionless processing, $S = \left(s_{ij}\right)_{m \times n}$. Then, S is normalized:

$$s'_{ij} = \frac{s_{ij}}{\sum_j \sum_i s_{ij}} \tag{3}$$

(3) Calculate the characteristic proportion f_{ij} of the characteristic object of the i-th evaluation object under the j-th index:

$$f_{ij} = \frac{s'_{ij}}{\sum_{i=1}^{m} s'_{ij}} \tag{4}$$

(4) Calculate the entropy value Hj of the j-th index:

$$H_j = k \sum_{i=1}^{m} f_{ij} ln f_{ij} \tag{5}$$

Among them, $k = \frac{1}{In(m)}$.

(5) Introduce the difference coefficient a_j and calculate the difference coefficient of the j-th index:

$$a_j = 1 - H_j, j = (1, 2, \ldots, n) \tag{6}$$

(6) Determine the entropy weight of the j-th index, that is, the weight of the j-th index:

$$x_j = \frac{a_j}{\sum_{j=1}^{n} a_j}, \; x_j \in [0,1], \; \sum_{j=1}^{n} x_i = 1 \tag{7}$$

(7) Calculate the weight of each index.

$$W = \begin{pmatrix} w_1 \\ \vdots \\ w_m \end{pmatrix} = S \cdot X = \begin{pmatrix} s_{11} & \cdots & s_{1n} \\ \vdots & \ddots & \vdots \\ s_{m1} & \cdots & s_{mn} \end{pmatrix} \begin{pmatrix} x_1 \\ \vdots \\ x_n \end{pmatrix}. \tag{8}$$

2.1.2. Weight Calculation of Analytic Hierarchy Process

The complete evaluation technique known as the analytic hierarchy process, developed by Saaty in the 1970s, combines qualitative and quantitative methodologies. It is frequently employed to find solutions to intricate issues with many goals. The basic idea is that after thoroughly examining the problem's nature and the overall goal that must be achieved, the problem is broken down into its component parts. The parts are then grouped and combined into a multi-level structural hierarchy model based on how closely they are related to one another and how much of a member they are. Finally, the weight of the decision-making scheme in relation to the overall goal is determined. The calculation stages will not be repeated in this study because the analytic hierarchy process is frequently utilized in the suitability assessment of engineering projects.

2.1.3. Combination Weight Calculation

The entropy weight method uses objective weighing, whereas the analytical hierarchy process uses subjective weighting. Both processes obtain the weight first, then mix it with the assessed object's original index data to produce the evaluation result. Based on the strengths of the two ways, the entropy weight–analytic hierarchy process weights and fuses the weights from the two methods to produce the combined weight.

The precise procedure is as follows:

Assume that the analytical hierarchy procedure gave the following weights to n evaluation indicators:

$$W_1 = [w_1, w_2, \ldots, w_n]^T \tag{9}$$

The following weights were given to n evaluation indicators using the entropy weight method:

$$W_2 = [w_1', w_2', \ldots, w_n']^T \tag{10}$$

The i-th assessment indicator's total weight is:

$$W = \frac{W_{1i} W_{2i}}{\sum_{i=1}^{n} W_{1i} W_{2i}} \tag{11}$$

2.2. Cloud Model

2.2.1. Cloud Model Definition

The cloud model is a mathematical way to express the mutual conversion between qualitative and quantitative information, reflecting the randomness and ambiguity of things. It is defined as follows: Let X be a set of exact values, X = {x}. X is called the universe of discourse, C is a qualitative concept on the set of X, and x is a random realization of the qualitative concept C. The membership degree $\mu(x)$ of x to C~[0, 1] and is a random number with a stable tendency. If $x \sim N\left(E_x, E_n'^2\right)$, where $E_n' \sim N\left(E_n, H_e^2\right)$, and the degree of membership of C satisfies $\mu(x) = \exp\left\{-(x - E_n)^2 / \left(2E_n'^2\right)\right\}$, then the distribution of x on the universe X is called a cloud or a normal cloud. Each x is called a cloud drop $(x, \mu(x))$.

In the above, $\mu(x)$ is the degree of membership, x is the original variable, E_x is the expectation of x, E_n is the entropy, and H_e is the hyper-entropy.

Expectation, entropy, and hyper-entropy H_e (Figure 2) are used to explain the numerical features of the cloud. The point that best embodies a qualitative concept is expectation, which also serves as the universe's focal point. Entropy is used to reflect the likelihood and ambiguity of qualitative concepts; the higher the entropy, the wider the range of qualitative concepts that can be accepted. Hyper-entropy He is the entropy of entropy, which symbolizes the thickness of the cloud. The degree of membership dispersion increases as the cloud drop thickness increases. The value is often based on empirical data; there is no precise guideline for choosing hyper-entropy.

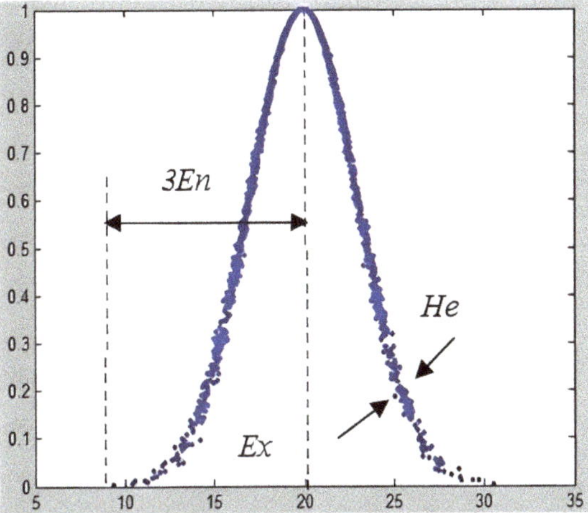

Figure 2. Digital characteristics of the cloud model.

2.2.2. Cloud Model Generator

The cloud model implements the reciprocal conversion between qualitative and the quantitative values using a cloud generator. The forward cloud generator (CG) and reverse cloud generator (CG-1) are two categories of cloud generator.

The forward cloud generator, which creates cloud droplets from the cloud digital feature (E_x, E_n, H_e), is a conversion model that translates the qualitative notion to the quantitative value (Figure 3).

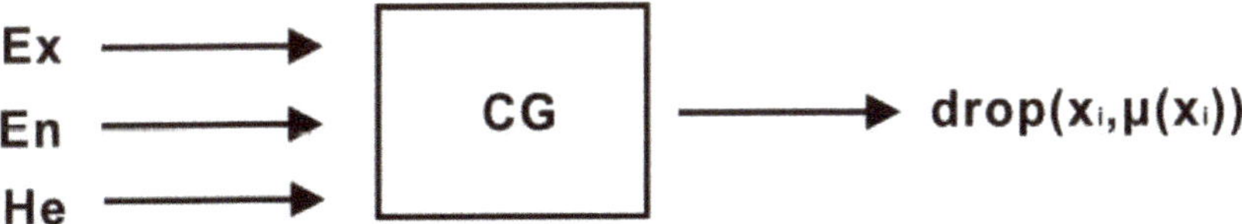

Figure 3. Forward cloud generator.

Figure 4 shows a conversion model called the reverse cloud generator that may transform a given quantity of precise data into a qualitative idea symbolized by the digital feature (E_x, E_n, H_e). It realizes the quantitative value to the qualitative concept.

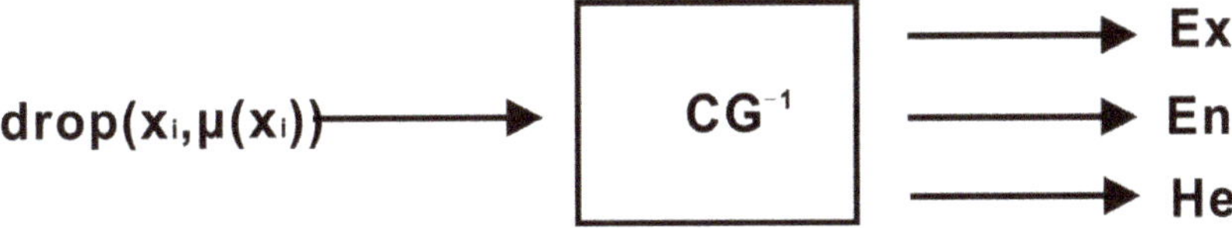

Figure 4. Backward cloud generator.

2.2.3. Constructing the Comprehensive Evaluation Cloud of Suitability

We first split the assessment area into t evaluation units, choose the comment set based on each secondary index system in Table 2, and then obtain the evaluation matrix in order to build a thorough evaluation cloud. The cloud generator calculates the secondary index cloud feature digital matrix, which is then combined with the secondary index weight matrix to produce the corresponding first-level evaluation index cloud feature digital matrix. The combined first-level evaluation index cloud feature digital matrix and weight matrix then produce the comprehensive cloud digital feature of the t-th evaluation. The suitability level of the t-th unit is then determined via comparing the complete cloud map of the t-th unit's suitability that is produced by the forward cloud generator (CG) with the standard assessment cloud map. Mapping software (https://www.esri.com/en-us/arcgis/products/mapping/overview, accessed on 2 August 2023) is used to create the complete suitability assessment zoning map of Xiongan New Area after repeating this procedure to collect the suitability evaluation grades of all t units.

$$C_t = \begin{pmatrix} W_1 \\ \vdots \\ W_t \end{pmatrix}^T \begin{pmatrix} E_{x1} & E_{n1} & H_{e1} \\ \vdots & \vdots & \vdots \\ E_{xt} & E_{nt} & H_{et} \end{pmatrix} = (E_{xt}^*,\ E_{nt}^*,\ H_{et}^*) \tag{12}$$

$$E_{xt}^* = \frac{E_{x1}E_{n1}W_1 + \cdots + E_{xt}E_{nt}W_t}{E_{n1}W_1 + \cdots + E_{nt}W_t} \tag{13}$$

$$E_{nt}^* = E_{n1}W_1 + \cdots + E_{nt}W_t \tag{14}$$

$$H_{et}^* = \frac{H_{e1}E_{n1}W_1 + \cdots + H_{et}E_{nt}W_t}{E_{n1}W_1 + \cdots + E_{nt}W_t} \tag{15}$$

Table 2. System of engineering construction suitability integrated evaluation indexes in Xiongan New Area.

First-Order Factor	Second-Order Factor	Evaluation Index Grading Standard			
		Unsuited	Less Suitable	Relatively Suitable	Suitability
		I	II	III	IV
Foundation soil condition	0–5 m foundation bearing capacity f_a	$f_a < 80$ kPa	80 kPa $\leq f_a < 100$ kPa	100 kPa $\leq f_a < 130$ kPa	$f_a \geq 130$ kPa
	5–10 m foundation bearing capacity f_a	$f_a < 100$ kPa	100 kPa $\leq f_a < 130$ kPa	130 kPa $\leq f_a < 160$ kPa	$f_a \geq 160$ kPa
	10–15 m foundation bearing capacity f_a	$f_a < 110$ kPa	110 kPa $\leq f_a < 130$ kPa	130 kPa $\leq f_a < 170$ kPa	$f_a \geq 170$ kPa
	15–30 m foundation bearing capacity f_a	$f_a < 130$ kPa	130 kPa $\leq f_a < 160$ kPa	160 kPa $\leq f_a < 200$ kPa	$f_a \geq 200$ kPa
	30–50 m foundation bearing capacity f_a	$f_a < 130$ kPa	130 kPa $\leq f_a < 160$ kPa	160 kPa $\leq f_a < 200$ kPa	$f_a \geq 200$ kPa
	Soil comprehensibility	Es > 15 MPa	11 MPa $\leq$ Es < 15 MPa	4 MPa $\leq$ Es < 11 MPa	Es < 4 MPa
Hydrogeology condition	Shallow groundwater depth	<1 m	1–3 m	3–6 m	>6 m
	Corrosion of soil and water	High corrosion	Medium corrosion	Low corrosion	Micro-corrosion
Environmental geological problem	Land subsidence(rate)	>50 mm/a	30–50 mm/a		<30 mm/a
	Sand liquefaction	Severe sand liquefaction	Medium sand liquefaction	Weak sand liquefaction	No sand liquefaction
	Flood inundation potential (the flood inundation depth or land elevation is lower than the flood prevention level)	>1.0 m	0.5–1.0 m	<0.5 m	No flood or land elevation higher than fortification height
Field stabilization	Seismic fortification intensity	$\geq$IX	VIII, VII		$\leq$VI

Among them, C_t is the comprehensive suitability cloud digital feature, for which E_{xt}^*, E_{nt}^*, H_{et}^* is the expectation, entropy, and super entropy of the higher-level index of the t-th unit, and W_t is the weight of the t-th index.

3. Engineering Example

3.1. Single-Factor Analysis

3.1.1. Foundation Soil Conditions

With the exception of Beiwangkou Town near Baiyangdian Duancun Town–Tongkou Town, Xiong County, the foundation soil conditions in Xiongan New Area are generally good, the soil mass is relatively uniform [18], there is only a small range of soft soil developed (Figure 5a), and the thickness is small (generally no more than 2 m). The stratum has a limited bearing capacity (95–115 kPa) in the 0–5 m depth range, making it unsuitable as the natural foundation–bearing layer for multi-story buildings. In most other places, the stratum below 100 m has a bearing capability of more than 120 kPa. It is appropriate for all types of engineering construction after treatment and may satisfy the criteria of pile foundation or pile end–bearing layer of multi-story structures (Figure 5b–f). The compressibility of soil in most areas of the new area is in the medium–low compression zone, and the soil is in the medium–high compression zone in the depth range of 0–5 m east of the first line of Zhanggang Township–Xiong County–Zhaobeikou Township–Yucun Township and west of the first line of Zhaili Township–Anzhou Township–Laohetou Township, as well as in the depth range of 30–50 m east of the first line of Nangzhang Township, southwest of Xiaoli Township, and Qiaogang Township–Zhugezhuang Township–Xiongzhou Township–Xinglonggong Township. The maximum weighted average compressive modulus is 27.17 MPa and 16.37 MPa, respectively (Figure 5g–k).

3.1.2. Hydrogeological Conditions

Only the surrounding region of Baiyangdian District has a shallow groundwater level below 5 m in most of Xiongan New Area (Figure 5l). The construction components of a foundation or pile foundation will corrode due to groundwater and soil, which will reduce the material's strength and influence the stability of engineering structures. The Liuli Zhuang Town–Tongkou Town region is where most of the places with moderate to severe groundwater corrosion are found. The main ions affecting groundwater erosion arc $SO_4{}^{2-}$ and Cl^-, and their concentrations range from 1290 to 4345 mg/L and 33 to 1727 mg/L, respectively (Figure 5m). The moderate and severe soil erosion areas are primarily located near the Baiyangdian area, which is impacted by evaporation and concentration and where salt is enriched, forming a salt accumulation zone (Figure 5n).

3.1.3. Environmental Geological Problems

According to Xie et al. [19], the yearly land subsidence rate of Xiongan New Area ranges from 0 to 90 mm. As of 2020, the area of the new district with an annual settlement rate greater than 50 mm accounts for about 12% of the total area of the new district. The new district is primarily located in Nanzhang Town, Luzhuang Township, Longhua Township, Laohetou Township, some areas of Anzhou Town, the junction of Santai Town and Xiaoli Town, the junction of Pingwang Township in Rongcheng County and Zhugezhuang Township in Xi. Urban subterranean pipes or piling foundations may be damaged by land subsidence, which might seriously jeopardize the security of project development and operation. The Xiongan New Area's central and southern regions are particularly susceptible to sand liquefaction, which is primarily dispersed 20 m below the surface. A tiny portion of Mozhou, Liuli Village, Anxin County, and other regions are dominated by medium–severe liquefaction, whereas the Anxin–Zhaobeikou area and the Anzhou–Qijianfang area in the south are dominated by minor liquefaction (Figure 5q).

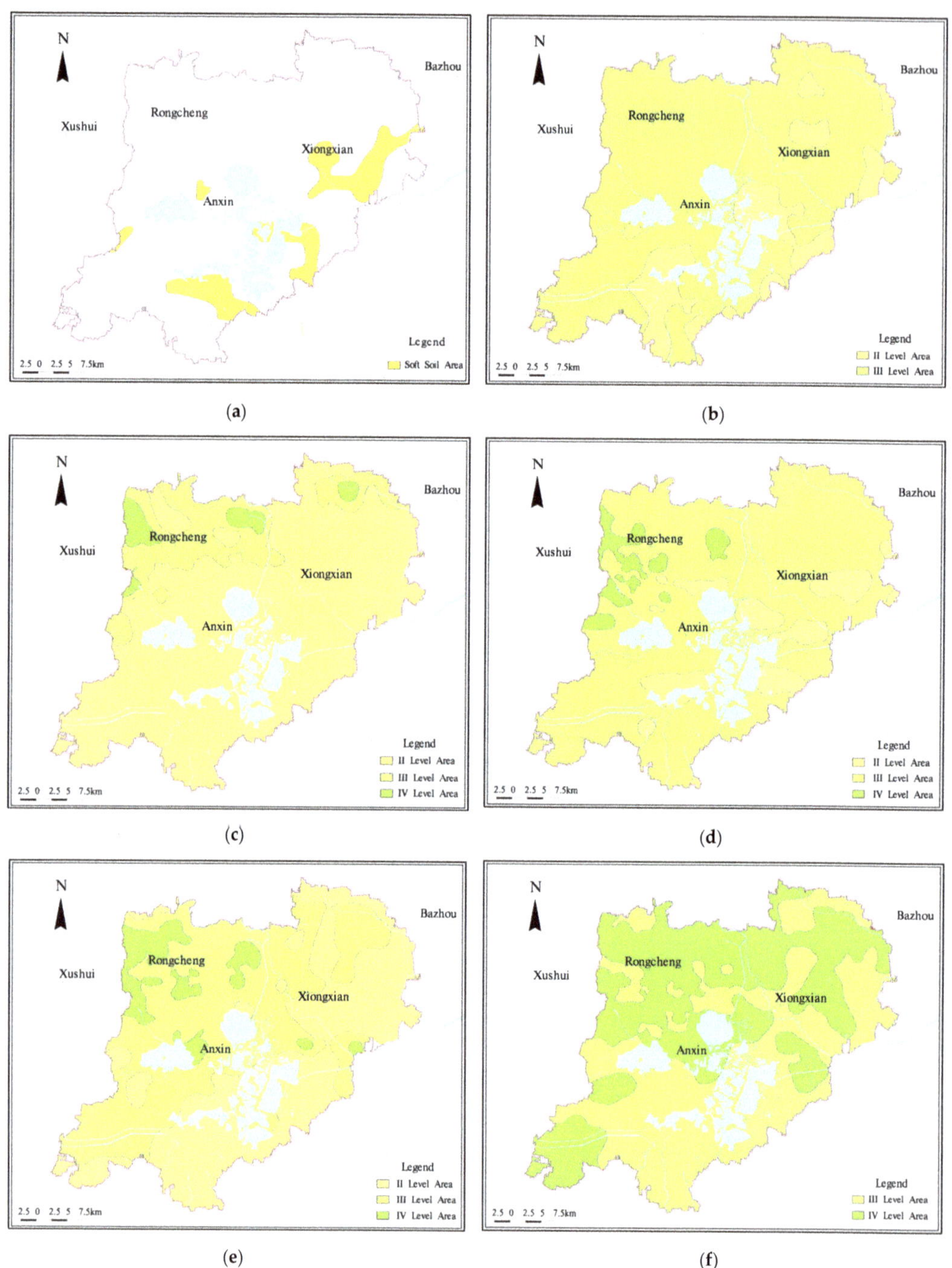

Figure 5. *Cont.*

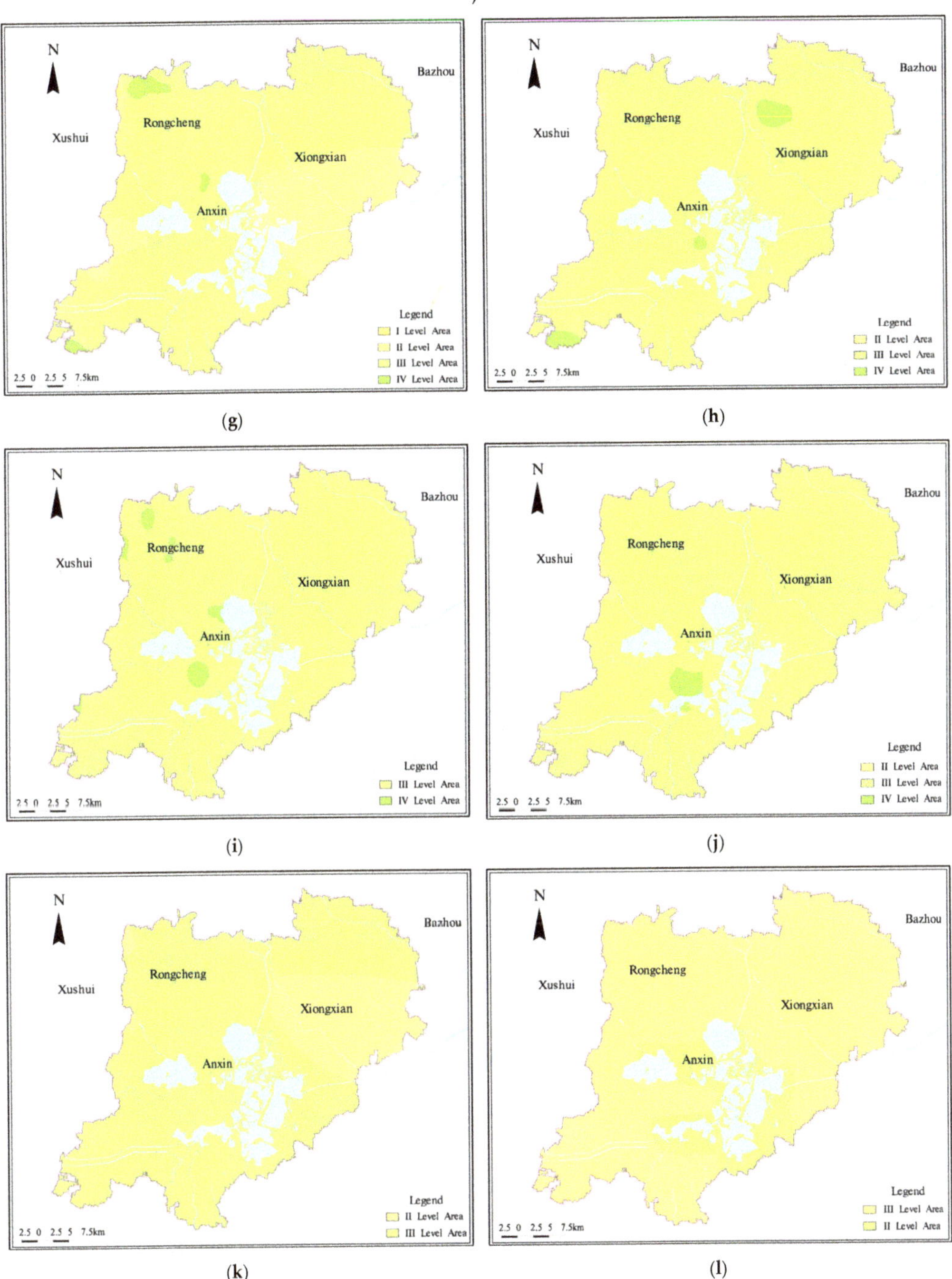

Figure 5. *Cont.*

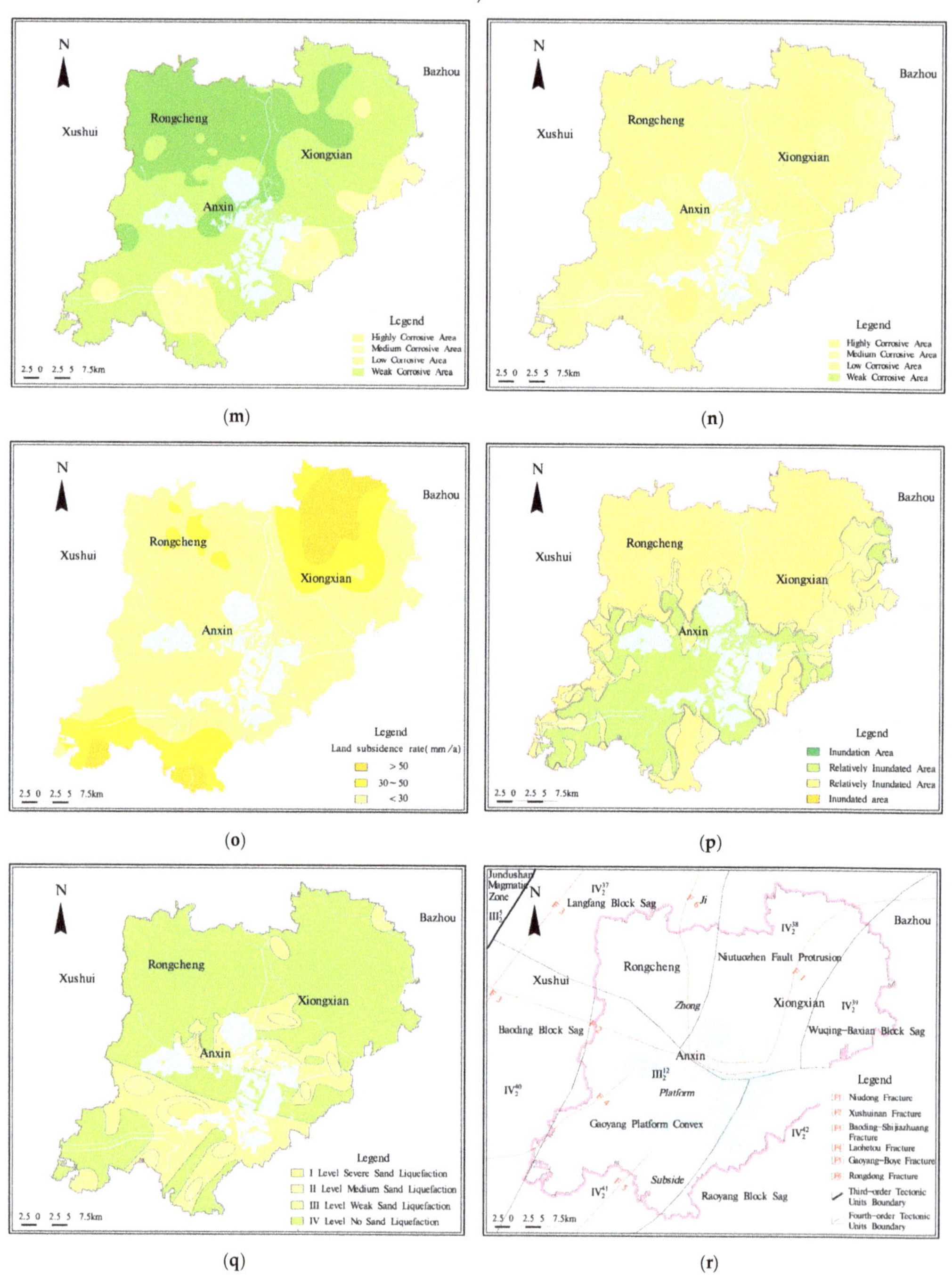

Figure 5. Evaluation results of single-factor analysis. (**a**) Soft soil distribution map in Xiongan New Area; (**b**) zoning map of 0–5 m foundation bearing capacity in Xiongan New Area; (**c**) zoning map of 5–10 m foundation bearing capacity in Xiongan New Area; (**d**) zoning map of 10–15 m foundation bearing capacity in Xiongan New Area; (**e**) zoning map of 15–30 m foundation bearing capacity in

Xiongan New Area; (**f**) zoning map of 30–50 m foundation bearing capacity in Xiongan New Area; (**g**) 0–5 m compressibility zoning diagram of soil in Xiongan New Area; (**h**) 5–10 m compressibility zoning diagram of soil in Xiongan New Area; (**i**) 10–15 m compressibility zoning diagram of soil in Xiongan New Area; (**j**) 15–30 m compressibility zoning diagram of soil in Xiongan New Area; (**k**) 30–50 m compressibility zoning diagram of soil in Xiongan New Area; (**l**) classification chart of shallow groundwater depth in Xiongan New Area; (**m**) zoning map of soil corrosion in Xiongan New Area; (**n**) zoning map of groundwater corrosion in Xiongan New Area; (**o**) gradation map of land subsidence rate in Xiongan New Area (2016); (**p**) zoning map of flood inundation potential in Xiongan New Area; (**q**) classification chart of the sand liquefaction in Xiongan New Area; (**r**) hidden fault structure distribution map in Xiongan New Area.

3.1.4. Dynamic Geological Action

The stability of the site is unaffected by any of the six developed faults in and around the Xiongan New Area, all of which are non-new active fault zones (Figure 5r). Earthquakes have extremely low magnitudes and frequency; their seismic fortification intensity is VIII degrees.

3.2. Index System and Weight Determination

This time, a number of second-level evaluation indicators are chosen underneath the first-level indicators to construct the evaluation index system. The foundation soil conditions, hydrogeological conditions, environmental geological problems, site stability, and underground space availability resources are chosen as the first-level evaluation indicators this time. This article divides the suitability of the environment and engineering construction and utilization in the new area into four grades, from poor suitability (a) to suitability (b), and analyzes the distribution characteristics of hydrogeological conditions, engineering geological conditions, and environmental geological problems in the study area and their impact on the development and utilization of above-ground and underground engineering in Xiongan New Area. Referring to pertinent data [2,20] and experts' recommendations that empirical values establish grading standards, there are two ways to determine the extent to which various types of geological environmental impact factors will have an impact on project construction. Table 2 displays the quantitative grading standards for the evaluation factors.

In this study, the subjective and objective integrated weighing approach and the entropy weight–analytic hierarchy process are used, which can effectively eliminate the drawbacks of employing the two methods independently. Through questionnaire surveys, in-person discussions, and other methods for weight evaluation and index assignment, as well as through the pertinent calculations to obtain the judgment matrix and the consistency test of the results, this evaluation arranges three groups of engineering construction suitability evaluation consultants, with each group consisting of ten people who are experienced engineering geology scholars. The evaluation index is weighted using the entropy approach. Five relevant experts from the consulting group create the initial matrix based on the scores of each index, and then the appropriate weighting computation is done. A calculation is made to determine the aggregate weight of each indicator, which displays how much of an impact each index has on the suitability of a project's construction (Table 3).

Table 3. Comprehensive weight table of engineering construction suitability evaluation index.

First-Order Factor	Weight	Second-Order Factor	Weight
Foundation soil condition	0.432	Foundation bearing capacity	0.242
		Soil comprehensibility	0.190
Hydrogeology condition	0.152	Shallow groundwater depth	0.120
		Corrosion of soil and water	0.032
Environmental geological problem	0.3	Land subsidence	0.159
		Sand liquefaction	0.113
		Flood inundation potential	0.028
Field stabilization	0.116	Seismic fortification intensity	0.116

3.3. Evaluation Cloud Construction

3.3.1. Standard Evaluation Cloud Construction

We initially develop a common assessment cloud before performing engineering, geological environment, and construction suitability evaluation. In accordance with the Xiongan New Area's established engineering, geological, and construction suitability evaluation system (Table 2), the new area's suitability evaluation grades are separated into four categories: suitable, more suitable, generally appropriate, and poorly suitable. The equivalent value range is set to [0, 100], with [0, 25) denoting the area that is most suitable, [25, 50) the area that is more suitable, [50, 75) the area that is generally suitable, and [75, 100] the area that is least suitable. Higher scores indicate a worse fit, whereas lower scores indicate greater suitability. The standard evaluation cloud model is then produced using the cloud model's inverse cloud generator (Figure 6). To ascertain the level of construction suitability for the project, the final results of the construction suitability evaluation are compared and examined. Table 4 displays the numerical properties of the derived standard cloud model.

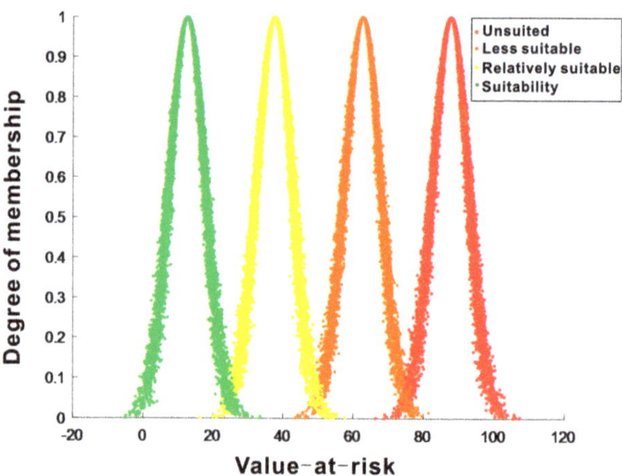

Figure 6. Cloud map of suitability evaluation standard for Xiongan New Area engineering construction.

Table 4. Engineering construction suitability classification and digital characteristics of standard cloud model.

Classification of Suitability	Degree of Suitability	Range	Digital Characteristics of Standard Cloud Model
I	Unsuited	[100, 75]	(87.5, 15.0, 0.5)
II	Less suitable	(75, 50]	(62.5, 15.0, 0.5)
III	Relatively suitable	(50, 25]	(37.5, 15.0, 0.5)
IV	Suitable	(25, 0]	(12.5, 15.0, 0.5)

3.3.2. Comprehensive Evaluation Cloud Construction

According to the construction suitability evaluation system in Table 2, Formulas (13)–(15) are used to determine the numerical characteristic values of each evaluation index (Tables 5 and 6). Since there is no exact guidance method for the selection of super entropy, this time, after consulting a large number of relevant materials, the value of super entropy He is set to 0.5. When generating random cloud drops, considering the computer operation ability and accuracy needs, the calculation is repeated 5000 times, that is, each normal cloud model generates 5000 cloud drops. The normal cloud model is generated for the selected 12 s level evaluation indicators (Figure 7) and 4 first level evaluation indicators (Figure 8) according to the four evaluation levels.

Table 5. Numerical characteristics of cloud model for suitability evaluation index of engineering construction in Xiongan New Area.

Assessment Index		Suitability			Relatively Suitable			Less Suitable			Unsuited		
		E_x	E_n	H_e	E_x	E_n	H_e	E_x	E_n	H_e	E_x	E_n	H_e
Foundation soil condition	0–5 m foundation bearing capacity f_a	155.000	131.644	0.500	115.000	97.672	0.500	90.000	76.439	0.500	40.000	33.972	0.500
	5–10 m foundation bearing capacity f_a	170.000	144.385	0.500	145.000	123.152	0.500	115.000	97.672	0.500	50.000	42.466	0.500
	10–15 m foundation bearing capacity f_a	310.000	263.290	0.500	150.000	127.399	0.500	115.000	97.670	0.500	55.000	46.713	0.500
	15–30 m foundation bearing capacity f_a	325.000	279.030	0.500	180.000	152.878	0.500	145.000	123.152	0.500	65.000	55.026	0.500
	30–50 m foundation bearing capacity f_a	325.000	276.030	0.500	180.000	152.878	0.500	145.000	123.152	0.500	65.000	55.206	0.500
	Soil comprehensibility	2.000	1.698	0.500	7.500	6.369	0.500	13.000	11.041	0.500	17.500	14.863	0.500
Hydrogeology condition	Shallow groundwater depth	8.000	6.794	0.500	4.500	3.821	0.500	2.000	1.698	0.500	0.500	0.424	0.500
	Corrosion of soil and water	150.000	1.274	0.500	900.00	1.274	0.500	2250.0	5.095	0.500	4000.0	8.917	0.500
Environmental geological problem	Land subsidence	15.000	12.739	0.500	40.000	33.972	0.500	70.000	9.452	0.500	70.000	59.452	0.500
	Sand liquefaction	0.050	0.042	0.500	3.050	2.590	0.500	12.000	10.191	0.500	21.500	18.260	0.500
	Flood inundation potential	0.050	0.042	0.500	0.300	0.254	0.500	0.750	0.637	0.500	1.500	1.274	0.500
Field stabilization	Seismic fortification intensity	3.000	2.548	0.500	7.500	6.370	0.500	7.500	6.370	0.500	10.500	8.9180	0.500

Table 6. Construction suitability of integrated cloud digital characteristics.

Assessment Index	Suitability			Relatively Suitable			Less Suitable			Unsuited		
	E_x	E_n	H_e	E_x	E_n	H_e	E_x	E_n	H_e	E_x	E_n	H_e
Foundation soil condition	293.584	56.286	0.500	156.478	33.732	0.500	120.529	27.950	0.500	50.246	14.476	0.500
Hydrogeology condition	126.300	4.892	0.500	883.500	24.919	0.500	2242.50	61.355	0.500	3998.1	108.764	0.500
Environmental geological problem	253.924	55.465	0.500	149.001	33.064	0.500	116.961	25.776	0.500	52.962	11.797	0.500
Field stabilization	3.000	2.548	0.500	7.500	6.369	0.500	7.500	6.369	0.500	10.5	8.917	0.500

The whole area of Xiongan New Area is divided into 100×150 evaluation units. The construction suitability cloud model is used to evaluate them. The digital characteristics of each respective evaluation unit are obtained and compared with the digital characteristics of the standard cloud. The engineering, geological environment, and construction suitability grid of Xiongan New Area is obtained (Figure 9).

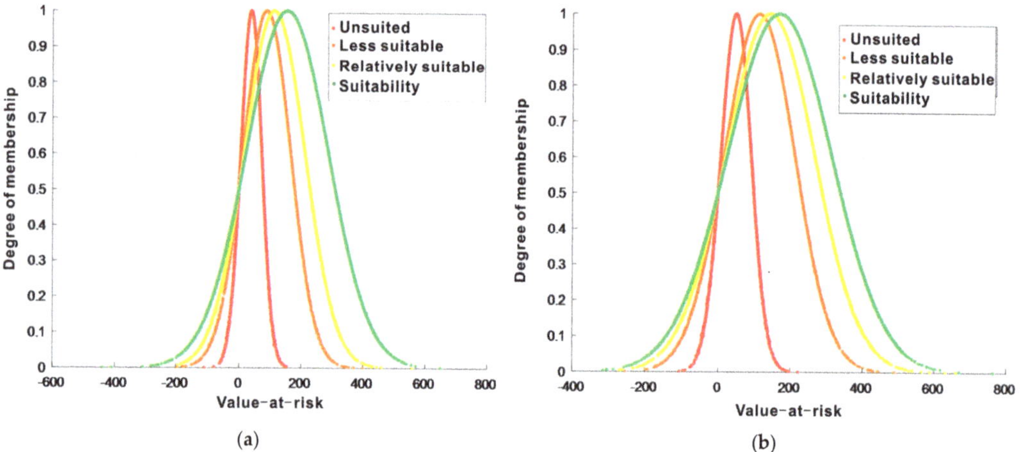

Figure 7. *Cont.*

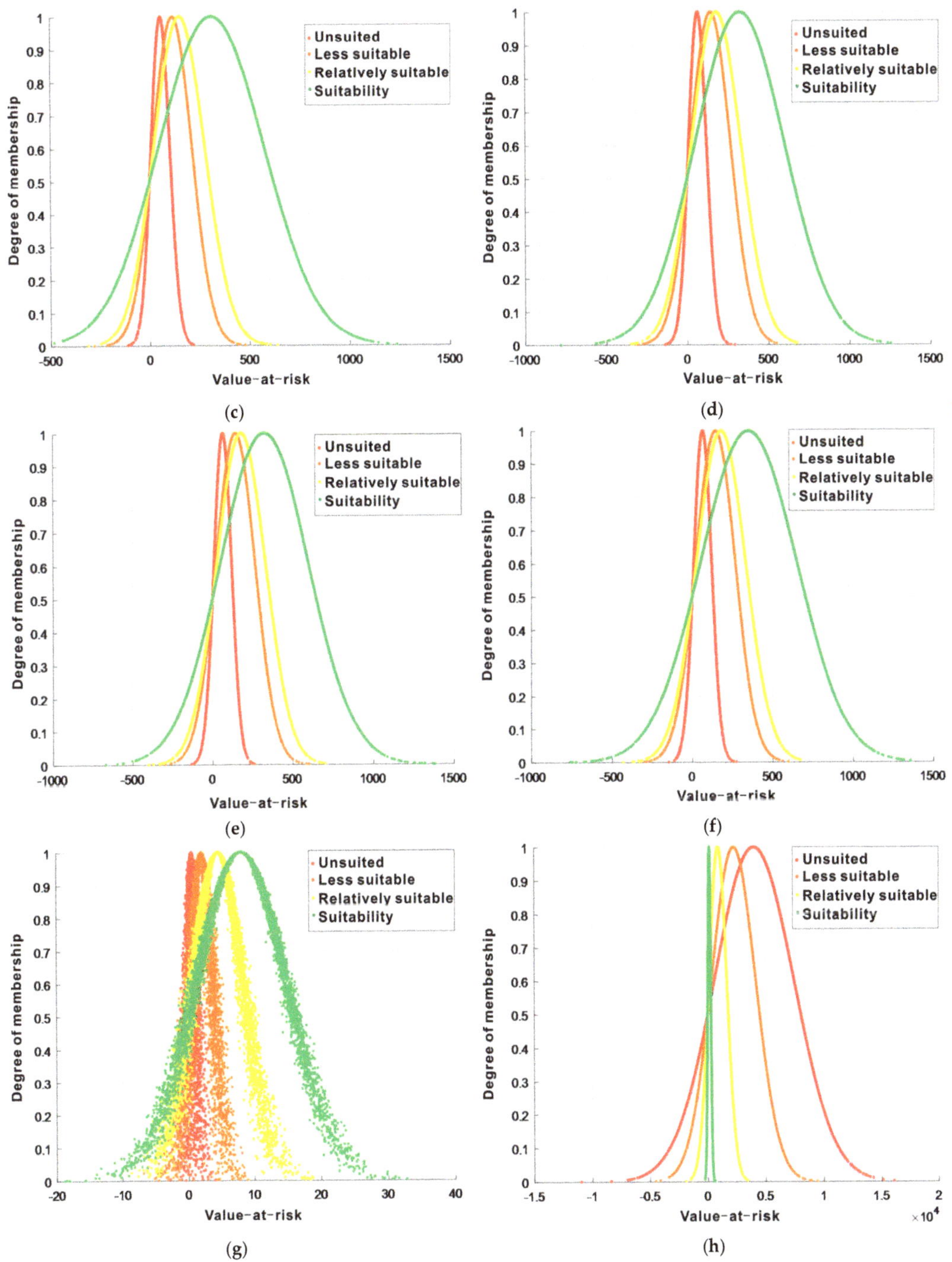

Figure 7. *Cont.*

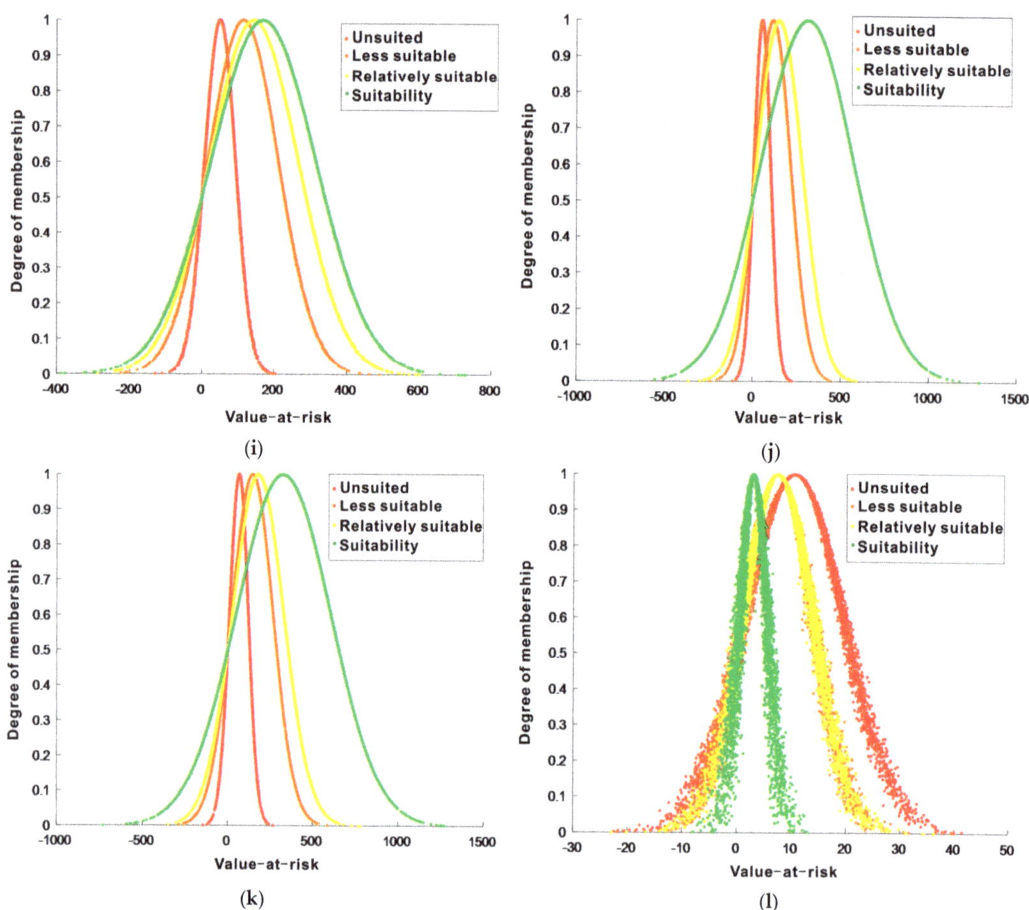

Figure 7. Cloud model of second-level index of suitability evaluation of engineering construction. (**a**) Cloud model of 0–5 m foundation bearing capacity; (**b**) cloud model of 5–10 m foundation bearing capacity; (**c**) cloud model of 10–15 m foundation bearing capacity; (**d**) cloud model of 15–30 m foundation bearing capacity; (**e**) cloud model of 30–50 m foundation bearing capacity; (**f**) cloud model of soil comprehensibility; (**g**) cloud model of shallow groundwater depth; (**h**) cloud model of corrosion of soil and water; (**i**) cloud model of land subsidence; (**j**) cloud model of sand liquefaction; (**k**) cloud model of flood inundation potential; (**l**) cloud model of seismic fortification intensity.

3.4. Results and Analysis

A partition diagram illustrating the applicability of cloud theory to building of the Xiongan New Area is shown in Figure 10. The chart shows that the majority of the above-ground regions in Xiongan New Area are appropriate and more suitable places for engineering building, with the suitable area accounting for 417.50 km², or 23.6% of the total area. This area has an excellent engineering geological environment that is suited for all types of engineering development and building, as well as general seismic fortification. 1022.89 km², or 57.8% of the total, is the area that is more appropriate. This region has better engineering geological conditions that are better suited for different engineering development and construction. Measures for liquefiable strata should be taken in accordance with standard seismic fortification. Only 8.9% of the available space, or 157.74 km², is generally appropriate. It is mostly dispersed in the areas around Baiyangdian.

Flooding is the main possibility. In addition, there are issues with sand liquefaction and land subsidence. To lessen or eliminate the risks of flooding, land subsidence, and sand liquefaction, appropriate precautions must be taken. Finally, 171.87 km^2—or 9.7% of the total area—is considered to be unsuitable. It is primarily found in the north of Xiong County, in the western parts of the towns of Duancun, Longhua, and Luzhuang. The area has substantial liquefaction of sandy soil and a ground settlement rate of more than 50 mm per year.

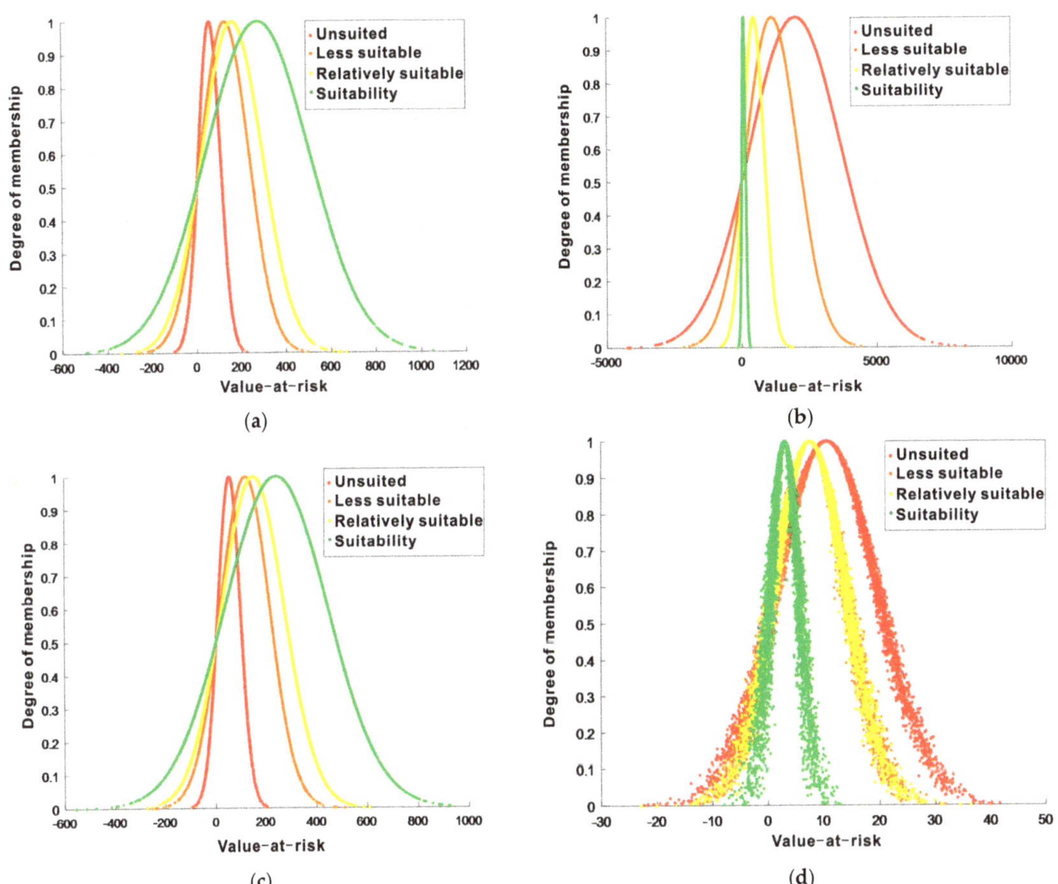

Figure 8. Cloud model of first level index of suitability evaluation of engineering construction. (**a**) Foundation soil condition; (**b**) hydrogeology condition; (**c**) environmental geological problem; (**d**) field stabilization.

This research also utilizes the average comprehensive index approach as a comparison reference to confirm the viability of the cloud model in the appropriateness evaluation of engineering construction (Figure 11). According to the results of the average comprehensive index method's suitability evaluation, the suitable area is 818.65 km^2, or 46.3% of the total area; the more suitable area is 673.87 km^2, or 38.1%; the generally suitable area is 163.18 km^2, or 9.2%; and the area of poor suitability is 114.30 km^2, or 6.4%. Comparing the two methods reveals that while the assessment outcomes of the two approaches are somewhat dissimilar, the overall outcomes are rather comparable (Table 7). The main distinction is that the comprehensive index method overemphasizes the importance of a single factor's maximum value in the evaluation process and ignores the weight of the evaluation index, whereas the cloud model overcomes these drawbacks and more accurately quantifies the weight of all evaluation factors' effects on the suitability of engineering construction. It is clear that the cloud model may provide a more accurate and thorough reflection of the evaluation outcomes.

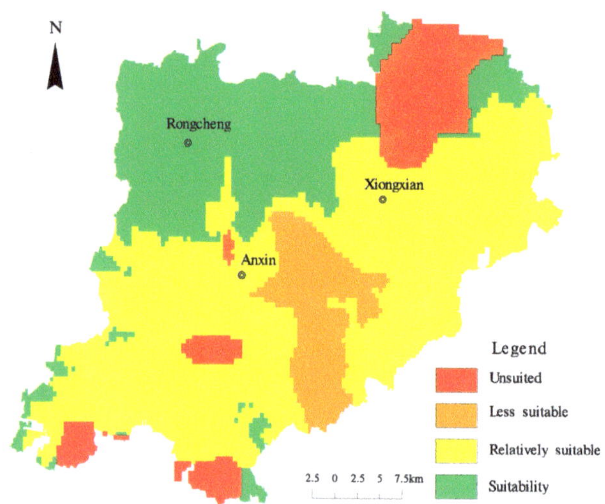

Figure 9. Grid profile of cloud model suitability for engineering construction in Xiongan New Area.

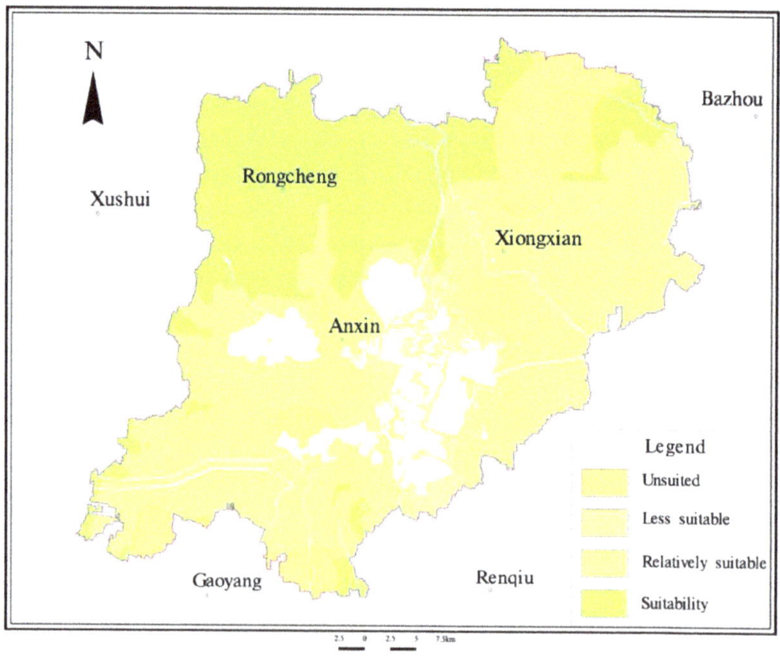

Figure 10. Cloud theory suitability zone map in Xiongan New Area.

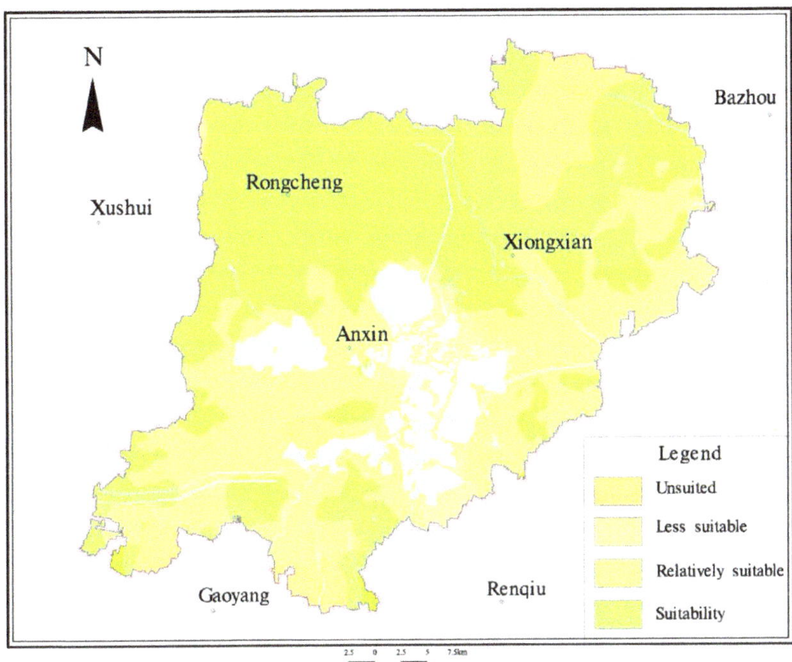

Figure 11. Average composite index method suitability zone map in Xiongan New Area.

Table 7. Comparison list of evaluation results of suitability of project development and construction in Xiongan New Area.

Cloud Theory				Average Composite Index Method			
Zoning of Suitability	Area Km^2	Proportion %	Distribution	Zoning of Suitability	Area Km^2	Proportion %	Distribution
Unsuited	171.87	9.7	Between Beishakou, Daying and Mijiawu; Duancun west side; Longhua and luzhuang	Unsuited	114.30	6.4	East of Beishakou and Daying; Duancun west side; Longhua and Luzhuang
Less suitable	157.74	8.9	Around Baiyangdian area	Less suitable	163.18	9.2	Around Baiyangdian area; Northwest of Nanzhang
Relatively suitable	1022.89	57.8	South of Santai, Dawang, Pingwang and Mijiawu	Relatively suitable	673.87	38.1	South of Zhaili, Anxin and Zhaobeikou
suitability	417.50	23.6	Areas other than those mentioned above	suitability	818.65	46.3	Areas other than those mentioned above

4. Conclusions

(1)   We created a thorough evaluation mechanism for the viability of project construction in Xiongan New Area. Eight second level evaluation indices of foundation bearing capacity at different depths, soil compressibility, groundwater depth, water and soil corrosivity, and land stability are screened out after the engineering geological

conditions and environmental geological conditions affecting the urban construction of the new district are taken into consideration.

(2) A thorough assessment of the appropriateness of engineering construction in Xiongan New Area was carried out. The suitability assessment grade standard was developed using a cloud model to assess the appropriateness of Xiongan New Area, and the index weight was determined using the analytical hierarchy process and the entropy weight technique. This study's findings indicate that Xiongan New Area's engineering construction conditions are generally good, that the suitable and more suitable areas make up more than 80% of the new area's total area, and that both the generally suitable area and the poorly suitable area have significant environmental geological problems like land subsidence, sand liquefaction, and flood inundation.

(3) Comparisons were made between the outcomes of the cloud model's evaluation and those of the conventional suitability evaluation model. Despite some discrepancies between the two, the general conclusions are the same, proving the viability and efficacy of the cloud model in the evaluation of engineering construction's appropriateness.

(4) The conceptual model of the evaluation index system must be established before the evaluation method based on cloud theory can be implemented. Through the cloud generator, this method realizes the organic combination of uncertainty and ambiguity among the multi-factor indicators of the suitability evaluation of urban engineering construction in comparison to the conventional assessment technique. The assessment index system is a crucial element in engineering practice that influences site selection for engineering projects as well as disaster prevention and mitigation. The choice of assessment indices in this work is mostly based on the most recent technical requirements of engineering and geological exploration. The consideration is not thorough enough, and the actual problems outside the framework are less complicated. It must be reinforced in a follow-up study to increase the research findings' maturity and applicability.

(5) This research was conducted for the Xiongan New Area's general planning, and as such, its evaluation accuracy does not satisfy the standards of regulatory detailed planning. The following phase is conducting a "targeted" project construction suitability review in accordance with the particular planning framework, taking into account the requirements of various project construction layouts, and undertaking various kinds of projects.

Author Contributions: Conceptualization, Y.-H.G. and B.H.; methodology, Y.-H.G.; software, Y.-H.G.; validation, B.H.; investigation, Y.-H.G., B.H. and S.J.; resources, B.H.; data curation, Y.-H.G. and B.H.; writing—original draft preparation, Y.-H.G.; writing—review and editing, B.H., J.-J.M. and H.-W.L.; visualization, Y.-H.G.; supervision, B.H. and H.-W.L.; project administration, B.H. and H.-W.L.; funding acquisition, B.H. All authors have read and agreed to the published version of the manuscript.

Funding: This research was funded by the following grants: "Monitoring and evaluation of resource and environment carrying capacity of the Beijing–Tianjin–Hebei Collaborative Development Area and Xiongan New Area" (Grant no. DD20221727); "Detailed investigation and risk control of geological disasters in Taihang Lvliang Mountain area" (Grant no. DD20230438).

Institutional Review Board Statement: Not applicable.

Informed Consent Statement: Not applicable.

Data Availability Statement: Not applicable.

Acknowledgments: Thanks to Jiang Wanjun for his guidance and suggestions on this article.

Conflicts of Interest: The authors declare that they have no conflict of interest regarding the publication of this paper.

References

1. Fang, W.; Lee, X.; Tian, D.; Wang, L. Comprehensive evaluation and analysis on the engineering geological stability of Tianjin urban construction. *Geol. Surv. Res.* **2016**, *39*, 64–70.
2. Hao, A.; Wu, A.; Ma, Z.; Liu, F.; Xia, Y.; Xie, H.; Lin, L.; Wang, T.; Bao, Y.; Zhang, J.; et al. A study of engineering construction suitability integrated evaluation of surface-underground space in Xiongan New Area. *Acta Geosci. Sin.* **2018**, *39*, 513–522.
3. Zhang, J.; Ma, Z.; Wu, A.; Bai, Y.; Xia, Y. A study of paleochannels interpretation by the spectrum of Lithology in Xiongan New Area. *Acta Geosci. Sin.* **2018**, *39*, 542–548.
4. Zhang, Y.; Li, X.; Sun, S. Evaluation and optimization of rural space suitability based on "the production, living and ecological space coordination" take beishakou township, the Xiongan New Area as an example. *Urban Dev. Stud.* **2019**, *26*, 116–124.
5. Lu, H.; Yan, Y.; Zhao, C.; Wu, G. Framework of constructing multi-scale ecological security pattern in Xiongan New Area. *Acta Ecol. Sin.* **2020**, *40*, 7105–7112.
6. Liu, H.; Wang, G.; Ma, C.; Gao, M.; Bai, Y.; Zhang, J.; Du, D. Study on evaluation index system of suitability for development and utilization of underground space in sedimentary plan: A Case Study of Tongzhou District in Beijing and Langfang north three counties in Hebei Province. *Geol. North China* **2022**, *45*, 68–74.
7. Bai, Y.; Ma, Z.; Zhang, J.; Guo, X.; Liu, H.; Miao, J.; Du, D. Suitability assessment of the engineering construction in the north-three-counties of Langfang. *Geol. Surv. Res.* **2019**, *42*, 117–122.
8. Sterling, R.L.; Nelson, S.R. *Planning for Underground Space: A Case Study for Minneapolis, Minnesota*; Underground Space Center, University of Minnesota: Minneapolis, MN, USA, 1982; pp. 21–40.
9. Mejia-Navarro, M.; Wohl, E.E.; Oaks, S.D. Geological hazards vulnerability and risk assessment using GIS: Model for Glenwood Springs, Colorado. *Geomorphology* **1994**, *10*, 331–354. [CrossRef]
10. Mejia-Navarro, M.; Garcia, L.A. Natural hazard and risk assessment using decision support systems, application: Glenwood Springs, Colorado. *Environ. Eng. Geosci.* **1996**, *2*, 299–324. [CrossRef]
11. Gan, X.; Lee, Y.; Xuan, Y. The building of fuzzy mathematic discriminant model of the plain area construction location suitability. *Geotech. Investig. Surv.* **2009**, *2*, 557–562.
12. Shan, M.; Hadidi, M.; Vessali, E.; Mosstafakhani, P.; Taheri, K.; Shahoie, S.; Khodamoradpour, M. Intelating multcriteria decision analysis for a GIS-based hazardous waste landfill sitting in Kurdistan Province, western Iran. *Waste Manag.* **2009**, *29*, 2740–2758.
13. Ahmed, Y.; Pradhan, B.; Tarabees, E. Integrated evaluation of urban development suitability based on remote sensing and GIS techniques: Contribution from the analytic hierarchy process. *Arab. J. Geosci.* **2011**, *4*, 463–473.
14. Liu, H.; Cao, H.; Fu, Y. Evaluation on the suitability of compound foundation for a city building land. In Proceedings of the 2018 National Engineering Survey Academic Conference, Xi'an, China, 21 June 2018.
15. Wang, W.; Guo, M. Engineering geological characteristics and suitability evaluation of site engineering construction in the Shangqiu urban planning area. *Miner. Explor.* **2022**, *13*, 130–138.
16. Lee, D.; Meng, H.; Shi, X. Membership clouds and membership cloud cenerators. *J. Comput. Res. Dev.* **1995**, *6*, 15–20.
17. Zhou, Z.; Shi, H.; Dong, Y.; Fan, W. Research on road collapse risk evaluation based on cloud model and AHP-EWM. *J. Saf. Environ.* **2023**, *23*, 1752–1761.
18. Ma, Z.; Xia, Y.; Wang, X.; Han, B.; Gao, Y. Integration of engineering geological investigation data and construction of a 3D geological structure model in the Xiongan New Area. *Geosci. Data Discov.* **2019**, *46*, 169–177.
19. Xie, H.; Xia, Y.; Meng, Q.; Zhao, C.; Ma, Z. Study on land subsidence assessment in evaluation of carrying capacity of geological environment. *Geol. Surv. Res.* **2019**, *42*, 104–108.
20. Ministry of Housing and Urban-Rural Development of the People's Republic of China (MOHURD). *Urban and Rural Planning Engineering Geological Survey Specifications (CJJ57-2012)*; China Building Industry Press: Beijing, China, 2013.

Article

Suitability Assessment of Multilayer Urban Underground Space Based on Entropy and CRITIC Combined Weighting Method: A Case Study in Xiong'an New Area, China

Hongwei Liu [1,2,3,*], Zhuang Li [2,3] and Qingcheng He [1]

[1] Chinese Academy of Geological Sciences, No. 26 Baiwanzhuang Street, Beijing 100037, China
[2] Tianjin Center, China Geological Survey, No. 4 Dazhigu 8th Road, Tianjin 300170, China
[3] Xiong'an Urban Geological Research Center, China Geological Survey, No. 4 Dazhigu 8th Road, Tianjin 300170, China
* Correspondence: liuhenry022@163.com; Tel.: +86-133-5209-9531

Abstract: Suitability assessment is an essential initial step in the scientific utilization of underground space. It plays a significant role in providing valuable insights for optimizing planning and utilization strategies. Utilizing urban underground space has the potential to enhance the capacity of urban infrastructure and public service facilities, as well as mitigate issues such as traffic congestion and land scarcity. To effectively plan and utilize urban underground space, it is crucial to conduct a suitability assessment. This assessment helps identify the factors that influence the utilization of underground space and their impacts, offering guidance on avoiding unfavorable conditions and ensuring the safety of planned underground facilities. To achieve objective and reasonable evaluation results, this paper proposed an assessment method that combines entropy and CRITIC (CRiteria Importance Through Intercriteria Correlation) weighting. Taking Xiong'an New Area as a study area, a suitability assessment indicator system for underground space was established. The system included criteria indicators and sub-criteria indicators. By analyzing the weights, the study identified the difference of suitability and critical affecting factors for shallow, sub-shallow, sub-deep, and deep underground space. The results showed that deep layers had better suitability than shallow layers in the study area. The regions with inferior and worse suitability were mostly located around Baiyangdian Lake, with proportions of acreage at 54.69% for shallow layer, 42.06% for sub-shallow layer, 41.69% for sub-deep layer, and 42.03% for deep layer. Additionally, the dominant affecting factors of suitability varied in different layers of underground space. These findings provide valuable evidence for the scientific planning and disaster prevention of underground space in Xiong'an New Area, and also serve as references for studying suitability in other areas.

Keywords: multilayer underground space; suitability assessment; entropy; CRITIC; Xiong'an New Area

Citation: Liu, H.; Li, Z.; He, Q. Suitability Assessment of Multilayer Urban Underground Space Based on Entropy and CRITIC Combined Weighting Method: A Case Study in Xiong'an New Area, China. *Appl. Sci.* **2023**, *13*, 10231. https://doi.org/10.3390/app131810231

Academic Editor: John Dodson

Received: 8 August 2023
Revised: 4 September 2023
Accepted: 7 September 2023
Published: 12 September 2023

1. Introduction

Due to rapid urbanization, many cities are facing saturation in the utilization of aboveground land, leading to problems such as traffic congestion and land resource shortage [1,2]. To accommodate urban sustainable development, the utilization of underground space has emerged as an efficient means to expand urban space [3–5]. Therefore, integrated planning that includes both aboveground and underground space is considered an optimal approach [6,7]. Unlike aboveground space, underground space is a non-renewable resource that is difficult to transform once built into some kind of infrastructure. As a result, conducting suitability assessment of underground space before urban planning becomes particularly significant.

Existing research shows that, as the carrier of underground space, the geological environment plays a crucial role in the utilization of underground space [8]. Understanding how the geological environment constrains the use of underground space and identifying the geological factors that affect it can help in avoiding or transforming unfavorable geological conditions. In particular, in newly building cities, the utilization of underground space is primarily influenced by geological factors rather than existing surface buildings and underground infrastructure, unlike in already built-up cities [9]. Assessing suitability can provide valuable insights for optimizing planning and utilization strategies of underground space. Therefore, studying the geological suitability of underground space holds significant theoretical and practical importance.

Existing studies on the suitability of underground space primarily rely on geological analysis, which is an essential and valid method [10,11]. The assessment system for the suitability of urban underground space (UUS) incorporates various geological factors, which differ depending on the region's geological characteristics [12–14]. These factors typically include active faults, land subsidence, groundwater level, bearing capacity of soils, compression of soils, ground elevation, sands liquefaction, soft soils [15–18]. However, while researchers generally agree on the geological factors that affect the suitability of underground space in the same study area, they may differ in the weights assigned to these factors. Therefore, the critical issue lies in selecting an assessment method that incorporates reasonable indicator weights. Various methods have been developed to analyze the suitability of underground space, including AHP (analytic hierarchy process) [19–21], entropy [22–24], neural network [25,26], fuzzy mathematics [27–29], TOPSIS (Technique for Order Preference by Similarity to Ideal Solution) [16,30], etc. The combined application of different methods has been shown to be more reasonable [31–33], considering that each method has its own advantages and disadvantages [34,35]. The critical issues in the selection of an assessment method are the objectivity of weights, correlation, discreteness, and comparative intensity of affecting factors. In this paper, an entropy and CRITIC (Criteria Importance Through Intercriteria Correlation) combined weighting method was applied to obtain reasonable weights. The entropy method, derived from information theory [36], was used to derive objective weights but focused on discreteness while ignoring correlation and comparative intensity of data [11,37]. However, the CRITIC method can effectively compensate for this deficiency [38–40]. Therefore, the combination of these two methods ensures the rationality of the objective weight of the selected indicators.

Consequently, taking Xiong'an New Area, a newly planned city, as a case study, this research established a multilayer assessment indicator system for the underground space. The study also identified the suitability and critical affecting factors of underground space. It is important to note that this paper specifically focused on the geological characteristics that affect the utilization of urban underground space (UUS), while other socioeconomic factors were not considered.

2. Study Area

The Xiong'an New Area, established as a national new district in 2017, is located in the eastern alluvial plain of the Taihang Mountains, at the central part of the North China Plain. It covers an area of 1770 km^2 and mainly consists of Rongcheng County, Anxin County, and Xiongxian County (Figure 1) [41]. Geographically, these counties are located within Hebei Province. The ground elevation in the area is less than 26 m, and the terrain is relatively flat. The study area is characterized by Quaternary deposits with a depth of over 100 m [42]. It is situated in the Jizhong platform depression, where buried faults are relatively well-developed, but modern activities are very weak [43]. The climate of the study area falls under the warm temperate continental monsoon zone, with an average annual temperature of 12.1 °C, average precipitation of 478 mm, and a maximum evaporation of 1762 mm [44,45]. Additionally, the area is home to Baiyangdian Lake, the largest freshwater wetland in the North China Plain [46].

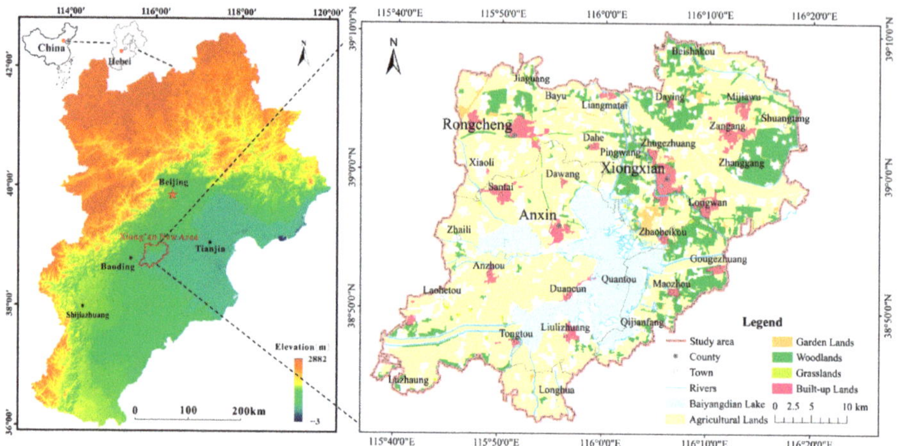

Figure 1. Location of the study area.

Furthermore, in accordance with the Planning Outline of Xiong'an New Area, the urban underground space will be utilized in an orderly manner based on different depth levels. These depth levels include the shallow layer (L_1: 0~−15 m), sub-shallow layer (L_2: −15~−30 m), sub-deep layer (L_3: −30~−50 m), and deep layer (L_4: −50~−100 m). The shallow and sub-shallow layers will be actively exploited, while the sub-deep and deep layers will be appropriately utilized.

3. Materials and Methods

3.1. Establishment of Suitability Assessment Frame

During the initial phase of assessing the suitability of urban underground spaces, a system of assessment indicators was developed by analyzing geological factors. Figure 2 illustrates four types of criteria indicators: topography, geotechnical characteristics, hydrogeological conditions, and adverse geological phenomena. These criteria indicators include various sub-criteria indicators and were considered in the suitability assessment. Some of these indicators are positive, such as geotechnical characteristics, where a larger value indicates better suitability of urban underground spaces. On the other hand, some indicators are negative, such as adverse geological phenomena, which indicate the opposite. To obtain comprehensive evaluation results, data on sub-criteria indicators from multiple layers should be obtained beforehand. Topography indicators, including ground elevation, were obtained from DEM (Digital Elevation Model) measurements. Geotechnical characteristics, such as bearing capacity and compression modulus of soils, were determined through standard penetration tests, shear tests, and geotechnical tests conducted in engineering geological boreholes. Hydrogeological conditions, such as groundwater buried depth, were measured in the field through groundwater level measurements, while aquifer thickness was inferred from lithology data obtained from geological boreholes. Adverse geological phenomena, such as land subsidence rate, were measured using InSAR (Interferometric Synthetic Aperture Radar) remote sensing. Sands liquefaction index was determined through standard penetration tests conducted in the engineering geological boreholes, and chemical corrosion of groundwater and soils was assessed through chemical experiments. All the aforementioned work was carried out by the China Geological Survey project.

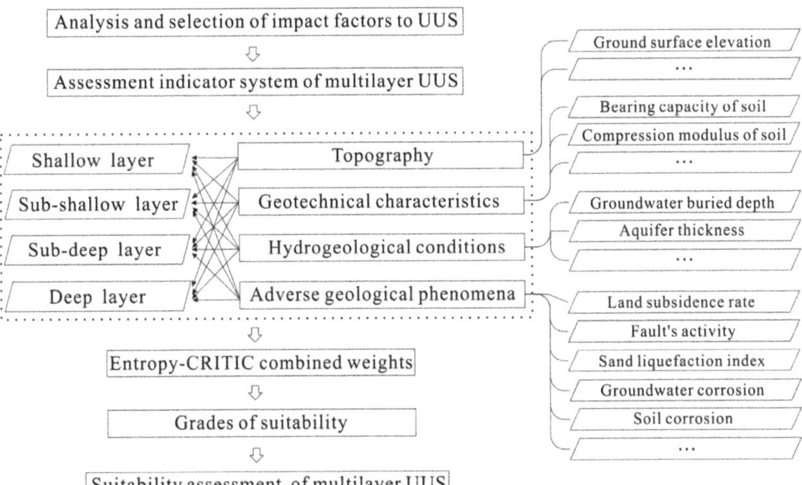

Figure 2. Frame flow chart of suitability assessment to UUS.

The relationships among various indicators were analyzed using entropy and CRITIC methods to determine the combined weights that represent the significance of different indicators. These indicator weights were then input into the evaluation model, which was the weighted average comprehensive index model, to calculate the comprehensive indexes of geological suitability. The final maps of grading evaluation for the multilayer UUS were generated using ArcGIS 10.8 software.

3.2. Establishment of Assessment Model

The suitability of underground space in Xiong'an New Area is evaluated using the weighted average comprehensive index model. The comprehensive index, *CI*, is obtained through the assignment of scores to indicators and calculation of indicator weights. The assessment model is as follows:

$$CI = \sum_{j=1}^{n} k_j \cdot w_j, j = 1, 2, \cdots, n$$

where *CI* is the comprehensive index, w_j is the indicator weight, k_j is the score assignment of the indicator, and *n* is the number of indicators.

3.3. Calculation of the Entropy-CRITIC Combined Weights

3.3.1. Entropy Weighting

(1)   Establish an initial assessment indicators matrix, *X*.

$$X = (X_{ij})_{m \times n} \tag{1}$$

where *m* is the number of evaluating objects, *n* is the number of indicators, *i* is the *i*-th object, and *j* is the *j*-th indicator.

(2)   Construct a normalization matrix, *Y*.

$$Y = (Y_{ij})_{m \times n} \tag{2}$$

where Y_{ij} satisfies the following Equations (3) or (4):

$$Y_{ij} = \frac{X_{ij} - minX_j}{maxX_j - minX_j}.$$ (3)

Here, j is a positive indicator.

$$Y_{ij} = \frac{maxX_j - X_{ij}}{maxX_j - minX_j}.$$ (4)

Here, j is a negative indicator.

(3)   Calculate the contribution of the i-th object to the j-th indicator, P_{ij}.

$$P_{ij} = \frac{Y_{ij}}{\sum_{i=1}^{m} Y_{ij}}$$ (5)

(4)   Calculate the entropy value of the j-th indicator, E_j.

$$E_j = -k\sum_{i=1}^{m} P_{ij} \ln P_{ij}$$ (6)

where $k = 1/\ln m$.

(5)   Calculate the otherness coefficient of the j-th indicator, G_j.

$$G_j = 1 - E_j$$ (7)

(6)   Calculate the entropy weight coefficient of the j-th indicator, v_j.

$$v_j = \frac{G_j}{\sum_{j=1}^{n} G_j}$$ (8)

3.3.2. CRITIC Weighting

(1)   Based on Equation (2), the standard deviation of the j-th indicator, S_j, is calculated.

$$S_j = \sqrt{\frac{\sum_{i=1}^{m} (Y_{ij} - \overline{Y}_j)^2}{n - 1}}$$ (9)

(2)   Based on Equation (2), the correlation coefficient between the b-th and j-th indicators, r_{bj}, is calculated.

$$r_{bj} = \frac{\sum_{b,j=1}^{n} (Y_{ib} - \overline{Y}_b)(Y_{ij} - \overline{Y}_j)}{\sqrt{\sum_{b=1}^{n} (Y_{ib} - \overline{Y}_b)^2 \sum_{j=1}^{n} (Y_{ij} - \overline{Y}_j)^2}}$$ (10)

(3)   Calculate the conflicting characteristic between the b-th and j-th indicators, A_j.

$$A_j = \sum_{b=1}^{n} (1 - r_{bj})$$ (11)

(4)   Calculate the quantity of information contained in j-th indicator, C_j.

$$C_j = S_j A_j$$ (12)

(5)   Calculate the weight coefficient of the j-th indicator, u_j.

$$u_j = \frac{C_j}{\sum_{j=1}^{n} C_j}$$ (13)

3.3.3. Combined Weighting

Based on v_j and u_j, the combination weight of j-th indicator, w_j, is calculated.

$$w_j = \frac{v_j u_j}{\sum_{j=1}^{n} v_j u_j} \tag{14}$$

4. Results

4.1. Assessment Indicator System

Based on the actual geological conditions of the study area, an assessment indicator system was established, comprising four kinds of criteria indicators: topography, geotechnical characteristics, hydrogeological conditions, and adverse geological phenomena (Table 1). Additionally, the assessment of four application layers of UUS (L_1: $0 \sim -15$ m, L_2: $-15 \sim -30$ m, L_3: $-30 \sim -50$ m, L_4: $-50 \sim -100$ m) involved the evaluation of different combinations of sub-criteria indicators. These sub-criteria indicators included ground elevation, bearing capacity of soils, compression modulus of soils, groundwater buried depth, aquifer thickness, land subsidence rate, sands liquefaction index, chemical corrosion of groundwater, and chemical corrosion of soils (Table 1).

Table 1. Assessment indicators of suitability to multilayer UUS in the study area.

Criteria Indicators	Sub-Criteria Indicators (Unit)	Grading Criteria of Suitability				Application Layer
		Grade I	Grade II	Grade III	Grade IV	
Topography	Ground elevation (m)	<8	8~8.5	8.5~9	>9	L_1~L_4
Geotechnical characteristics	Bearing capacity of soils ($0 \sim -5$ m) (kPa)	<80	80~100	100~130	>130	L_1
	Bearing capacity of soils ($-5 \sim -10$ m) (kPa)	<100	100~130	130~160	>160	L_1
	Bearing capacity of soils ($-10 \sim -15$ m) (kPa)	<110	110~130	130~170	>170	L_1
	Bearing capacity of soils ($-15 \sim -30$ m) (kPa)	<130	130~160	160~200	>200	L_2
	Bearing capacity of soils ($-30 \sim -50$ m) (kPa)	<130	130~160	160~200	>200	L_3
	Bearing capacity of soils ($-50 \sim -100$ m) (kPa)	<130	130~160	160~200	>200	L_4
	Compression modulus of soils ($0 \sim -5$ m) (MPa)	<4	4~11	11~15	>15	L_1
	Compression modulus of soils ($-5 \sim -10$ m) (MPa)	<4	4~11	11~15	>15	L_1
	Compression modulus of soils ($-10 \sim -15$ m) (MPa)	<4	4~11	11~15	>15	L_1
	Compression modulus of soils ($-15 \sim -30$ m) (MPa)	<4	4~11	11~15	>15	L_2
	Compression modulus of soils ($-30 \sim -50$ m) (MPa)	<4	4~11	11~15	>15	L_3
	Compression modulus of soils ($-50 \sim -100$ m) (MPa)	<4	4~11	11~15	>15	L_4

Table 1. *Cont.*

Criteria Indicators	Sub-Criteria Indicators (Unit)	Grading Criteria of Suitability				Application Layer
		Grade I	Grade II	Grade III	Grade IV	
Hydrogeological conditions	Groundwater buried depth (m)	<5	5~10	10~15	>15	L_1~L_4
	Aquifer thickness (0~−15 m) (m)	>7.5	5~7.5	2.5~5	<2.5	L_1
	Aquifer thickness (−15~−30 m) (m)	>7.5	5~7.5	2.5~5	<2.5	L_2
	Aquifer thickness (−30~−50 m) (m)	>7.5	5~7.5	2.5~5	<2.5	L_3
	Aquifer thickness (−50~−65 m) (m)	>7.5	5~7.5	2.5~5	<2.5	L_4
	Aquifer thickness (−65~−80 m) (m)	>7.5	5~7.5	2.5~5	<2.5	L_4
	Aquifer thickness (−80~−100 m) (m)	>7.5	5~7.5	2.5~5	<2.5	L_4
Adverse geological phenomena	Land subsidence rate (mm/a)	>50	30~50	10~30	<10	L_1~L_4
	Sands liquefaction index	>18	6~18	<6	0	L_1
	Chemical corrosion of groundwater	Strong	Medium	Weak	Micro	L_1~L_4
	Chemical corrosion of soils	Strong	Medium	Weak	Micro	L_1~L_4

The suitability of underground space in the study area was categorized into four grades: grade I (Worse suitability), grade II (Inferior suitability), grade III (Moderate suitability), and grade IV (Good suitability) [15,18]. The classification of ground elevation, chemical corrosion of groundwater, and chemical corrosion of soils was based on the grading criteria outlined in the 'Code for geo-engineering site investigation and evaluation of urban and rural planning (CJJ57-2012)'. The classification of bearing capacity of soils and compression modulus of soils was referenced from Gao's grading criteria [47]. The land subsidence rate was classified according to the grading criteria specified in the 'Specifications for risk assessment of geological hazard (GB/T 40112-2021)', while the sands liquefaction index was classified based on the grading criteria provided in the 'Code for seismic design of buildings (GB 50011-2010)'. The classification of groundwater buried depth and aquifer thickness was determined in accordance with the recommendations of local geological experts.

4.2. Characteristics of Geological Indicators

4.2.1. Topography

The study area exhibited distinct sedimentary facies characteristics, with an alluvial-proluvial facies observed in the northwestern region and an alluvial-lacustrine facies observed in the southeastern region. This differentiation in sedimentary facies was roughly delineated by the connection line between Rongcheng urban districts and northern Anxin [48].

Based on the DEM elevation measuring data in 2021 (Figures 3 and 4), the study area exhibited a relatively flat topography with a gradual decrease in elevation from northwest to southeast, characterized by a gradient of less than 2‰. The ground elevation mostly ranged between 6 m and 10 m, with the highest point reaching 26 m in Jiaguang town, Rongcheng County. However, around Baiyangdian Lake, the ground elevation was less than 5 m. It is important to note that when the ground elevation drops below 9 m, there is a potential risk of inundation for underground space, as indicated by the warning water level of Baiyangdian Lake.

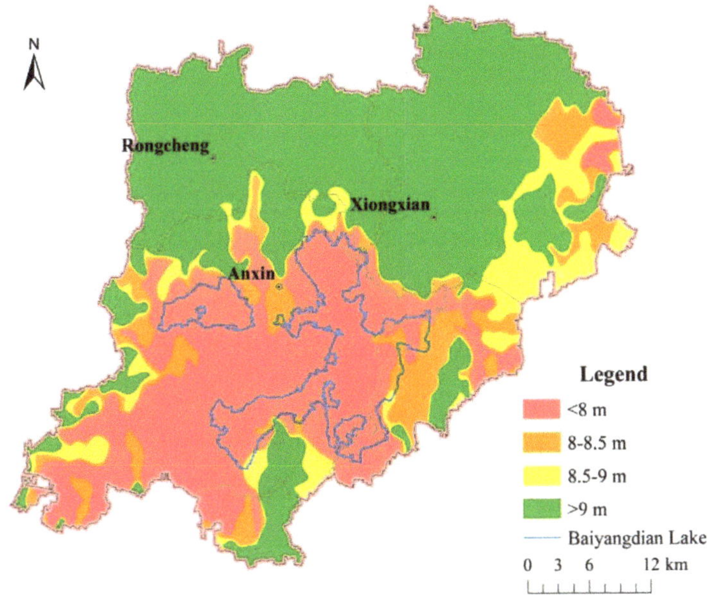

Figure 3. Zoning map of ground elevation.

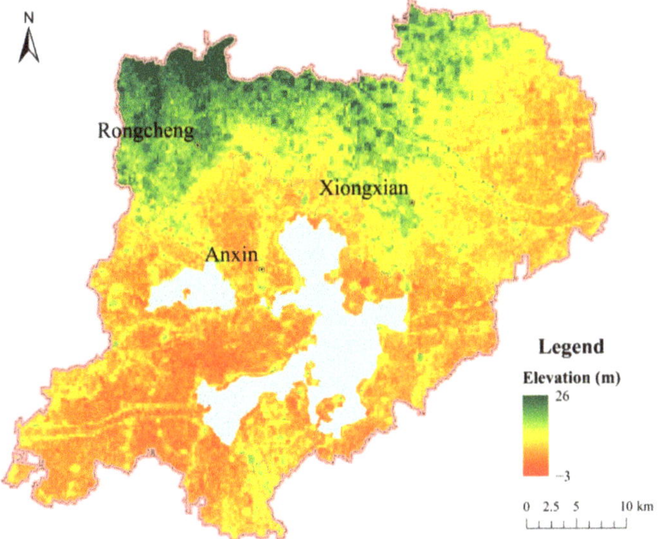

Figure 4. DEM map of ground elevation.

4.2.2. Geotechnical Characteristics

The planning areas of underground space were found to be deposited with Quaternary soils, which consisted of sands, clays, and silts. Additionally, the main soil layers were evenly distributed to a certain extent, and their sedimentary rhythm was relatively stable. The geotechnical test data, including natural void ratio, liquidity index, and organic matter content, did not indicate the presence of soft soils.

The bearing capacity and compression modulus of the soils varied with depth. Based on data from standard penetration tests, shear tests, and geotechnical tests conducted in engineering geological boreholes, the bearing capacity of each individual soil layer was determined. The bearing capacity of the foundation soils at different depths was then calculated by integrating the data of the individual soil layers using the weighted average method. The bearing capacity ranges for depths of $0\sim-5$ m, $-5\sim-10$ m, $-10\sim-15$ m, $-15\sim-30$ m, $-30\sim50$ m, and $-50\sim-100$ m were found to be 95~130 kPa, 100~180 kPa, 110~250 kPa, 145~240 kPa, 180~280 kPa, and 180~280 kPa, respectively (Figure 5). Additionally, the compression modulus of the soils at different depths was determined based on lateral confined test data. The majority of the compression modulus values for depths of $0\sim-5$ m, $-5\sim-10$ m, $-10\sim-15$ m, $-15\sim-30$ m, $-30\sim-50$ m, and $-50\sim-100$ m ranged from 4 MPa to 15 MPa (Figure 6).

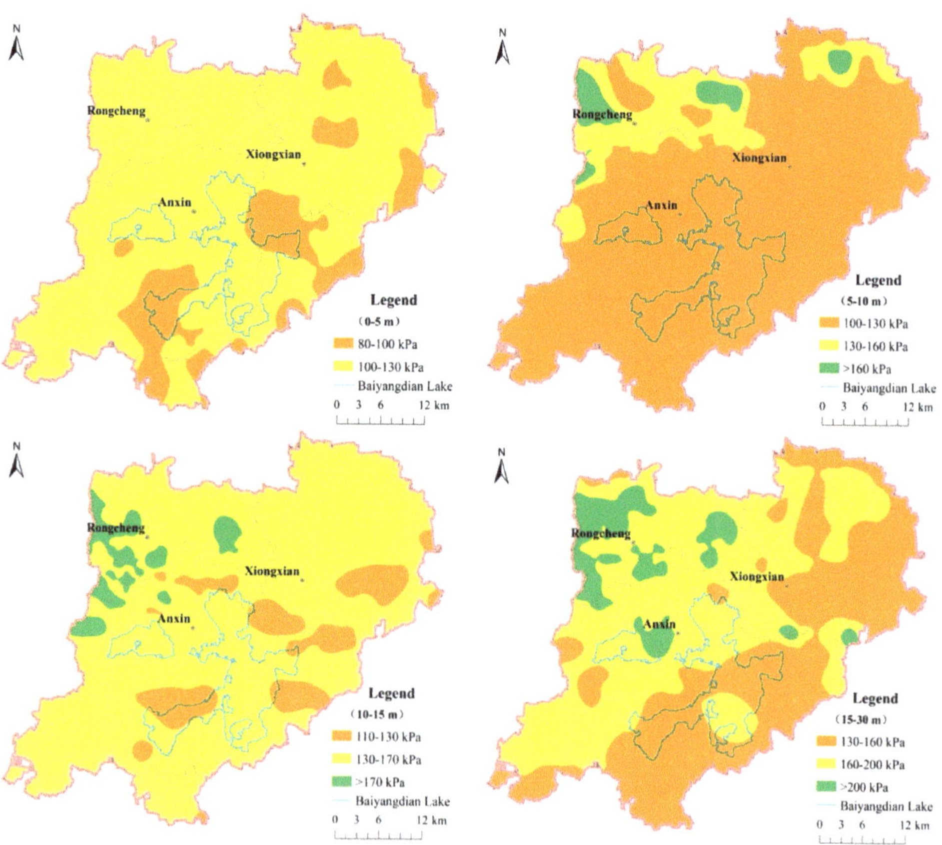

Figure 5. *Cont.*

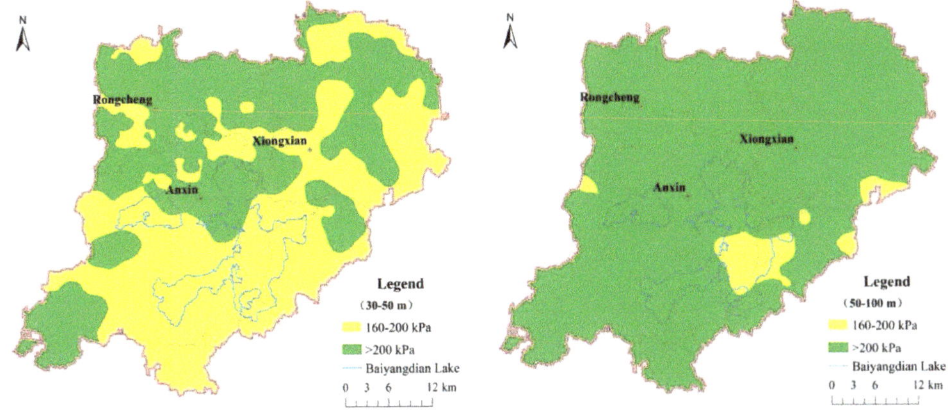

Figure 5. Bearing capacity zoning map of soils within different buried depths.

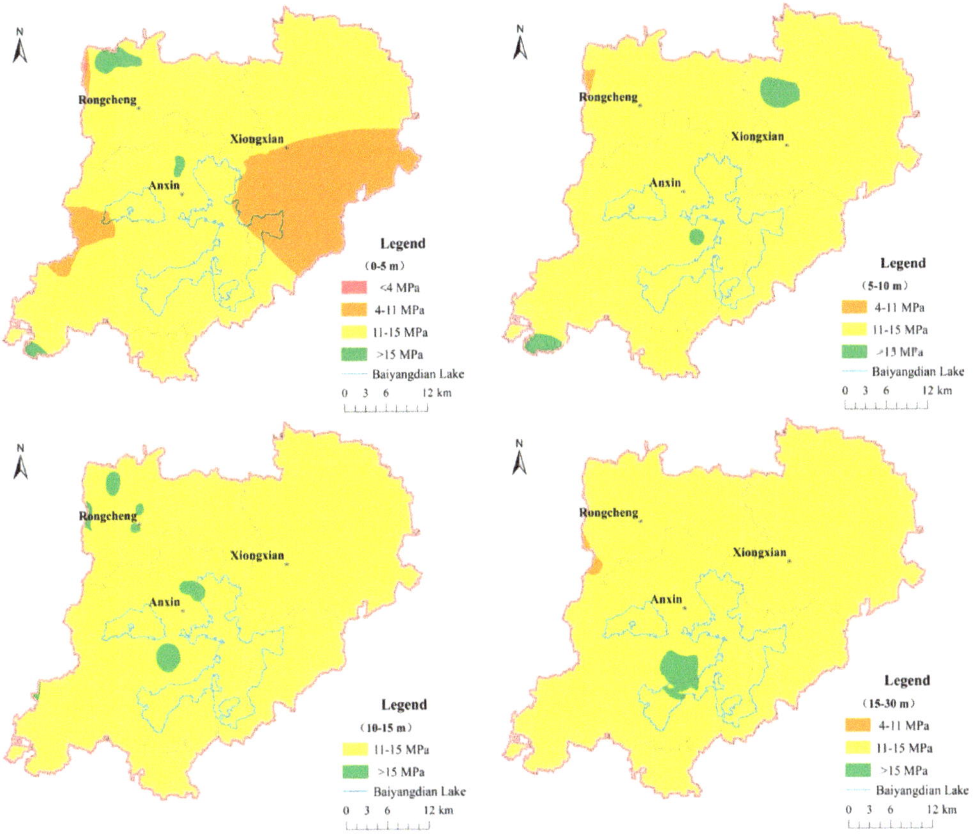

Figure 6. *Cont.*

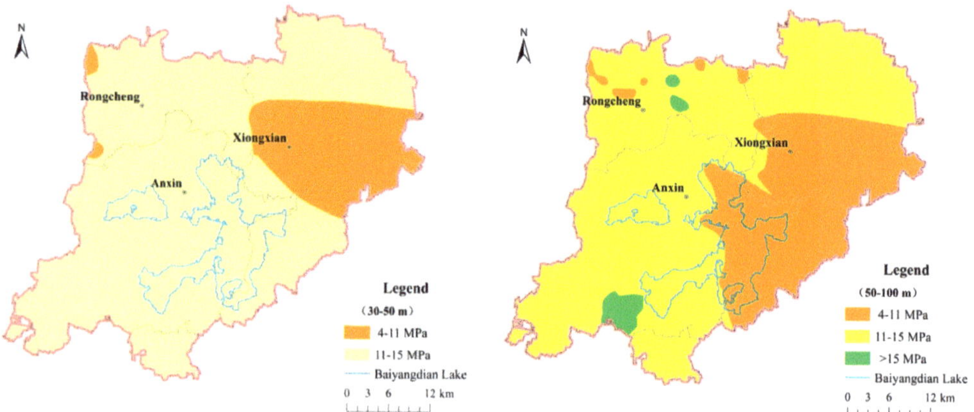

Figure 6. Compression modulus zoning map of soils within different buried depths.

4.2.3. Hydrogeological Conditions

The regional aquifer structure was determined by generalizing the lithology data obtained from geological boreholes in the Quaternary strata. Within a depth of 100 m, there were six relatively continuous aquifers located at different intervals: 0~−15 m, −15~−30 m, −30~−50 m, −50~−65 m, −65~−80 m, and −80~−100 m. Although the thickness of each aquifer varied regionally, the average thickness in most areas was less than 2.5 m (Figure 7).

Based on the measurement data of groundwater level in 2021, it was observed that the predominant buried depth of mixed groundwater in six aquifers was between 5 m and 20 m (Figure 8). However, the surrounding zone of Baiyangdian Lake exhibited a lesser depth of less than 5 m. In general, the groundwater flow direction under current conditions was observed to be from northwest to southeast [49].

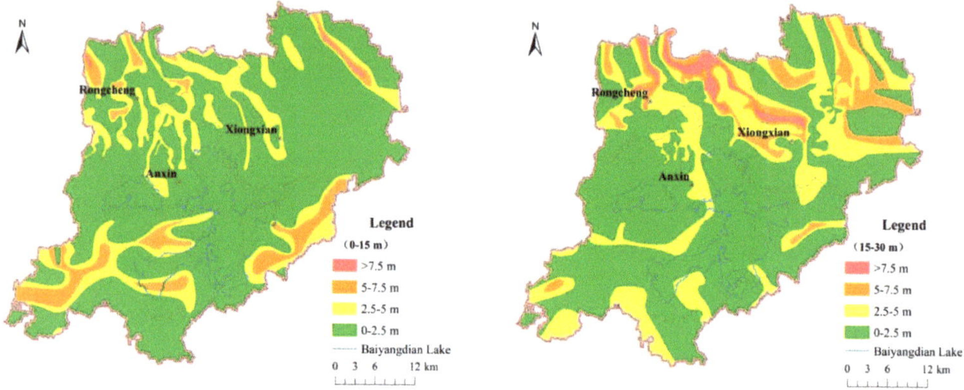

Figure 7. *Cont.*

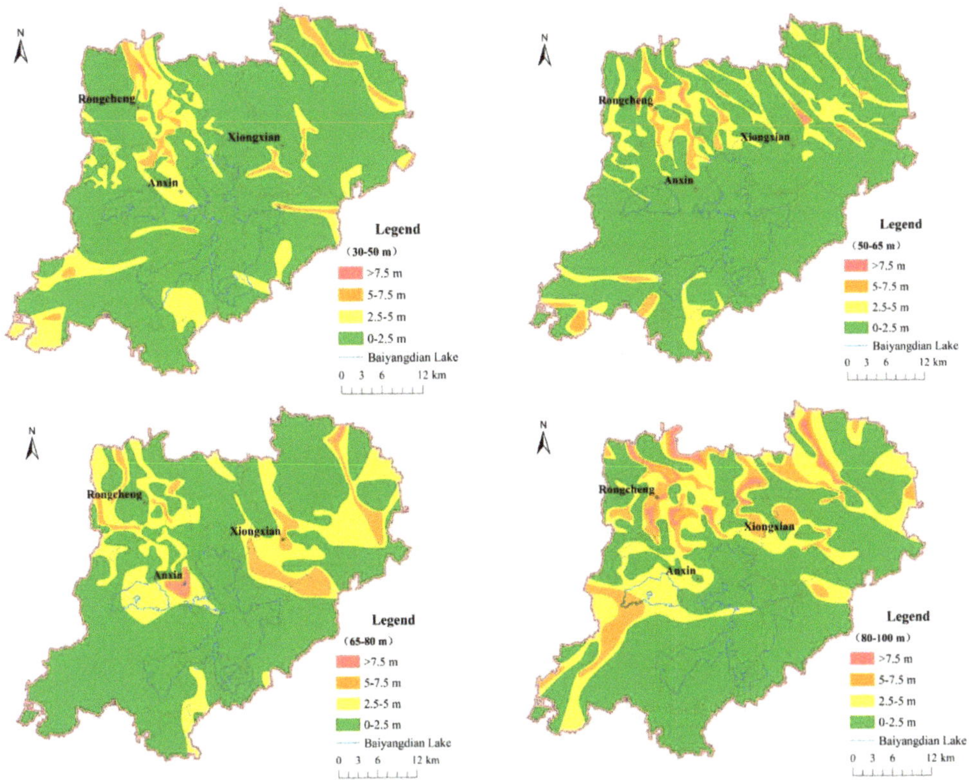

Figure 7. Aquifer thickness zoning map of soils within different buried depths.

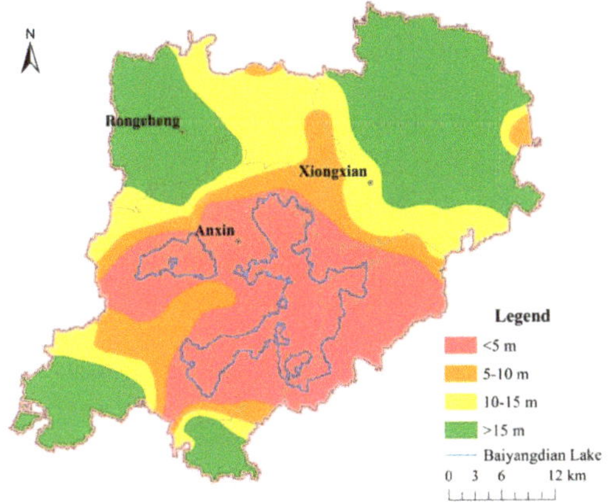

Figure 8. Zoning map of groundwater buried depth.

4.2.4. Adverse Geological Phenomena

Based on the comprehensive geophysical prospecting and borehole stratigraphic dislocation records, multiple faults were detected in Xiong'an New Area, such as the Niudong fault, Rongdong fault [50]. However, these faults have shown inactive features since the

Late Pleistocene, suggesting a relatively stable geological structure [43,51]. Nonetheless, there have been some adverse geological phenomena observed [52], including land subsidence, liquefaction of sands, chemical corrosion of groundwater, and chemical corrosion of soils.

According to the PS-InSAR (Persistent Scatterer Interferometric Synthetic Aperture Radar) remote sensing data during 2021, the land subsidence rate exhibited spatial variation (Figure 9). Most regions showed a rate lower than 10 mm/a, while several northern sites displayed higher rates of more than 30 mm/a. The sands liquefaction index was calculated based on the data of standard penetration tests conducted in the engineering geological boreholes. According to the index, potential sands liquefaction would primarily occur in the central and southern parts of the study area (Figure 10), considering the current groundwater buried depth. The corrosion intensity of both groundwater and soils, as observed from the data of groundwater chemical and soil chemical experiments, showed distribution characteristics of being higher in the southeast and lower in the northwest (Figures 11 and 12).

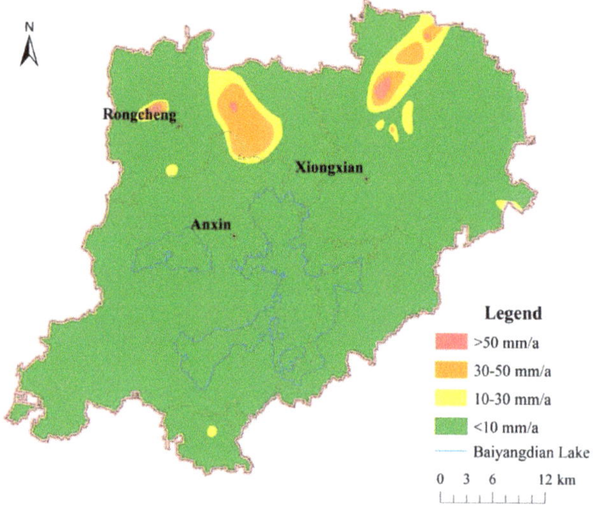

Figure 9. Zoning map of land subsidence rate.

4.3. Weights of Indicators

In this study, the study area was divided into multiple cells using a grid of 500 m by 500 m. These cells were selected as assessment samples for four layers of UUS, namely, L_1, L_2, L_3, and L_4. An initial assessment indicator matrix was created using Equation (1). Entropy and CRITIC weights were then calculated based on Equations (2)–(13). Using Equation (14), combined weights were calculated based on entropy and CRITIC weighting methods (Figure 13). The L_1 layer of UUS was taken as an example to illustrate the calculating processes and outcomes; the information is presented in Tables 2 and 3.

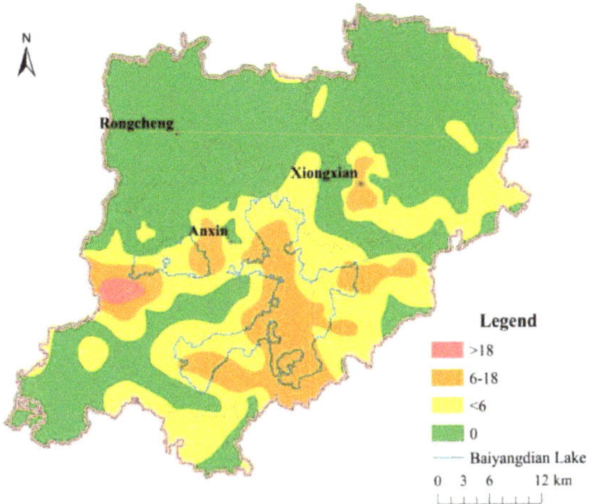

Figure 10. Zoning map of sands liquefaction index.

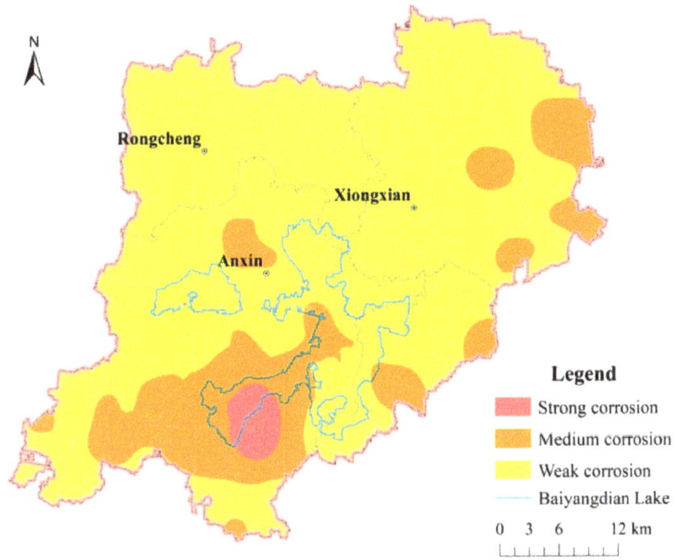

Figure 11. Zoning map of groundwater chemical corrosion intensity.

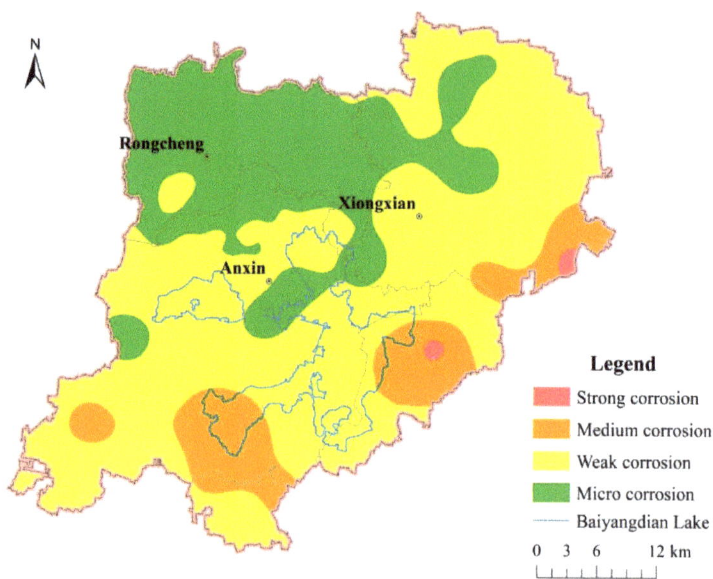

Figure 12. Zoning map of soil chemical corrosion intensity.

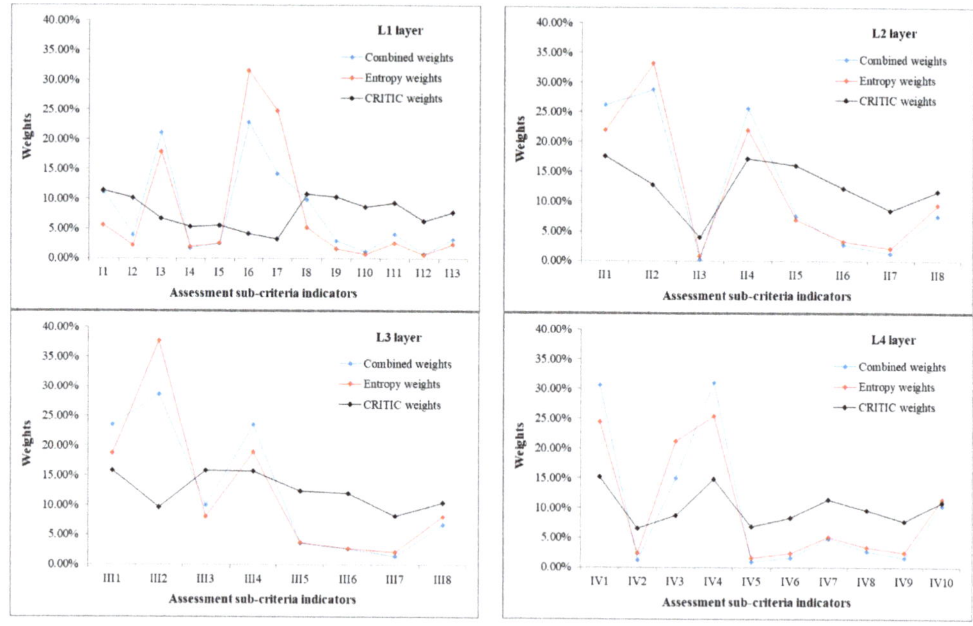

Figure 13. Weights curve of sub-criteria indicators for L_1~L_4 layers.

Table 2. Processes and outcomes information with entropy weighting method for L_1 layer.

Criteria Indicators	Sub-Criteria Indicators	Entropy Values, E	Otherness Coefficients, G	Entropy Weights, v
Topography	Ground elevation, I_1	0.9281	0.0719	5.61%
Geotechnical characteristics	Bearing capacity of soils ($0\sim-5$ m), I_2	0.9710	0.0290	2.26%
	Bearing capacity of soils ($-5\sim-10$ m), I_3	0.7707	0.2293	17.90%
	Bearing capacity of soils ($-10\sim-15$ m), I_4	0.9759	0.0241	1.88%
	Compression modulus of soils ($0\sim-5$ m), I_5	0.9660	0.0340	2.65%
	Compression modulus of soils ($-5\sim-10$ m), I_6	0.5948	0.4052	31.62%
	Compression modulus of soils ($-10\sim-15$ m), I_7	0.6823	0.3177	24.80%
Hydrogeological conditions	Groundwater buried depth, I_8	0.9330	0.0670	5.23%
	Aquifer thickness ($0\sim-15$ m), I_9	0.9788	0.0212	1.65%
Adverse geological phenomena	Land subsidence rate, I_{10}	0.9903	0.0097	0.76%
	Sands liquefaction index, I_{11}	0.9671	0.0329	2.57%
	Chemical corrosion of groundwater, I_{12}	0.9913	0.0087	0.68%
	Chemical corrosion of soils, I_{13}	0.9694	0.0306	2.39%

Table 3. Processes and outcomes information with CRITIC weighting method for L_1 layer.

Criteria Indicators	Sub-Criteria Indicators	Deviation Values, S	Conflicting Values, A	Quantity of Information, C	CRITIC Weights, u
Topography	Ground elevation, I_1	0.439	9.508	4.173	11.45%
Geotechnical characteristics	Bearing capacity of soils ($0\sim-5$ m), I_2	0.345	10.747	3.711	10.18%
	Bearing capacity of soils ($-5\sim-10$ m), I_3	0.255	9.640	2.459	6.75%
	Bearing capacity of soils ($-10\sim-15$ m), I_4	0.191	10.258	1.963	5.39%
	Compression modulus of soils ($0\sim-5$ m), I_5	0.192	10.464	2.013	5.52%
	Compression modulus of soils ($-5\sim-10$ m), I_6	0.129	11.674	1.507	4.14%
	Compression modulus of soils ($-10\sim-15$ m), I_7	0.092	13.087	1.200	3.29%
Hydrogeological conditions	Groundwater buried depth, I_8	0.415	9.535	3.957	10.86%
	Aquifer thickness ($0\sim-15$ m), I_9	0.304	12.356	3.761	10.32%
Adverse geological phenomena	Land subsidence rate, I_{10}	0.228	13.771	3.145	8.63%
	Sands liquefaction index, I_{11}	0.350	9.728	3.403	9.34%
	Chemical corrosion of groundwater, I_{12}	0.225	10.259	2.305	6.32%
	Chemical corrosion of soils, I_{13}	0.298	9.557	2.846	7.81%

4.4. Assessment Results

The suitability of urban underground space at different depths was evaluated based on the distribution of geological indicators in the study area, as shown in Figures 14–17.

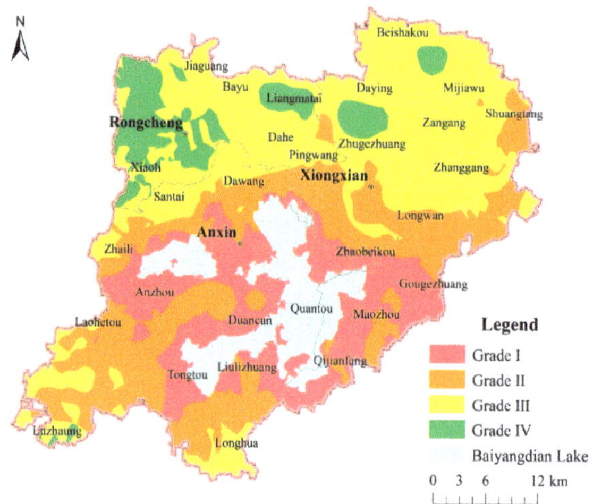

Figure 14. Suitability zoning map of L_1 layer.

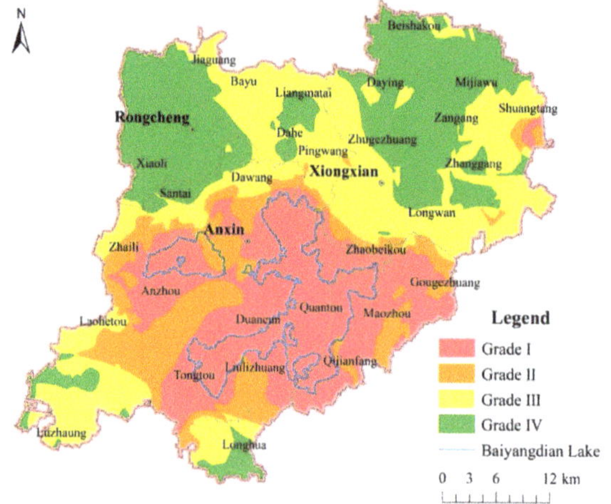

Figure 15. Suitability zoning map of L_2 layer.

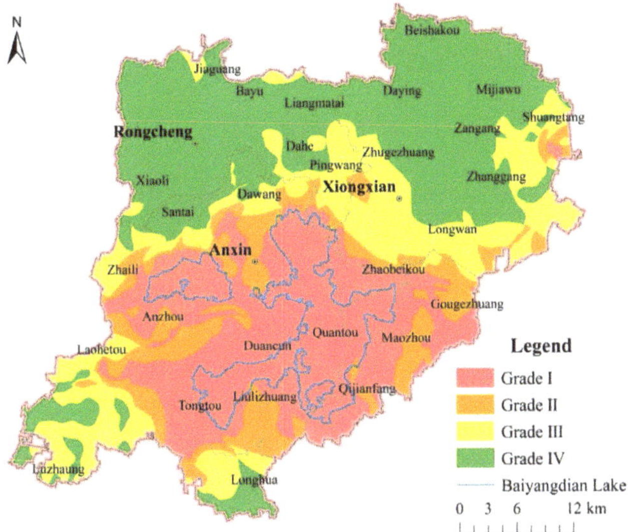

Figure 16. Suitability zoning map of L_3 layer.

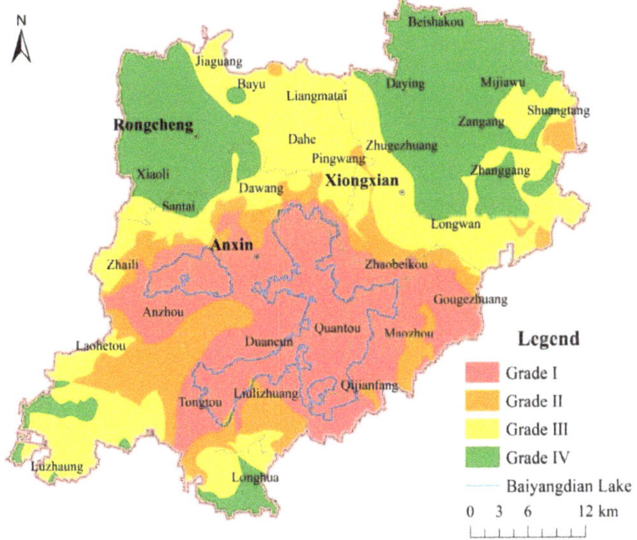

Figure 17. Suitability zoning map of L_4 layer.

For the shallow layer (L_1) within a depth of 0~−15 m, grade I, grade II, grade III, and grade IV accounted for 28.87%, 25.82%, 37.59%, and 7.72% of the study area, respectively (Figure 14). It was important to note that the Baiyangdian Lake region occupied 34.72% of the space classified as grade I and 0.81% of the space classified as grade II. Some scholars have suggested that larger lakes could have an impact on the use of shallow underground space [17]. In order to account for this, Baiyangdian Lake, including its wetland area, was used as a sensitive indicator to adjust the assessment result. Accordingly, all regions of Baiyangdian Lake were adjusted to grade I. The revised acreages with grade I, grade II, grade III, and grade IV accounted for 29.08%, 25.61%, 37.59%, and 7.72%, respectively. Based on the evaluation result, it was found that most of Rongcheng and Xiongxian, particularly the Rongcheng regions, were considered good or moderately suitable for underground space utilization. On the other hand, the regions around the central-eastern areas of Anxin were found to be less suitable due to issues with the bearing capacity and compression modulus of soils. As a result, the compression modulus of soils ($-5\sim-10$ m), bearing capacity of soils ($-5\sim-10$ m), and compression modulus of soils ($-10\sim-15$ m) were given higher weights in assessing underground space suitability compared to other sub-indicators.

For the deep layer (L_2) within a depth of $-15\sim-30$ m, grade I accounted for about 28.27% of the total acreage, grade II accounted for 13.79%, grade III accounted for 29.91%, and grade IV accounted for 28.03% (Figure 15). The analysis showed that the most suitable regions were mainly located near the west of Rongcheng and the central-north of Xiongxian. On the other hand, the regions around Baiyangdian Lake were found to be less suitable due to factors such as soil bearing capacity, ground elevation, and groundwater buried depth. Among these factors, the weights assigned to bearing capacity of soils ($-15\sim-30$ m), ground elevation, and groundwater buried depth were relatively high, indicating their significant influence on the suitability of underground space, accounting for 28.89%, 26.22%, and 25.60%, respectively, compared to other sub-indicators.

For the sub-deep layer (L_3), which was within a depth of $-30\sim-50$ m, the area occupied by grade I, grade II, grade III, and grade IV was 29.32%, 12.37%, 20.64%, and 37.67%, respectively. This indicated that the majority of the central-eastern regions of Rongcheng and Xiongxian were highly suitable for underground space utilization due to factors such as soil bearing capacity, groundwater buried depth, and ground elevation (Figure 16). Among these factors, the weights assigned to soil bearing capacity ($-30\sim-50$ m), groundwater buried depth, and ground elevation were relatively higher at 28.62%, 23.57%, and 23.55%, respectively, compared to other sub-indicators.

For the deep layer (L_4) within a depth of $-50\sim-100$ m, approximately 28.46% of the total area was classified as grade I, 13.57% as grade II, 29.52% as grade III, and 28.45% as grade IV (Figure 17). The assessment result revealed that the regions with good suitability were primarily located near the west of Rongcheng and the central-north of Xiongxian. On the other hand, the regions with poorer suitability were found around the central-east of Anxin, mainly due to factors such as groundwater buried depth, ground elevation, and compression modulus of soils. These factors, groundwater buried depth, ground elevation, and compression modulus of soils ($-50\sim-100$ m), were assigned higher weights, indicating their significance in determining underground space suitability, accounting for 31.02%, 30.66%, and 15.09%, respectively, compared to other sub-indicators.

5. Discussion

Firstly, in the suitability assessment of UUS, faults are typically an important geological factor to consider [53]. However, in this case study, faults were not included in the assessment indicator system due to their weak activity and minimal impact on the planning and construction of Xiong'an New Area [43,51]. The study area did not show other adverse geological phenomena, such as collapsible loess and gravel. However, it is important to note that the river flow in the study area is generally small and seasonal. Since these rivers have limited impact, they have not been included in the assessment indicator system. Additionally, based on the evaluation results of the four layers, the areas that were deemed good and moderate in suitability were mainly located in the north, south, and southwest of the study area. On the other hand, the inferior and worse suitable regions were mostly found around Baiyangdian Lake (Figure 18). This finding aligns well with the characteristics of stratigraphic sedimentary facies.

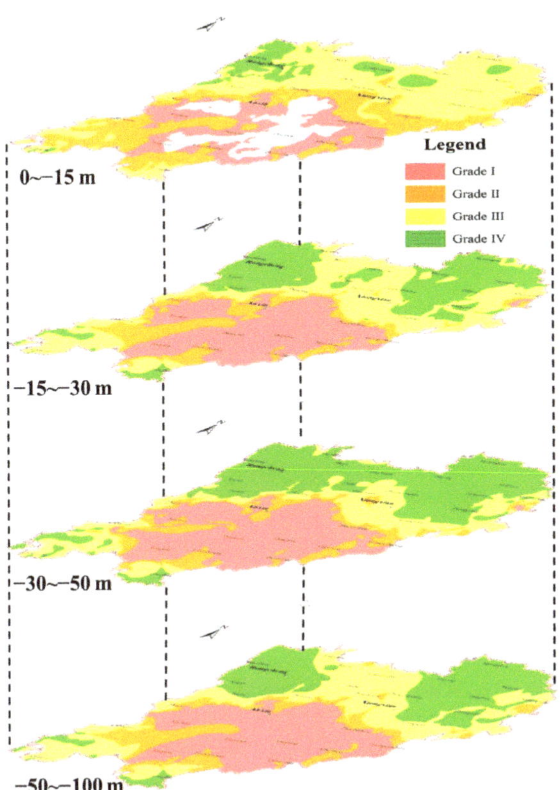

Figure 18. The integrated suitability assessment results of multilayer urban underground space.

Secondly, when the study area is divided into smaller evaluation units, such as smaller grids and more cells, the calculated weights are theoretically more accurate. However, in this paper, the grid size of 500 m by 500 m was temporarily set without further subdivision, in order to focus on the research method and reduce the amount of calculation. Subsequent researchers can refine the segmentation based on the acreage of the selected study area.

Thirdly, the planning and construction period of a city, especially for newly building cities such as Xiong'an New Area with a 'millennium plan' [54], is generally long. The utilization of urban underground space is a gradual process that may be also influenced by various socioeconomic conditions. In this study, we focused on examining the geological factors that affect the suitability of multilayer underground urban spaces (UUS), without considering socioeconomic factors and excavation types of underground facilities. However, it is important to note that future research should also investigate these aspects. This is also the limitation in this paper. If data on socioeconomic factors and excavation types could be obtained, they would provide more valuable insights for planning managers in optimizing planning and utilization strategies of underground space.

Finally, up to now, there have been limited studies on the suitability of underground space in Xiong'an New Area, as revealed by the literature review. Only Gao [47] has conducted a study on this topic. In contrast to this paper, Gao classified the underground space of Xiong'an New Area into three layers: shallow layer (0~−30 m), sub-deep layer (−30~−50 m), and deep layer (−50~−70 m). The evaluation indicators used in Gao's study included compression modulus of soils, groundwater depth, land subsidence rate, and aquifer thickness. However, the bearing capacity of soils was not considered. The evaluation method employed was the entropy-cloud model. The evaluation results differed from those of this paper. The areas with the worst grades were primarily located in the northern part of Xiongxian County. Similar to the findings of this paper, the Baiyangdian area generally exhibited a lower grade compared to its surrounding areas. After consulting the opinions of local geological experts, it is believed that the weights of indicators determined in this paper are more reasonable. This is because the evaluation takes into account the correlation, discreteness, and comparative intensity of the affecting indicators, making the results more valuable for reference.

6. Conclusions

(1)　A process was proposed to assess the suitability of multilayer UUS based on the impacts of geological environments. The case study of Xiong'an New Area was used to establish an assessment indicator system, including four types of criteria indicators: topography, geotechnical characteristics, hydrogeological conditions, and adverse geological phenomena. The suitability of four vertical layers within a depth of 100 m was identified, providing guidance for urban planners and constructors. To ensure an objective and rational assessment, the entropy and CRITIC combined method was employed to determine the weights of sub-criteria indicators, considering the discreteness, correlation, and comparative intensity of data. Compared with previously used combination by other researcher, this paper demonstrated a better method. This combined weighting method can be applied to similar regions for the suitability assessment of UUS.

(2)　The evaluation results indicated that the suitability of UUS in Xiong'an New Area varied depending on the depth of strata. Deeper layers showed better suitability, while shallower layers showed relatively worse suitability. The proportions of acreage with good and moderate suitability were as follows: 45.31% for shallow layer, 57.94% for sub-shallow layer, 58.31% for sub-deep layer, and 57.97% for deep layer. When considering the plane distribution state, the good and moderate suitable zones were predominantly located in the north, south, and southwest of the study area. Conversely, the inferior and worse suitable zones were mainly found around Baiyangdian Lake. Although the factors influencing the suitability of the four layers differed, the main indicators were the bearing capacity of soils, compression modulus of soils, groundwater buried depth, and ground elevation, which carried higher weights. Groundwater buried depth was significantly affected by artificial exploitation of groundwater, while the other three indicators are primarily influenced by the sedimentary process of strata.

(3)　The paper focused on the geological indicators affecting the utilization of UUS in Xiong'an New Area. However, it did not include socioeconomic factors and excavation types of underground facilities. Although the UUS utilization in Xiong'an New Area is still in the initial stage, future studies should consider socioeconomic progress, existing underground infrastructures, and the difficulty of excavation based on the evaluation results.

Author Contributions: Conceptualization, H.L. and Q.H.; methodology, H.L.; software, H.L. and Z.L.; validation, H.L. and Z.L.; investigation, H.L. and Z.L.; data curation, H.L. and Z.L.; writing—original draft preparation, H.L.; writing—review and editing, H.L. and Q.H.; funding acquisition, H.L. All authors have read and agreed to the published version of the manuscript.

Funding: This research was funded by China Geological Survey project (Grant No. DD20221727).

Institutional Review Board Statement: Not applicable.

Informed Consent Statement: Not applicable.

Data Availability Statement: Not applicable.

Acknowledgments: The authors would like to thank Bo Han and Yihang Gao for their valuable suggestions.

Conflicts of Interest: The authors declare no conflict of interest.

References

1. Romano, B.; Zullo, F.; Fiorini, L.; Marucci, A.; Ciabò, S. Land transformation of Italy due to half a century of urbanization. *Land Use Policy* **2017**, *67*, 387–400. [CrossRef]
2. Zhang, Y.H.; Zhu, J.B.; Liao, Z.Y.; Guo, J.; Xie, H.P.; Peng, Q. An intelligent planning model for the development and utilization of urban underground space with an application to the Luohu District in Shenzhen. *Tunn. Undergr. Space Technol.* **2021**, *112*, 103933. [CrossRef]
3. Bobylev, N.; Sterling, R. Urban underground space: A growing imperative. Perspectives and current research in planning and design for underground space use. *Tunn. Undergr. Space Technol.* **2016**, *55*, 1–4. [CrossRef]
4. Chen, Z.L.; Chen, J.Y.; Liu, H.; Zhang, Z.F. Present status and development trends of underground space in Chinese cities: Evaluation and analysis. *Tunn. Undergr. Space Technol.* **2018**, *71*, 253–270. [CrossRef]
5. Tann, L.; Ritter, S.; Hale, S.; Langford, J.; Salazar, S. From urban underground space (UUS) to sustainable underground urbanism (SUU): Shifting the focus in urban underground scholarship. *Land Use Policy* **2021**, *109*, 105650. [CrossRef]
6. Admiraal, H.; Cornaro, A. *Underground Spaces Unveiled: Planning and Creating the Cities for the Future*; ICE Publishing: London, UK, 2018.
7. Cui, J.Q.; Broere, W.; Lin, D. Underground space utilisation for urban renewal. *Tunn. Undergr. Space Technol.* **2021**, *108*, 103726. [CrossRef]
8. Li, X.Z.; Li, C.C.; Parriaux, A.; Wu, W.B.; Li, H.Q.; Sun, L.P.; Liu, C. Multiple resources and their sustainable development in Urban Underground Space. *Tunn. Undergr. Space Technol.* **2016**, *55*, 59–66. [CrossRef]
9. Lin, D.; Broere, W.; Cui, J.Q. Underground space utilisation and new town development: Experiences, lessons and implications. *Tunn. Undergr. Space Technol.* **2022**, *119*, 104204. [CrossRef]
10. Admiraal, H.; Cornaro, A. Future cities, resilient cities-The role of underground space in achieving urban resilience. *Undergr. Space* **2020**, *5*, 223–228. [CrossRef]
11. Tan, F.; Wang, J.; Jiao, Y.Y.; Ma, B.; He, L. Suitability evaluation of underground space based on finite interval cloud model and genetic algorithm combination weighting. *Tunn. Undergr. Space Technol.* **2021**, *108*, 103743. [CrossRef]
12. Šipetić, N.; Kuzmić, P. Design the Future Urban Plan of the Underground Construction from the Aspect of Geological and Geotechnical Features of Belgrade's Inner City Area. *Procedia Eng.* **2016**, *165*, 641–648. [CrossRef]
13. Tong, D.; Tan, F.; Ma, B.; Jiao, Y.Y.; Wang, J. A Suitability Evaluation Method of Urban Underground Space Based on Rough Set Theory and Conditional Entropy: A Case Study in Wuhan Changjiang New Town. *Appl. Sci.* **2022**, *12*, 1347. [CrossRef]
14. Peng, Z.L.; Zhang, Y.; Tan, F.; Lv, J.H.; Li, L.H. Variable-Weight Suitability Evaluation of Underground Space Development Considering Socioeconomic Factors. *Sustainability* **2023**, *15*, 3574. [CrossRef]
15. Hou, W.; Yang, L.; Deng, D.C.; Ye, J.; Clarke, K.; Yang, Z.J.; Zhuang, W.; Liu, J.X.; Huang, J.C. Assessing quality of urban underground spaces by coupling 3D geological models: The case study of Foshan city, South China. *Comput. Geosci.* **2016**, *89*, 1–11. [CrossRef]
16. Lu, Z.L.; Wu, L.; Zhuang, X.Y.; Rabczuk, T. Quantitative assessment of engineering geological suitability for multilayer Urban Underground Space. *Tunn. Undergr. Space Technol.* **2016**, *59*, 65–76. [CrossRef]
17. Zhang, M.S.; Wang, H.Q.; Dong, Y.; Li, L.; Sun, P.P.; Zhang, G. Evaluation of urban underground space resources using a negative list method: Taking Xi'an City as an example in China. *China Geol.* **2020**, *3*, 124–136. [CrossRef]
18. Deng, F.; Pu, J.; Huang, Y.; Han, Q.D. 3D geological suitability evaluation for underground space based on the AHP-cloud model. *Undergr. Space* **2023**, *8*, 109–122. [CrossRef]
19. Youssef, A.M.; Pradhan, B.; Tarabees, E. Integrated evaluation of urban development suitability based on remote sensing and GIS techniques: Contribution from the analytic hierarchy process. *Arab. J. Geosci.* **2011**, *4*, 463–473. [CrossRef]
20. Zhang, X.B.; Wang, C.S.; Fan, J.; Wang, H.J.; Li, H.L. Optimizing the analytic hierarchy process through a suitability evaluation of underground space development in Tonghu District, Huizhou City. *Energies* **2020**, *13*, 742. [CrossRef]
21. Duan, Y.; Xie, Z.; Zhao, F.; Zeng, H.; Lin, M.; Chen, H.; Zuo, X.; He, J.; Hou, Z. Suitability of underground space development in plateau cities based on geological environment analysis: Case study in Kunming, China. *J. Urban Plan. Dev.* **2021**, *147*, 05021014. [CrossRef]
22. Dai, S.; Niu, D. Comprehensive evaluation of the sustainable development of power grid enterprises based on the model of fuzzy group ideal point method and combination weighting method with improved group order relation method and entropy weight method. *Sustainability* **2017**, *9*, 1900. [CrossRef]
23. Guo, J.H.; Liu, K.; Deng, Y.F.; Yu, C.H. Geological evaluation of underground space resources based on the entropy weight optimization method. *Geol. Bull. China* **2023**, *42*, 385–396.
24. Zhang, Y.B.; Xie, Z.Q.; Jiang, F.S.; Xu, T.; Yang, S.Q.; Yin, S.Q.; Zhao, F.; He, J.L.; Chen, Y.B.; Hou, Z.Q.; et al. Evaluating the Socioeconomic Value of Urban Underground Space in Kunming, China, Using the Entropy Method and Exponential Smoothing Prediction. *J. Urban Plan. Dev.* **2023**, *149*, 05023001. [CrossRef]

25. Durmisevic, S.; Sariyildiz, S. A systematic quality assessment of underground spaces–public transport stations. *Cities* **2001**, *18*, 13–23. [CrossRef]
26. Xu, K.; Kong, C.; Li, J.; Zhang, L.; Wu, C. Suitability evaluation of urban construction land based on geo-environmental factors of Hangzhou, China. *Comput. Geosci.* **2011**, *37*, 992–1002. [CrossRef]
27. Bobylev, N. Sustainability and vulnerability analysis of critical underground infrastructure. In *Managing Critical Infrastructure Risks: Decision Tools and Applications for Port Security*; Springer: Berlin/Heidelberg, Germany, 2007; pp. 445–469.
28. Costa, A.N.; Polivanov, H.; Alves, M.G.; Ramos, D.P. Multicriterial analysis in the investigation of favorable areas for edifications with shallow and deep foundations in the Municipality of Campos dos Goytacazes-Rio de Janeiro, Brazil. *Eng. Geol.* **2011**, *123*, 149–165. [CrossRef]
29. Zhou, C.A.; Ren, H.; Liu, G.; Chen, C. Comprehensive evaluation and case study of urban underground space development under multiple constraints. *J. Bulg. Chem. Commun.* **2017**, *49*, 90–97.
30. Dou, F.F.; Xing, H.X.; Li, X.H.; Yuan, F.; Lu, Z.T.; Li, X.L.; Ge, W.Y. 3D Geological suitability evaluation for urban underground space development based on combined weighting and improved TOPSIS. *Nat. Resour. Res.* **2022**, *31*, 693–711. [CrossRef]
31. Cao, X.T.; Wei, C.F.; Xie, D.T. Evaluation of scale management suitability based on the entropy-TOPSIS method. *Land* **2021**, *10*, 416. [CrossRef]
32. Sałabun, W.; Wątróbski, J.; Shekhovtsov, A. Are MCDA methods benchmarkable? A comparative study of TOPSIS, VIKOR, COPRAS, and PROMETHEE II methods. *Symmetry* **2020**, *12*, 1549. [CrossRef]
33. Chen, P.Y. Effects of normalization on the entropy-based TOPSIS method. *Expert Syst. Appl.* **2019**, *136*, 33–41. [CrossRef]
34. Mahmoud, P.H.A.; Morteza, N.H.; Behnam, M.I.; Heresh, S. A hybrid genetic particle swarm optimization for distributed generation allocation in power distribution networks. *Energy* **2020**, *209*, 118218.
35. Zeleny, M.; Cochrane, J.L. *Multiple Criteria Decision Making*; McGraw-Hill: New York, NY, USA, 1982; p. 34.
36. Shannon, C.E. A mathematical theory of communication. *Bell Syst. Technol. J.* **1948**, *27*, 379–423. [CrossRef]
37. Aomar, R.A. A combined ahp-entropy method for deriving subjective and objective criteria weights. *Int. J. Ind. Eng.* **2010**, *17*, 12–24.
38. Diakoulaki, D.; Mavrotas, G.; Papayannakis, L. Determining objective weights in multiple criteria problems: The critic method. *Comput. Oper. Res.* **1995**, *22*, 763–770. [CrossRef]
39. Madić, M.; Radovanović, M. Ranking of some most commonly used nontraditional machining processes using ROV and CRITIC methods. *UPB Sci. Bull. Ser. D* **2015**, *77*, 193–204.
40. Anath, R.K.; Maznah, M.K.; Hamid, R.; Mohd, F.G. A modified CRITIC method to estimate the objective weights of decision criteria. *Symmetry* **2021**, *13*, 973.
41. Cao, Y.C.; Yang, J.; Chen, X.L.; Yang, X.J. A Beacon of Space and Time: Detailed Depiction of Human Space in the Xiong'an New Area Guided by Material Cultural Heritages. *Discret. Dyn. Nat. Soc.* **2021**, *2021*, 7746556. [CrossRef]
42. Liu, K.M.; Xu, Q.M.; Duan, L.F.; Niu, W.C.; Teng, F.; Wang, X.D.; Zhang, W.; Dong, J. Quaternary stratigraphic architecture and sedimentary evolution from borehole GB014 in the western Xiong'an New Area. *Sci. China Press* **2020**, *65*, 2145–2160.
43. He, D.F.; Shan, S.Q.; Zhang, Y.Y.; Lu, R.Q.; Zhang, R.F.; Cui, Y.Q. 3-D geologic architecture of Xiong'an New Area: Constraints from seismic reflection data. *Sci. China* **2018**, *61*, 1007–1022. [CrossRef]
44. Zhao, K.; Qi, J.X.; Chen, Y.; Ma, B.H.; Yi, L.; Guo, H.M.; Wang, X.Z.; Wang, L.Y.; Li, H.T. Hydrogeochemical characteristics of groundwater and pore-water and the paleoenvironmental evolution in the past 3.10 Ma in the Xiong'an New Area, North China. *China Geol.* **2021**, *4*, 476–486. [CrossRef]
45. Gao, Y.H.; Shen, J.H.; Chen, L.; Li, X.; Jin, S.; Ma, Z.; Meng, Q.H. Influence of underground space development mode on the groundwater flow field in Xiong'an new area. *J. Groundw. Sci. Eng.* **2023**, *11*, 68–80. [CrossRef]
46. Wang, Y.J.; Song, L.C.; Han, Z.Y.; Liao, Y.M.; Xu, H.M.; Zhai, J.Q.; Zhu, R. Climate-related risks in the construction of Xiongan New Area, China. *Theor. Appl. Clim.* **2020**, *141*, 1301–1311. [CrossRef]
47. Gao, Y.H. Study on Engineering Geological Environment and Construction Suitability of Xiong'an New Area. Ph.D. Dissertation, Chengdu University of Technology, Chengdu, China, 2023.
48. Xia, Y.B.; Li, H.T.; Wang, B.; Ma, Z.; Guo, X.; Zhao, K.; Zhao, C.R. Characterization of Shallow Groundwater Circulation Based on Chemical Kinetics: A Case Study of Xiong'an New Area, China. *Water* **2022**, *14*, 1880. [CrossRef]
49. Li, H.T.; Feng, W.; Wang, K.L.; Zhao, K.; Li, G.; Zhang, Y.; Li, M.Z.; Sun, L.; Chen, Y.C.; You, B. Groundwater resources in Xiong'an New Area and its exploitation potential. *Geol. China* **2021**, *4*, 1112–1126.
50. Shang, S.J.; Feng, C.J.; Tan, C.X.; Qi, B.S.; Zhang, P.; Meng, J.; Wang, M.M.; Sun, M.Q.; Wan, J.W.; Wang, H.J.; et al. Quaternary activity study of major buried faults near Xiongan New Area. *Acta Geosci. Sin.* **2019**, *40*, 836–846.
51. Yue, G.F.; Wang, G.L.; Ma, F.; Zhu, X.; Zhang, H.X.; Zhou, J.W.; Na, J. Fracture Characteristics and Reservoir Inhomogeneity Prediction of the Gaoyuzhuang Formation in the Xiong'an New Area: Insights From a 3D Discrete Fracture Network Model. *Front. Earth Sci.* **2022**, *10*, 849361. [CrossRef]
52. Ma, Z.; Huang, Q.B.; Lin, L.J.; Zhang, X.; Han, B.; Xia, Y.B.; Guo, X. Practice and application of multi-factor urban geological survey in Xiong'an New Area. *North China Geol.* **2022**, *45*, 58–68.

53. Liu, H.W.; Wang, G.M.; Ma, C.M.; Gao, M.D.; Bai, Y.N.; Zhang, J.; Du, D. Study on evaluation index system of suitability for development and utilization of underground space in sedimentary plain: A Case Study of Tongzhou District in Beijing and Langfang north three counties in Hebei Province. *North China Geol.* **2022**, *45*, 68–78.
54. Zou, Y.H.; Zhao, W.X. Making a new area in Xiong'an: Incentives and challenges of China's "Millennium Plan". *Geoforum* **2018**, *88*, 45–48. [CrossRef]

Article

Geo-Environment Suitability Evaluation for Urban Construction in Rongcheng District of Xiong'an New Area, China

Hongwei Liu [1,2,3,*] and Bo Han [2,3]

1 Chinese Academy of Geological Sciences, No. 26 Baiwanzhuang Street, Beijing 100037, China
2 Tianjin Center, China Geological Survey, No. 4 Dazhigu 8th Road, Tianjin 300170, China
3 Xiong'an Urban Geological Research Center, China Geological Survey, No. 4 Dazhigu 8th Road, Tianjin 300170, China
* Correspondence: liuhenry022@163.com

Abstract: Xiong'an New Area is a national event and a project planned for a millennium of China. Its high-quality construction is of great significance to easing the noncapital functions of Beijing and the coordinated development of the Beijing-Tianjin-Hebei region. As an emerging city, the development and construction of Xiong'an New Area is bound to be restricted by geological and resource conditions. Therefore, geo-environment suitability analysis is the necessary basis of urban development and construction. Geo-environment suitability analysis of urban construction is a complex process that requires various geological indicator information, and relevant expertise to analyze their relevance. This paper focuses on the analytic hierarchy process (AHP) for the assessment of geo-environment suitability for urban construction in Rongcheng district, which is a Start Construction Region in Xiong'an New Area. Multiple factors, including the characteristic value of bearing capacity of foundation soil, land subsidence rate, geological faults, ground fissures, potential liquefied sands, quality of groundwater chemistry, quality of soil chemistry, chemical corrosion of concrete by groundwater, chemical corrosion of steel by groundwater, and enrichment of deep groundwater and geothermal resource, were used for the suitability assessments. From the evaluation achievements, the high and very high suitable lands for urban construction, with an acreage percentage of 89.2%, were located in most parts of the study area. Meanwhile, for another 9.1% of the land, the impacts of geological faults, land subsidence, and potential liquefied sands needed to be noted preferentially for urban construction.

Keywords: suitability; urban construction; geo-environment; Rongcheng; Xiong'an new area

Citation: Liu, H.; Han, B. Geo-Environment Suitability Evaluation for Urban Construction in Rongcheng District of Xiong'an New Area, China. *Appl. Sci.* **2023**, *13*, 9981. https://doi.org/10.3390/app13179981

Academic Editor: Paulo Santos

Received: 28 July 2023
Revised: 27 August 2023
Accepted: 28 August 2023
Published: 4 September 2023

1. Introduction

With the acceleration of urban construction, the increasing demand for construction lands has become the key factor restricting planning development. At the same time, inappropriate utilization of the geo-environment and irrational development of geological resources were becoming increasingly significant, directly restricting urban construction [1–3]. Therefore, how to maximize the optimal allocation of urban construction and geological environment, and explore the evaluation method to effectively solve the practical dilemma, is particularly important [4–6].

Multicriteria analysis was a common tool used for complicated decision-making questions [7–11]. A pivotal step of geo-environment suitability analysis of urban construction was used to confirm the weight of each criterion or indicator [12]. For the application of weight confirming methods, scholars had not formed a unified scientific understanding. Generally speaking, various means widely used at present could be roughly divided into the following five categories: geographic information system (GIS) spatial data superposition analysis method [13], artificial neural network method [14], analytic hierarchy process (AHP) [15], grey comprehensive evaluation method [16], and fuzzy comprehensive evaluation method [17]. For examples, based on the GIS platform, Mario Mejia-Navarro et al. [18,19] established a geological disaster risk assessment system for Glenwood Springs, Colorado, evaluated the risk of geological disasters, and provided a basis for regional urban planning and construction. Mozafar et al. [20] took a waste treatment plant in Kurdistan Province of Iran as the research object, and systematically analyzed its location suitability by AHP. Liu et al. [21] evaluated the suitability of composite foundation for construction land through a fuzzy synthetic evaluation model in a city. Wang et al. [22] evaluated the suitability of urban construction land using multifactor grading weighted index in the alluvial plain of the Yellow River. Hu et al. [23] carried out land use zoning based on the principle of ecological priority and geo-environmental suitability using the AHP method in Weifang North Plain. Das et al. [24] finished the landslide susceptibility zonation mapping in and around the Kalimpong region by applying AHP method integrated with fifteen factors such as slope, lithology, elevation, thrust, and faults. Wang et al. [25] carried out the geological and ecological bearing capacity evaluation from three aspects of geological, ecological, and social attributes based on the GIS platform and evaluation index system, and determined the factor weights by using the AHP method.

In spite of the existence of various methods to identify weights of the selected criteria [26–29], the analytic hierarchy process (AHP) integrated GIS overlay analysis was regarded as one of the excellent multicriteria decision-making means [30–32]. Before that, Saaty [33] introduced AHP and how to use it, and provided many study examples.

Consequently, the objectives of this paper were (1) to establish a comprehensive evaluation frame for evaluating the geo-environmental suitability for urban construction land based on geo-environmental factors in Rongcheng district of Xiong'an New Area; (2) to identify the relationship and its contribution of geological indicators, including the characteristic value of the bearing capacity of foundation soil, land subsidence rate, geological faults, ground fissures, potential liquefied sands, quality of groundwater chemistry, quality of soil chemistry, chemical corrosion of concrete by groundwater, chemical corrosion of steel by groundwater, and enrichment of deep groundwater and geothermal resource, to urban construction land with the AHP method; and (3) to provide some significant information to improve decision-making for urban land planning according to the assessment results.

2. Study Area

The Rongcheng district, with an acreage of 314 km^2, is located northwest of Xiong'an New Area in North China [34], which is part of the alluvial plain of the Taihang Mountains [35] (Figure 1). The surface ground elevation displays a characteristic of decreasing from the northwest to the southeast, and varies from 5 m to 26 m with a gradient lower than 2‰ [36]. The research district belongs to the warm temperate zone with a semiarid climate, and the average annual precipitation is 482.7 mm. Meanwhile, it is close to the North China Plain's largest freshwater wetland, named Baiyangdian Lake [37]. Quaternary sediments are widely distributed in the surface ground, where rich geothermal and groundwater resources occur underground [38].

The recharging sources of shallow Quaternary groundwater are from precipitation, river and lake infiltration, farmland irrigation and underground lateral runoff, while the main discharging modes are artificial abstraction and underground flow. Furthermore, the deep groundwater in Quaternary and bedrock stratum has a certain hydraulic connection with shallow groundwater. Moreover, there are different geological problems, such as land subsidence, geological faults, ground fissures, potential liquefied sands, and so on.

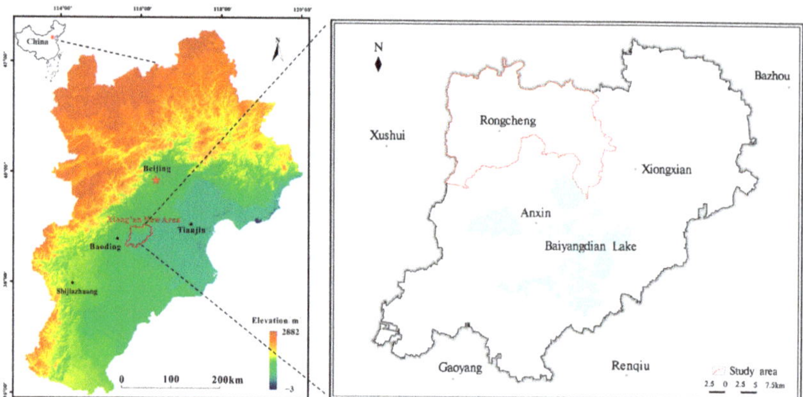

Figure 1. The study area of the Rongcheng district in Xiong'an New Area, China.

3. Methods

3.1. Field Survey and Data Collection

Geological factors play a significant role in urban construction. One such example is the bearing capacity of foundation soils, which determines the natural load-bearing capacity for buildings. Other factors such as land subsidence, geological faults, ground fissures, potential liquefaction of sands, and chemical corrosion of concrete and steel by groundwater can increase the risk of building deformation. On the other hand, the enrichment of deep groundwater and geothermal resources can provide residents in buildings with drinking water and heating. According to the data of the standard penetration test, shear test, and geotechnical test from engineering geological boreholes, the characteristic value of the bearing capacity of a single soil layer was determined. Characteristic values of the bearing capacity of foundation soils at different depths were calculated, including 0–5 m (meter), 5–10 m, 10–15 m, 15–30 m, and 30–50 m, through the data integration of single soil layers by the weighted average method. At the same time, the land subsidence rate was measured with the PS-InSAR remote sensing method, while ground fissures were investigated by high-density resistivity prospecting to depth, manual measurement to length, and compass measurement to direction. Enrichment of deep groundwater and geothermal resource distribution was analyzed by using the collected data. Meanwhile, geological faults were measured by multigeophysical exploration with controlled source audio-frequency magnetotelluric and resistivity tomography methods. Potential liquefied sands were identified on the basis of the sand liquefaction index calculated from the standard penetration test. Based on groundwater chemical and soil chemical experiments, the quality of groundwater chemistry, quality of soil chemistry, chemical corrosion of concrete by groundwater and chemical corrosion of steel by groundwater were evaluated.

3.2. Comprehensive Evaluation Frame

In view of the characteristics of the geological environment in Rongcheng district, based on suggestions from local geologists, the indicators closely related to geo-environment suitability for urban construction were chosen, and a comprehensive evaluation index system was established. As exhibited in Figure 2, two criteria, consisting of geological conditions and resource conditions, were taken into consideration for the suitability evaluation, including four subcriteria, i.e., engineering geological status, environmental geological status, hydrogeological status, and resource guarantee status. A total of 15 indicators were involved in the system. The engineering geological status included five indicators, such as the bearing capacity of the foundation soils at depths of 0~5 m, 5~10 m, 10~15 m, 15~30 m, and 30~50 m. The environmental geological status included six indicators, such as land subsidence rate, geological faults, ground fissures, potential liquefied sands, quality of groundwater chemistry, and quality of soil chemistry. The hydrogeological status included two indicators, such as chemical corrosion of concrete by groundwater and chemical corrosion of steel by groundwater. The resource guarantee status included two indicators, such as enrichment of deep groundwater and geothermal resource/geothermal gradient. Furthermore, the analytic hierarchy process (AHP), which was a decision-making method combining qualitative and quantitative analysis, was employed to identify the relations among various indicators or criteria, and to obtain final evaluation results. Obviously, the grading and weights of the abovementioned 15 indicators should be defined before evaluation, where the weights displayed the importance of different indicators. Furthermore, the comprehensive suitability index was calculated. Afterwards, final grading evaluation of geo-environment suitability for urban construction was achieved with ArcGIS 10.8 software.

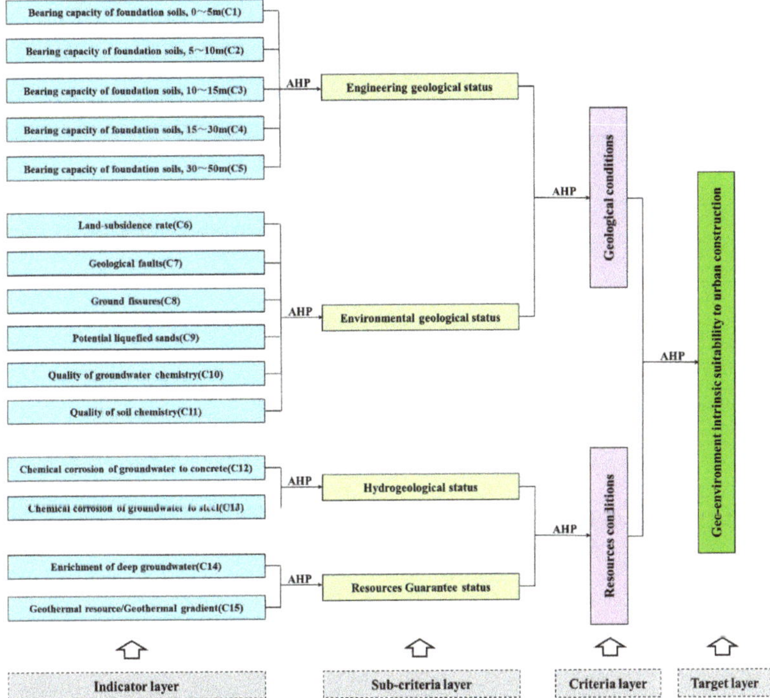

Figure 2. Comprehensive evaluation frame of geo-environment suitability.

3.3. Evaluation Method

By constructing the hierarchical structure model and its importance judgment matrix, the weight value of each evaluation indicator was obtained with the modified scaling method, and then the comprehensive index was calculated.

(1)   Establish an importance matrix, *A*.

$$A = \begin{bmatrix} C_{11} & \cdots & C_{1n} \\ \vdots & \ddots & \vdots \\ C_{n1} & \cdots & C_{nn} \end{bmatrix}$$

where *n* is the number of indicators, and the relative importance is checked from Table 1.

Table 1. Scale of relative importance between different indicators.

Scale	Meaning of the Scale
Scale = 1	Equal importance, two indicators contribute equally to the object
Scale = 1/9	Extreme unimportance, the evidence favoring one indicator over another is of the lowest possible order of affirmation
1/9 < Scale < 1, 1 < Scale < 9	More and more importance, judgment more and more strongly favors one indicator over another
Scale = 9	Extreme importance, the evidence favoring one indicator over another is of the highest possible order of affirmation

Annotation: the relative importance of the indicators being compared is closer together when the scale is equal to 1 [33].

(2)   Identify the weights

Based on the importance judgment matrix, the maximum eigenvalue and eigenvector were obtained [39], and then the eigenvector was normalized to calculate the weight value of different indicators [40,41]. Moreover, a consistency test of the judgment matrix should be carried out.

The calculation equation of the product of each row element (M_i) is:

$$M_i = \Pi_{j=1}^{n} C_{ij}$$

The calculation equation of normalized eigenvector (W_i) is:

$$W_i = \frac{\sqrt[n]{M_i}}{\sum_{i=1}^{n} \sqrt[n]{M_i}}$$

The calculation equation of eigenvalue (λ_i) is:

$$\lambda_i = \sum_{j=1}^{n} C_{ij} W_j$$

The calculation equation of maximum eigenvalue (λ_{max}) is:

$$\lambda_{max} = \sum_{i=1}^{n} \frac{\lambda_i}{n W_i}$$

The calculation equation of consistency ratio (CR) is:

$$CR = \frac{(\lambda_{max} - n)/(n-1)}{RI} \tag{1}$$

where RI is the mean random consistency index, which can be checked from Table 2. Meanwhile, CR needs to be less than 0.1.

Table 2. Mean random consistency index (RI) values of 11–15 order judgment matrix.

Order-Number	11	12	13	14	15
RI value	1.51	1.48	1.56	1.57	1.59

(3)   Calculate the suitability comprehensive index, *SI*.

$$SI = \sum_{i=1}^{n} u_i \cdot w_i, i = 1, 2, \ldots, n \tag{2}$$

where u_i is the score value of each indicator, w_i is the weight of each indicator, n is total number of indicators.

4. Results and Discussion

4.1. Results and Discussion of Geo-Environment Indicator Distribution

4.1.1. Bearing Capacity of Foundation Soils

The bearing capacity of foundation soils in different depths showed dissimilar characteristic values, varying from 105 kpa to 280 kpa, with an increasing trend from lower to deeper layers in Rongcheng County (Figure 3).

The bearing capacity of foundation soils for suitability for urban construction between 0–5 m, making 115 kpa and 125 kpa as the grading standards, could be divided into three grades, corresponding to very high (125–130 kpa), high (115–125 kpa), and moderate (105–115 kpa) (Table 3). The bearing capacity of foundation soils between 5–10 m, making 120 kpa, 130 kpa, and 140 kpa as the grading standards, could be divided into four grades, corresponding to very high (140–180 kpa), high (130–140 kpa), moderate (120–130 kpa), and low (110–120 kpa) (Table 3). The bearing capacity of foundation soils between 10–15 m, making 150 kpa, 170 kpa, and 190 kpa as the grading standards, could be divided into four grades, corresponding to very high (190–250 kpa), high (170–190 kpa), moderate (150–170 kpa), and low (110–150 kpa) (Table 3). The bearing capacity of foundation soils between 15–30 m, making 165 kpa, 175 kpa, and 185 kpa as the grading standards, could be divided into four grades, corresponding to very high (185–240 kpa), high (175–185 kpa), moderate (165–175 kpa), and low (155–165 kpa) (Table 3). The bearing capacity of foundation soils between 30–50 m, making 200 kpa, 210 kpa, and 220 kpa as the grading standards, could be divided into four grades, corresponding to very high (220–280 kpa), high (210–220 kpa), moderate (200–210 kpa), and low (190–200 kpa) (Table 3).

Figure 3. *Cont.*

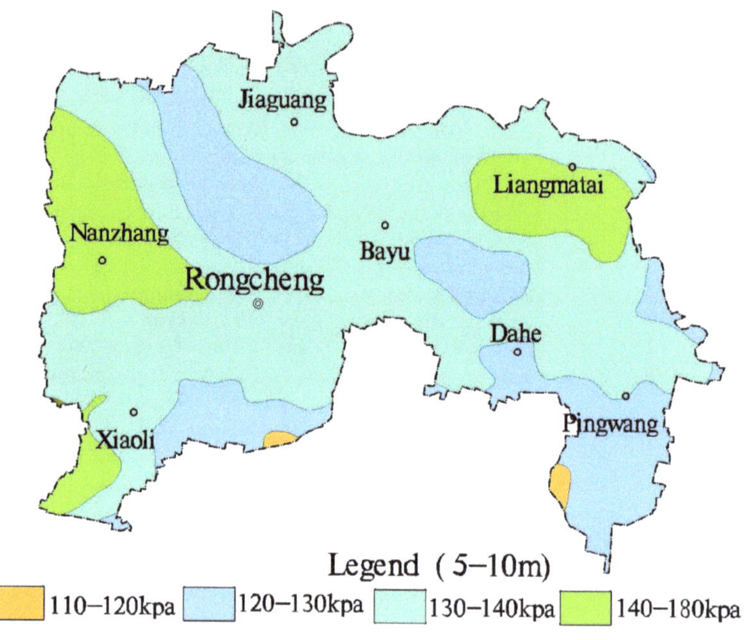

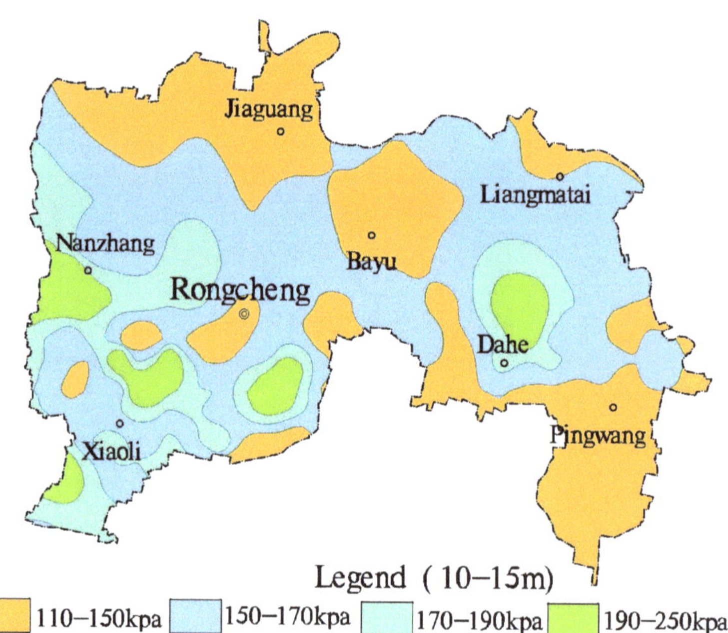

Figure 3. *Cont.*

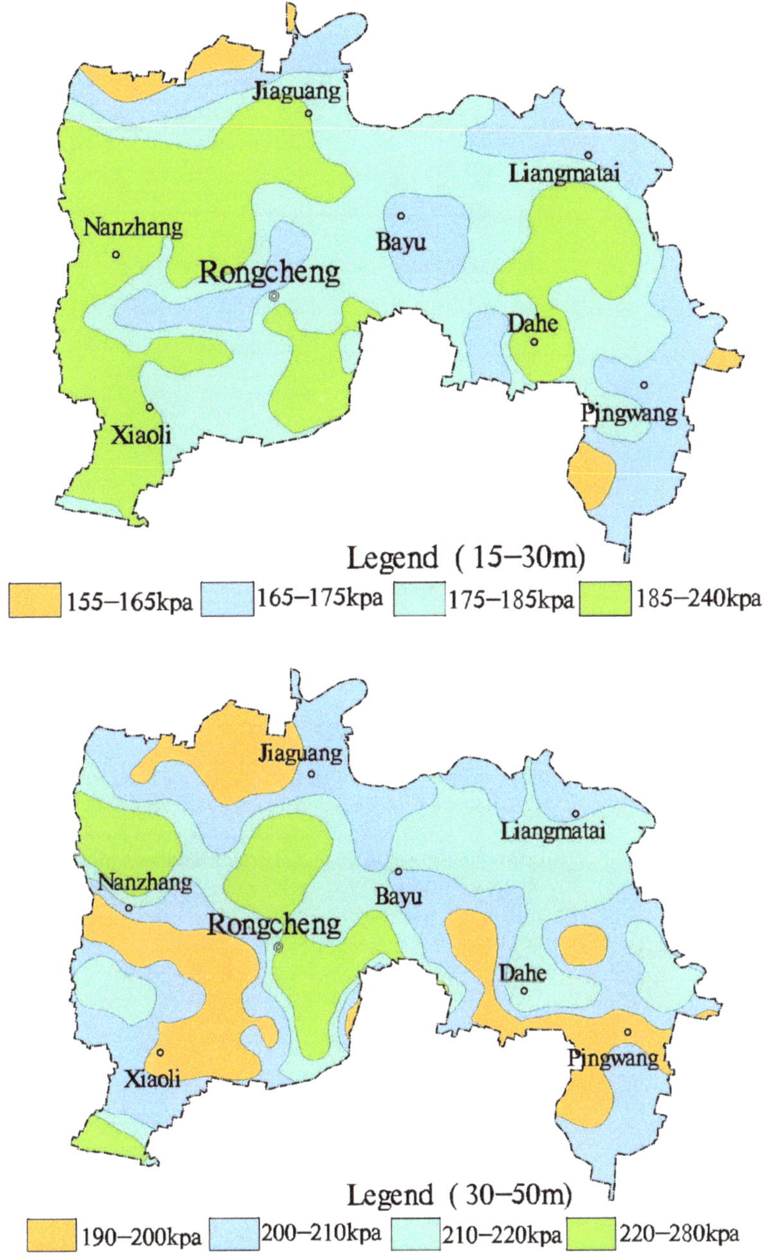

Figure 3. Spatial distributions of characteristic values of bearing capacity of foundation soils at different depths.

Table 3. The evaluation criteria for geo-environment suitability for urban construction.

Criteria Layer	Subcriteria Layer	Indicator Layer		Grading Criteria of Suitability			
				Very High	High	Moderate	Low
Geological conditions	Engineering geological status	Bearing capacity of foundation soils	0~5 m (C1)	125~130 kpa	115~125 kpa	105~115 kpa	/
			5~10 m (C2)	140~180 kpa	130~140 kpa	120~130 kpa	110~120 kpa
			10~15 m (C3)	190~250 kpa	170~190 kpa	150~170 kpa	110~150 kpa
			15~30 m (C4)	185~240 kpa	175~185 kpa	165~175 kpa	155~165 kpa
			30~50 m (C5)	220~280 kpa	210~220 kpa	200~210 kpa	190~200 kpa
	Environmental geological status	Land subsidence rate (C6)		<0 mm/a	0~10 mm/a	10~30 mm/a	>30 mm/a
		Geological faults (C7)		None	Away	Near	Existing
		Ground fissures (C8)		None	Away	Near	Existing
		Potential liquefied sands (C9)		None		Slight	Moderate
		Quality of groundwater chemistry (C10)		Can be used as a source of drinking water		Can be used as drinking water after proper treatment	Not suitable to be a source of drinking water
		Quality of soil chemistry (C11)		Very clean	Clean	Mildly polluted	Serious polluted
Resource conditions	Hydrogeological status	Chemical corrosion of concrete by groundwater (C12)		Slight			
		Chemical corrosion of steel by groundwater (C13)		Slight		A little	
	Resource Guarantee status	Enrichment of deep groundwater (C14)		>5000 m^3/d	3000–5000 m^3/d	1000–3000 m^3/d	<1000 m^3/d
		Geothermal resource/Geothermal gradient (C15)		$\geq$6 °C/100 m	$\geq$5 °C/100 m	$\geq$3 °C/100 m	<3 °C/100 m

4.1.2. Land Subsidence

According to the statistics data with PS-InSAR measurements from January to December in 2016, the land subsidence rate in most areas was between 30 mm/a and 10 mm/a, except the urban district and the northern area of Rongcheng County, Jiaguang, Bayu, and the western area of Dahe, with a rate of 30–40 mm/a, and the southern area of Pingwang, with a rate of less than 10 mm/a (Figure 4). The land subsidence rate for suitability for urban construction, making 0 mm/a, 10mm/a, and 30 mm/a as grading standards [42], could be divided into four grades, corresponding to very high (<0 mm/a), high (0–10 mm/a), moderate (10–30 mm/a), and low (>30 mm/a) (Table 3).

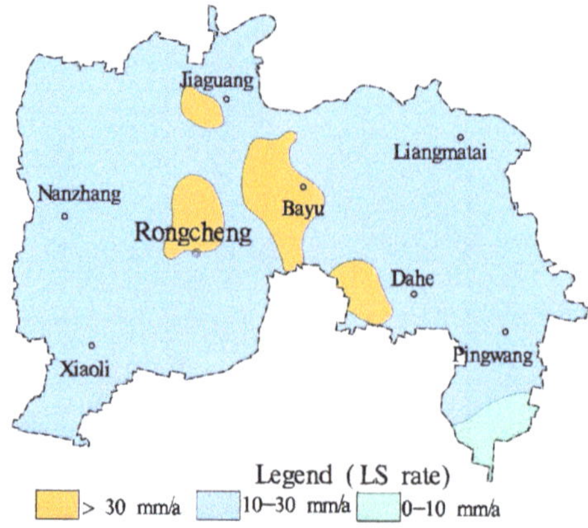

Figure 4. Zonal distributions of land subsidence rate.

4.1.3. Geological Faults

On the basis of geophysical exploration data, there were four geological faults, mainly distributed in Rongcheng, named Rongdong (RD) fault, Shunyi-Gaobeidian (SG) fault, Xushui-Anxin (XA) fault, and Qianxi-Jixian-Baoding-Shijiazhuang (QJBS) fault, which are inactive faults (Figure 5). It is worth noting that although the geological faults are currently inactive, they would lose stability with the reinjection of groundwater during deep geothermal resource exploitation in Rongcheng district of Xiong'an New Area [43]. Therefore, geological faults were also selected as an evaluation indicator for suitability evaluation. Based on the influence degree of distance to faults [44], geological faults for suitability for urban construction could be divided into four grades, corresponding to very high (none), high (away), moderate (near), and low (existing) (Table 3).

Figure 5. Distributions of geological faults.

4.1.4. Ground Fissures

There were about 20 discovered ground fissures, with a general distribution along the NW–SE direction. Most of the fissures appeared in forests and farmland, without endangering or damaging lives or property. Moreover, these fissures appeared in short lengths, most of which were less than 1000 m. The buried depth of the surface cracks were shallow, with depths of less than 20 m (Figure 6). Depending on the influence degree of distance to fissures [45], ground fissures for suitability for urban construction could be divided into four grades, corresponding to very high (none), high (away), moderate (near), and low (existing) (Table 3).

4.1.5. Potential Liquefied Sands

There were silty and fine sand layers distributed within 20 m of depth underground, which might result in liquefaction of seismic sands. According to the relevant provisions of the Code for Seismic Design of Buildings, the seismic intensity in this area was 7 degrees, the basic seismic acceleration was 0.10 g, and belonged to the second seismic group. According to the requirements of the general planning of this region, this evaluation of sand liquefaction was made according to the seismic intensity of 7.5 degrees, the designed basic seismic acceleration of 0.15 g, and 2 m of groundwater level, within a depth of 20 m.

Depending on the risk of liquefaction, potential liquefied sands for suitability for urban construction could be divided into four grades, corresponding to very high and high (none), moderate (slight liquefied), and low (moderate liquefied) (Figure 6, Table 3).

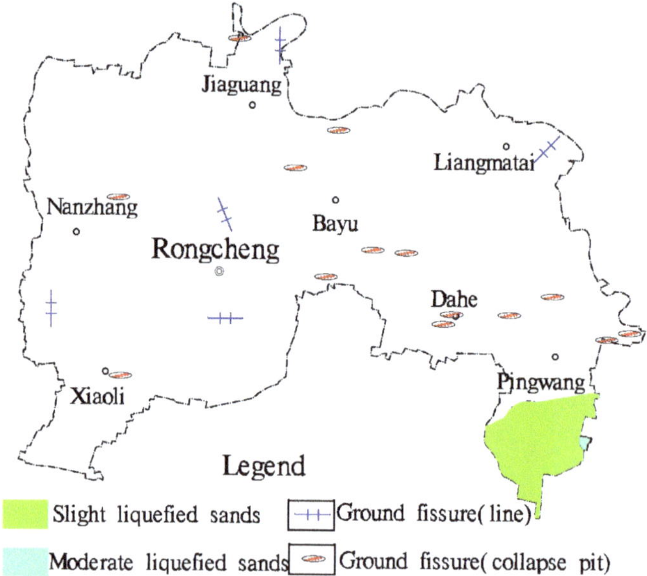

Figure 6. Distributions of ground fissures and potential liquefied sands.

4.1.6. Qualities of Groundwater and Soil Chemistry

Based on the qualities of groundwater and soil chemistry in 2017, groundwater in most areas could be used as a source of drinking water directly or after proper treatment, except the area of northeast Liangmatai (Figure 7). In addition, soils in most districts were clean and very clean, except areas such as Dongniubei, Xujiayuan, Wufangdong, Dongli, and Zanzhuang.

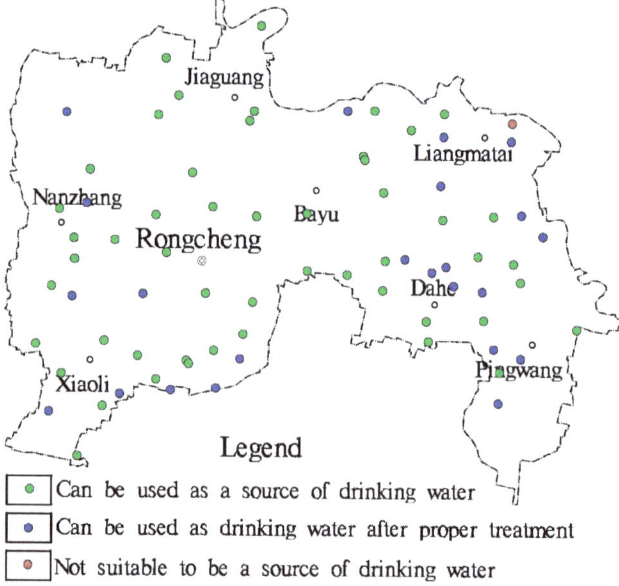

Figure 7. Quality distribution of groundwater chemistry samples.

Depending on the difference of categories, the quality of groundwater chemistry for suitability for urban construction could be divided into four grades, corresponding to very high, high, moderate, and low (Table 3). Meanwhile, the quality of soil chemistry could be divided into four grades, corresponding to very high (very clean), high (clean), moderate (mildly polluted), and low (serious polluted) [46] (Table 3).

4.1.7. Chemical Corrosion of Concrete and Steel by Groundwater

The chemical corrosion of concrete by groundwater in Rongcheng all belonged to a slight grade, and the chemical corrosion of steel by groundwater in most areas was in the slight category, except northeastern Xiaoli and southern Pingwang (Figure 8).

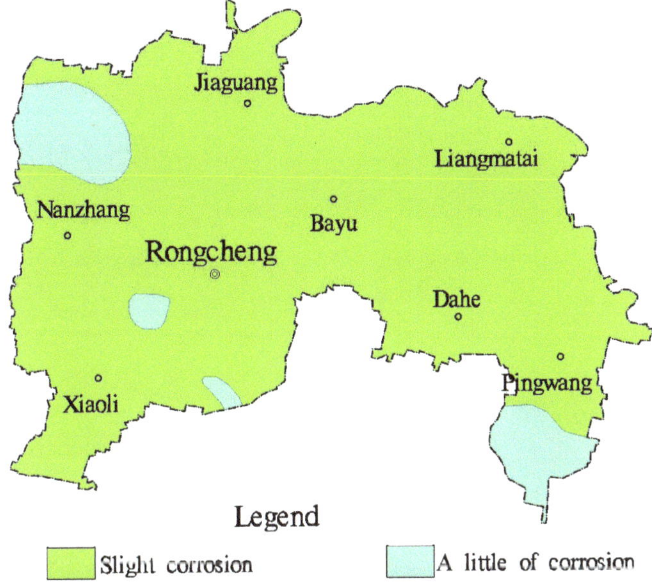

Figure 8. Zonal distributions of chemical corrosion of steel by groundwater.

According to the difference of categories, the chemical corrosion of concrete by groundwater was suitable for urban construction (Table 1). Simultaneously, chemical corrosion of steel by groundwater could be divided into four grades, corresponding to very high and high (slight corrosion), moderate, and low (a little corrosion) (Table 3).

4.1.8. Enrichment of Deep Groundwater and Geothermal Resource

Based on the available data, groundwater and geothermal resources were relatively abundant in Rongcheng, and showed certain zonation characteristics (Figures 9 and 10).

The enrichment of deep groundwater for suitability for urban construction, making 1000 m^3/d, 3000 m^3/d, and 5000 m^3/d as grading standards [47], could be divided into four grades, corresponding to very high (>5000 m^3/d), high (3000–5000 m^3/d), moderate (1000–3000 m^3/d), and low (<1000 m^3/d) (Figure 9) (Table 3). Meanwhile, geothermal resources, according to the difference of geothermal gradient, could be divided into four grades, corresponding to very high ($\geq$6 °C/100 m), high ($\geq$5 °C/100 m), moderate ($\geq$3 °C/100 m), and low (<3 °C/100 m) (Figure 10) (Table 3).

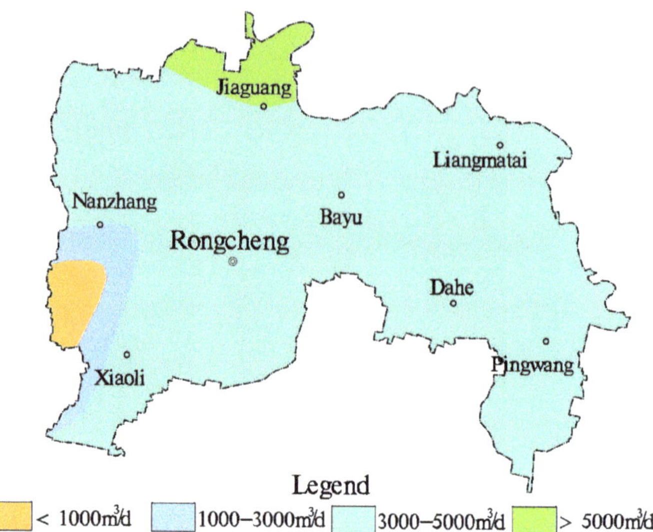

Figure 9. Zonal distributions of enrichment of deep groundwater.

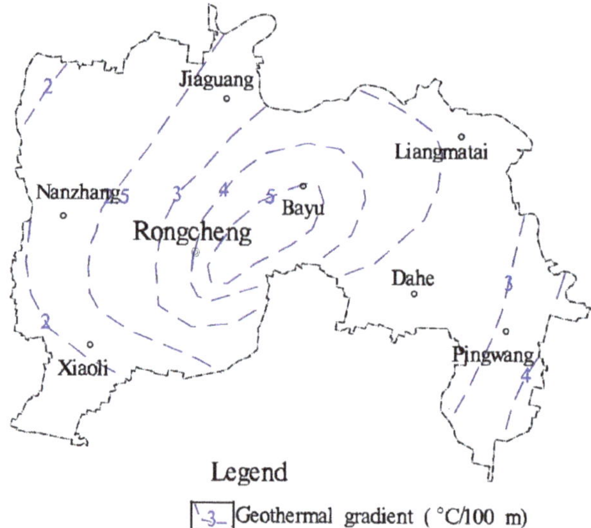

Figure 10. Distributions of geothermal gradient.

4.2. Results and Discussion of Geo-Environment Suitability for Urban Construction

According to the judgment matrix and weights of geo-environment indicators (Tables 3 and 4), the comprehensive evaluation indexes were calculated. The evaluation results showed that geo-environment suitability for urban construction in most areas of Rongcheng were in the high and very high grades, of which the very high zone covered an area of about 98 km^2, and the high zone was nearly 182 km^2 (Figure 11). The acreage of the moderate grade was approximately 5.5 km^2, and the low grade was close to 28.5 km^2. Meanwhile, the main affecting factors were dissimilar (Table 5); the impacts of geological faults, land subsidence rate, and potential liquefied sands should be noted preferentially for urban construction.

Table 4. Index judgment matrix and weights of geo-environment indicators.

	C1	C2	C3	C4	C5	C6	C7	C8	C9	C10	C11	C12	C13	C14	C15	Weights
C1	1	1	1	5/4	5/4	5/8	5/9	5/4	5/7	5/3	5/1	5/4	5/4	6/5	5/4	0.07
C2		1	1	5/4	5/4	5/8	5/9	5/4	5/7	5/3	5/1	5/4	5/4	6/5	5/4	0.07
C3			1	5/4	5/4	5/8	5/9	5/4	5/7	5/3	5/1	5/4	5/4	6/5	5/4	0.07
C4				1	1	1/2	4/9	6/5	2/3	3/2	4/1	6/5	6/5	7/6	6/5	0.06
C5					1	1/2	4/9	6/5	2/3	3/2	4/1	6/5	6/5	7/6	6/5	0.06
C6						1	7/9	2/1	8/7	4/1	5/1	8/3	8/3	8/3	4/1	0.12
C7							1	9/4	9/7	9/4	6/1	3/1	8/3	9/4	3/1	0.13
C8								1	4/7	2/1	2/1	4/5	4/5	5/2	5/3	0.06
C9									1	7/3	7/3	7/4	7/4	2/1	7/3	0.09
C10										1	8/7	1/2	1/2	7/9	7/8	0.04
C11											1	1/4	1/4	1/3	1/2	0.02
C12												1	1	5/4	5/3	0.06
C13													1	5/4	5/3	0.06
C14														1	4/3	0.05
C15															1	0.04

Annotation: the consistency ratio is 0.02.

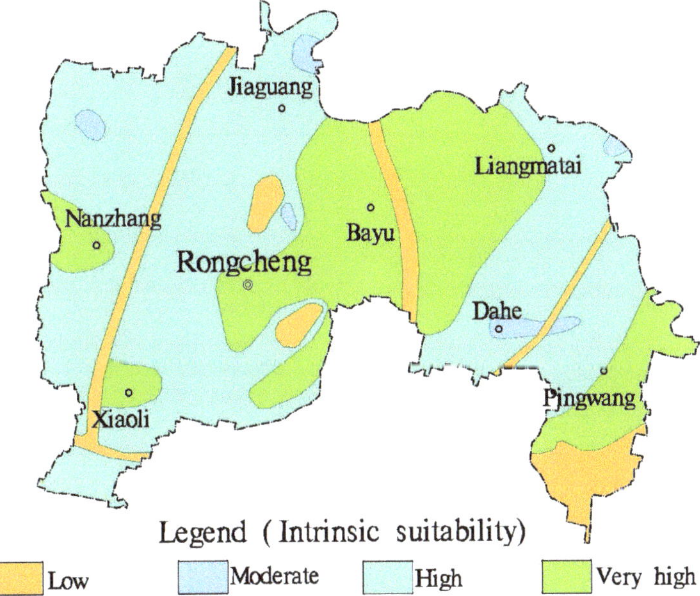

Figure 11. Zonal distributions of geo-environment suitability for urban construction.

Generally, it is a gradual process for the planning and construction of Rongcheng as a Start Construction Region in Xiong'an New Area; however, the geo-environment suitability evaluation for urban construction should be regarded as preliminary work. This paper selected as many geological indicators as possible to analyze the geo-environment suitability for urban construction. Nevertheless, it focused on geo-environment characteristics, and other socioeconomic features were not included. In future studies, population quantity and industrial structure should be considered, in order to improve decision-making for precise urban land planning.

Table 5. The evaluation results for geo-environment suitability for urban construction in Rongcheng.

Grade	Acreage (sq.km.)	Percentage	Main Affecting Factors
Very high	98	31.2%	Geothermal resource
			Enrichment of deep groundwater
			Characteristic value of bearing capacity of foundation soil
High	182	58%	Characteristic value of bearing capacity of foundation soil
			Quality of groundwater chemistry
Moderate	5.5	1.7%	Ground fissures
			Chemical corrosion of concrete by groundwater
			Chemical corrosion of steel by groundwater
			Quality of soil chemistry
Low	28.5	9.1%	Geological faults
			Land subsidence rate
			Potential liquefied sands

5. Conclusions

(1) In order to evaluate the geo-environment suitability for urban construction in Rongcheng district of Xiong'an New Area, the analytic hierarchy process (AHP) integrated GIS overlay analysis was used, based on the construction of a comprehensive evaluation frame. Moreover, two criteria, consisting of geological conditions and resource conditions, were taken into consideration for suitability evaluation, including four subcriteria, i.e., engineering geological status, environmental geological status, hydrogeological status, and resource guarantee status, which involved 15 indicators. Regrettably, the evaluation did not include the compressibility indicator of foundation soils due to a lack of data. When evaluating the suitability of the geo-environment for urban construction in other areas with the AHP method, more indicators of foundation soils could be taken into consideration. Furthermore, the analytic hierarchy process has certain advantages compared to other methods, such as the artificial neural network method and grey comprehensive evaluation method. It not only provides a quantitative mathematical calculation, but also incorporates the comparative judgment of geological experts regarding the importance of different geological indicators.

(2) The evaluation results showed that the geo-environment suitability for urban construction in most areas was in high and very high grades, of which, the very high zone covered an area of about 98 km^2, and the high zone was nearly 182 km^2. The acreage of the moderate grade was approximately 5.5 km^2, and the low grade was close to 28.5 km^2. The most suitable areas for urban construction, with an acreage percentage of 31.2%, were mainly located in the central parts of the study area. In the meantime, the least suitable areas, with an acreage percentage of 9.1%, were situated in the southeast corner and three linear belts.

(3) It is crucial to emphasize that faults, land subsidence rate, and potential liquefied sands are the primary factors that influence decision-making regarding future construction activities. When urban construction takes place in areas close to faults, buildings should maintain a certain distance from them, and these areas should be designated as green spaces. In regions experiencing a land subsidence rate of more than 30 mm/a, it is advisable to reduce groundwater extraction and lower the height of planned buildings. Additionally, engineering protection measures should be implemented in areas with potential liquefied sands. By addressing these issues, the study area can reduce infrastructure construction costs, and minimize the risk of geological disasters.

Author Contributions: Conceptualization and methodology, H.L.; validation, B.H.; investigation, H.L. and B.H.; writing—original draft preparation, H.L.; writing—review and editing, H.L. and B.H.; funding acquisition, H.L. and B.H. All authors have read and agreed to the published version of the manuscript.

Funding: This research was funded by China Geological Survey project of the Monitoring and Evaluation of Resource and Environment Carrying Capacity of the Beijing-Tianjin-Hebei Collaborative Development Area and Xiongan New Area (Grant no. DD20221727).

Institutional Review Board Statement: Not applicable.

Informed Consent Statement: Not applicable.

Data Availability Statement: Not applicable.

Conflicts of Interest: The authors declare no conflict of interest.

References

1. Xu, K.; Kong, C.F.; Li, J.F.; Zhang, L.; Wu, C. Suitability evaluation of urban construction land based on geo-environmental factors of Hangzhou, China. *Comput. Geosci.* **2011**, *37*, 992–1002. [CrossRef]
2. Zhou, X.N.; Wang, Z.X.; Miao, Q.Z.; Zhang, B. Study the shallow groundwater chemical characteristics in the typical area of Zhanghe catchment basin. *Geol. Surv. Res.* **2020**, *43*, 265–270.
3. Tann, L.; Ritter, S.; Hale, S.; Langford, J.; Salazar, S. From urban underground space (UUS) to sustainable underground urbanism (SUU): Shifting the focus in urban underground scholarship. *Land Use Policy* **2021**, *109*, 105650. [CrossRef]
4. Svoray, T.; Bar, P.; Bannet, T. Urban land-use allocation in a Mediterranean ecotone: Habitat heterogeneity model incorporated in a GIS using a multi-criteria mechanism. *Landsc. Urban Plan.* **2005**, *72*, 337–351. [CrossRef]
5. Liu, H.; Chen, F.C.; Chen, G. Fuzzy evaluation of construction land suitability based on the geological environment. *Resour. Environ. Eng.* **2016**, *30*, 645–653.
6. Dou, F.; Xing, H.; Li, X.; Yuan, F.; Lu, Z.; Li, X.; Ge, W. 3D Geological suitability evaluation for urban underground space development based on combined weighting and improved TOPSIS. *Nat. Resour. Res.* **2022**, *31*, 693–711. [CrossRef]
7. Feick, R.; Hall, B. A method for examining the spatial dimension of multi-criteria weight sensitivity. *Int. J. Geogr. Inf. Sci.* **2004**, *18*, 815–840. [CrossRef]
8. Ananda, J.; Herath, G. A critical review of multi-criteria decision making methods with special reference to forest management and planning. *Ecol. Econ.* **2009**, *68*, 2535–2548. [CrossRef]
9. Mosadeghi, R.; Warnken, J.; Tomlinson, R.; Mirfenderesk, H. Comparison of fuzzy-AHP and AHP in a spatial multi-criteria decision making model for urban land-use planning. *Comput. Environ. Urban Syst.* **2015**, *49*, 54–65. [CrossRef]
10. Romano, G.; Dal Sasso, P.; Trisorio Liuzzi, G.; Gentile, F. Multi-criteria decision analysis for land suitability mapping in a rural area of Southern Italy. *Land Use Policy* **2015**, *48*, 131–143. [CrossRef]
11. Ustaoglua, E.; Aydınoglu, A.C. Suitability evaluation of urban construction land in Pendik district of Istanbul, Turkey. *Land Use Policy* **2020**, *99*, 104783. [CrossRef]
12. Zhang, J.; Su, Y.; Wu, J.; Liang, H. GIS based land suitability assessment for tobacco production using AHP and fuzzy set in shandong province of China. *Comput. Electron. Agric.* **2015**, *114*, 202–211. [CrossRef]
13. Youssef, A.M.; Pradhan, B.; Tarabees, E. Integrated evaluation of urban development suitability based on remote sensing and GIS techniques: Contribution from the analytic hierarchy process. *Arab. J. Geosci.* **2011**, *4*, 463–473. [CrossRef]
14. Durmisevic, S.; Sariyildiz, S. A systematic quality assessment of underground spaces–public transport stations. *Cities* **2001**, *18*, 13–23. [CrossRef]
15. Duan, Y.; Xie, Z.; Zhao, F.; Zeng, H.; Lin, M.; Chen, H.; Hou, Z. Suitability of underground space development in plateau cities based on geological environment analysis: Case study in Kunming, China. *J. Urban Plan. Dev.* **2021**, *147*, 05021014. [CrossRef]
16. Xu, G.; Yang, Y.P.; Lu, S.Y.; Li, L.; Song, X. Comprehensive evaluation of coal-fired power plants based on grey relational analysis and analytic hierarchy process. *Energy Policy* **2011**, *39*, 2343–2351. [CrossRef]
17. Costa, A.N.; Polivanov, H.; Alves, M.G.; Ramos, D.P. Multicriterial analysis in the investigation of favorable areas for edifications with shallow and deep foundations in the Municipality of Campos dos Goytacazes-Rio de Janeiro, Brazil. *Eng. Geol.* **2011**, *123*, 149–165. [CrossRef]
18. Mejía-Navarro, M.; Wohl, E.E.; Oaks, S.D. Geological hazards vulnerability and risk assessment using GIS: Model for Glenwood Springs, Colorado. *Geomorphology* **1994**, *10*, 331–354. [CrossRef]
19. Mejía-Navarro, M.; Garcia, L.A. Natural hazard and risk assessment using decision support systems, application: Glenwood Springs, Colorado. *Environ. Eng. Geosci.* **1996**, *2*, 299–324. [CrossRef]
20. Shan, M.; Hadidi, M.; Vessali, E. Intelating multcriteria decision analysis for a GIS-based hazardous waste landfill sitting in Kurdistan Province, western Iran. *Waste Manag.* **2009**, *29*, 2740–2758.
21. Liu, H.; Cao, H.; Fu, Y. Fuzzy comprehensive evaluation of adaptability of natural foundation in a city. *Eng. Technol.* **2016**, *7*, 269–270.

22. Wang, W.T.; Guo, M.W. Engineering geological zoning and suitability evaluation of engineering construction in Shangqiu planning area. *Miner. Explor.* **2022**, *13*, 130–138.
23. Hu, X.J.; Gao, L.; Ma, C.M.; Hu, X.J. Land use zoning of Weifang North Plain based on ecological function and geo-environmental suitability. *Bull. Eng. Geol. Environ.* **2020**, *79*, 2697–2719. [CrossRef]
24. Das, S.; Sarkar, S.; Kanungo, D.P. GIS-based landslide susceptibility zonation mapping using the analytic hierarchy process (AHP) method in parts of Kalimpong Region of Darjeeling Himalaya. *Environ. Monit. Assess.* **2022**, *194*, 234. [CrossRef] [PubMed]
25. Wang, Z.F.; He, X.Q.; Zhang, C.; Xu, J.W.; Wang, Y.J. Evaluation of geological and ecological bearing capacity and spatial pattern along du-wen road based on the analytic hierarchy process (AHP) and the technique for order of preference by similarity to an ideal solution (TOPSIS) method. *ISPRS Int. J. Geo-Inf.* **2020**, *9*, 237. [CrossRef]
26. Banai-Kashani, A.R. A new method for sitesuitability analysis: Theanalytic hierarchy process. *Environ. Manag.* **1989**, *13*, 685–693. [CrossRef]
27. Kalogirou, S. Expert systems and GIS: An application of land suitability suitability. *Comput. Environ. Urban Syst.* **2002**, *26*, 89–112. [CrossRef]
28. Marinoni, O. Implementation of the analytical hierarchy process with VBA in ArcGIS. *Comput. Geosci.* **2004**, *30*, 637–646. [CrossRef]
29. Wang, Q.; Xu, J.G.; Xu, W.W. A GIS approach to the urban land suitability evaluation: A case study of gaochun in Nanjing. *Prog. Geophy.* **2005**, *20*, 877–880.
30. Bandyopadhyay, S.; Jaiswal, R.K.; Hegde, V.S.; Jayaraman, V. Assessment of land suitability potentials for agriculture using a remote sensing and GIS based approach. *Int. J. Remote Sens.* **2009**, *30*, 879–895. [CrossRef]
31. Seyedmohammadi, J.; Sarmadian, F.; Jafarzadeh, A.A.; McDowell, R.W. Development of a model using matter element, using AHP and GIS techniques to assess the suitability of land for agriculture. *Geoderma* **2019**, *352*, 80–95. [CrossRef]
32. Rasli, F.N.; Kanniah, K.D.; Muthuveerappan, C.; Ho, C.S. An integrated approach of analytical hierarchy process. *Int. J. Geoinformatics* **2016**, *12*, 67–77.
33. Saaty, R.W. The analytic hierarchy process—What it is and how it is used. *Math. Model.* **1987**, *9*, 161–176. [CrossRef]
34. Gao, Y.H.; Shen, J.H.; Chen, L.; Li, X.; Jin, S.; Ma, Z.; Meng, Q.H. Influence of underground space development mode on the groundwater flow field in Xiong'an new area. *J. Groundw. Sci. Eng.* **2023**, *11*, 68–80. [CrossRef]
35. Zhao, K.; Qi, J.X.; Chen, Y.; Ma, B.H.; Yi, L.; Guo, H.M.; Li, H.T. Hydrogeochemical characteristics of groundwater and pore-water and the paleoenvironmental evolution in the past 3.10 Ma in the Xiong'an New Area, North China. *China Geol.* **2021**, *4*, 476–486. [CrossRef]
36. Xia, Y.; Li, H.; Wang, B.; Ma, Z.; Guo, X.; Zhao, K.; Zhao, C. Characterization of Shallow Groundwater Circulation Based on Chemical Kinetics: A Case Study of Xiong'an New Area, China. *Water* **2022**, *14*, 1880. [CrossRef]
37. Wang, Y.; Song, L.; Han, Z.; Liao, Y.; Xu, H.; Zhai, J.; Zhu, R. Climate-related risks in the construction of Xiongan New Area, China. *Theor. Appl. Clim.* **2020**, *141*, 1301–1311. [CrossRef]
38. Ma, Z.; Huang, Q.B.; Lin, L.J. Practice and application of multi-factor urban geological survey in Xiong'an New Area. *North China Geol.* **2022**, *45*, 58–68.
39. Saaty, T. A scaling method for priorities in hierarchical structures. *J. Math. Psycholgy* **1977**, *15*, 234–281. [CrossRef]
40. Uyan, M. GIS-based solar farms site selection using analytic hierarchy process (AHP) in Karapinar region, Konya/Turkey. *Renew. Sustain. Energy Rev.* **2013**, *28*, 11–17. [CrossRef]
41. Gompf, K.; Traverso, M.; Hetterich, J. Using analytical hierarchy process (AHP) to introduce weights to social life cycle assessment of mobility services. *Sustainability* **2021**, *13*, 1258. [CrossRef]
42. Li, M.; Zhang, L.; Ge, D.; Liu, B.; Wang, Y.; Guo, X. PSInSAR technique to monitor coastal lowland subsidence along the Eastern Coast of China-a case study in Zhejiang coast. In Proceedings of the 2016 IEEE International Geoscience and Remote Sensing Symposium (IGARSS), Beijing, China, 10–15 July 2016; pp. 5955–5958.
43. Zhu, S.; Feng, C.; Tan, C.; Ma, X.; Meng, J.; Qi, B.; Zhang, Z. Fault slip potential induced by water injection in the Rongcheng deep-seated geothermal reservoir, Xiong'an New Area. *Chin. J. Rock Mech. Eng.* **2022**, *41* (Suppl. 1), 2735–2756.
44. He, X.; Xu, C.; Xu, X.; Yang, Y. Advances on the avoidance zone and buffer zone of active faults. *Nat. Hazards Res.* **2022**, *2*, 62–74. [CrossRef]
45. Deng, Y.; Chang, J.; Lu, Q.; Li, L.; Mu, H.; Feng, L. Disaster characteristics and influence range of earth fissure on nearby structures. *Arab. J. Geosci.* **2022**, *15*, 47. [CrossRef]
46. Wang, Y.; Sun, W.; Zhao, Y.; He, P.; Wang, L.; Nguyen, L.T.T. Assessment of Heavy Metal Pollution Characteristics and Ecological Risk in Soils around a Rare Earth Mine in Gannan. *Sci. Program.* **2022**, *2022*, 5873919.
47. Li, Y.Z.; Zhan, S.Q.; Wu, X. Geological environment vulnerability evaluation of the economic zone in south Shandong. *Geoscience* **2014**, *28*, 1096–1102.

MDPI AG
Grosspeteranlage 5
4052 Basel
Switzerland
Tel.: +41 61 683 77 34

Applied Sciences Editorial Office
E-mail: applsci@mdpi.com
www.mdpi.com/journal/applsci